Cells, Gels and the Engines of Life

Cells, Gels and the Engines of Life

A New, Unifying Approach to Cell Function

Ebner & Sons
Seattle WA, USA

Gerald H. Pollack

Ebner and Sons Publishers
3714 48th Avenue NE
Seattle Washington 98105

For more information about this book, including quantity orders, visit:
www.cellsandgels.com

Library of Congress Cataloging-in-Publication Data

Pollack, Gerald H., Ph.D.
Cells, gels and the engines of life; a new, unifying approach to cell function.
p. 320, cm.
Includes biographical references and index.
ISBN: Casebound: 0-9626895-1-3
Paperback: 0-9626895-2-1
1. Bioengineering. 2 Bioengineering -- cell function. 3. Cell function. 4. Cell physiology. I. Pollack, Gerald H. II. Title
574.6 P78 00-13784
947--dc CIP

9 8 7 6 5 4 3 2 1
Printed in South Korea

Design and illustrations by David Olsen.

Cover image by Jakob Zbaeren, Dept. of Klinische Forschung, Inselspital, Bern, Switzerland. Fluoresence micrograph of HUVEC culture. F-actin: red; von Willebrand factor: green; nuclei: blue.

for Seth, Ethan and Mia

.....who made it all worthwhile

ACKNOWLEDGEMENTS

Producing a book of this sort owes debt to numerous people. It is difficult to know where to begin.

From the foundational perspective, I warmly acknowledge the influence of Gilbert Ling. I met Gilbert at a meeting in Hungary in the mid-1980s. There I discovered a new conceptual world. It became clear that an approach to cell physiology orthogonal to current wisdom had considerable merit and enjoyed appreciable experimental support from a cadre of intellectually independent scientists. Ling has been a continuing force not only among that group, but also for myself. Without his influence, this work would never have begun.

Next, I acknowledge the intellectual stimulation provided by students and fellows in my laboratory, many of whom have taken keen interest in this subject area. Prime among them were Marc Bartoo, whose influence helped shape the book's early chapters, and Fritz Reitz, whose creative suggestions and sharp-eyed skepticism helped throughout.

Students and fellows also helped with chapter-by-chapter critiques. At times, I must admit that I was less than thrilled with their brutal honesty. But in the end, these critiques led to a product that is a good deal more lucid and less invested with flabby logic than had been the case initially. Helpful critiques were also provided by students in my introductory cellular and molecular biomechanics class, whom I thank. Among fellows and students who contributed well beyond the call of duty in this vein, I particularly thank: Kevin Anderson, Marc Bartoo, Felix Blyakhman, Dwayne Dunaway, Mark Fauver, Outi Hyyti, Fettah Kosar, Justin Mih, Thomas Neumann, Megan Peach, Brian Rabkin, Fritz Reitz, Nate Segerson, Betty Sindelar, and Paul Yang.

A book whose scope is as broad as this one inevitably touches on subjects beyond the sphere of expertise of any one author. I therefore sent drafts to colleagues whom I regard as experts in these various subjects. One or two responded with hostility, and I acknowledge their willingness to answer follow-up questions that helped me understand why this approach might be seen by some as unpalatable. Mainly, the responses were filled with encouragement and rich with helpful feedback. This constructive criticism has led to a book that is considerably freer of error and misinterpretation than had been the case initially. It goes without saying that any lingering errors are my responsibility, not theirs.

Among colleagues whose help has made a real difference, I would like to mention: Joe Andrade, Shu Chien, Wei-chun Chin, Jim Clegg, Rainer Gülch, Harold Hillman, Linda Hufnagel, Sir Andrew Huxley, Hiromasa Ishiwatari, Paul Janmey, Ed Korn, Gilbert Ling, Pascale Mentré, Charlotte Omoto, Seth Pollack, Fred Sachs, Ichiji Tasaki, Lars Thuneberg, Pedro Verdugo, Erwin Vogler, and Maish Yarmush. I cannot thank these busy people enough for sharing their wisdom with me.

The artwork that fills this volume was created entirely by David Olsen. I knew David was the right man when he first smiled at my doorstep. David was one of those rare students of art who was also a student of biochemistry. By the time he began, he had already illustrated a book of his own, on the subject of teaching English to Chinese children. Often times I could merely outline a concept and David would appear in a few days with an insightful first draft design. I hope the reader will agree that David's artwork is imaginative, occasionally whimsical, and consistently clear. David has been a pleasure to work with—fully understanding of deadlines, unusually tolerant of my excessive demands for revision, and pleasantly agreeable to the end. I thank him warmly.

I also thank Don Scott for a thoroughgoing job of editing the manuscript. Don's astute eye helped minimize awkward phraseology, clarify my intent, and un-mix my metaphors. Heidi Mair was helpful, accurate, and reassuring in securing permissions and gathering obscure references.

Finally, I thank my life-partner, Emily Freedman, and the rest of my family for putting up with me during the several years it took to prepare this book. During early mornings and late evenings, I could be found wielding the word processor instead of the vacuum cleaner. Emi has been extremely understanding and encouraging during this extended process and I can only promise her that the next book will be shorter.

CONTENTS

PREFACE

This book is about the cell and how it functions. Books about cell function have become commonplace, but the approach taken here is not at all common. It builds on fundamental principles of physics and chemistry rather than on accepted higher-level constructs. The premise is that nature's functional paradigms are elegant, simple, and probably far less convoluted than current wisdom purports them to be. The book's goal is to identify these simple paradigms from physical chemical basics.

The story begins with a curious scene described in the standard cell-biology textbook by Alberts *et al.* (first edition, p. 604). A crawling cell somehow manages to get stuck onto the underlying surface. The leading edge continues to press on, tears free, and continues its journey onward as though oblivious to the loss of its rear end.

Something about this scenario does not seem right. We have been taught that cell function depends on the integrity of the cell membrane; yet the membrane of this cell has been ripped asunder and the cell continues to function as though none of this matters. It's business as usual—even though the cell has effectively been decapitated.

When first exposed to this scenario, I reacted the way you may be reacting—the cell must somehow reseal. The membrane must spread over the jagged surfaces, covering the wound and restoring the cell's integrity. I was inclined to accept this seemingly flimsy account even though I could not understand how a bilayer membrane built of molecules with fixed lateral packing could spread over half again as much area. Nevertheless, it seemed expedient to shelve this anomaly and press on with more immediate matters; someone else could figure out what might be going on.

With time, it became clear that the survival of the decapitated cell was but one example of a set of related anomalies. Cells survive sundry insults equivalent to being guillotined, drawn and quartered, and shot full of holes with electrical bullets (Chapter 2). If the ruptured membrane does not reseal—and no evidence I've seen convinces me that "resealing" is any more than a conveniently invoked expedient—the implication is that membrane integrity may be less consequential than presumed.

From this realization springs the early material of this book. If membrane integrity is not essential, then what holds the contents inside the cell? If you are willing to consider the cytoplasm as a gel instead of an ordinary aqueous solution, then an answer may be at hand, for gels don't necessarily disintegrate when sliced; the contents will not spill out.

The gel-like nature of the cytoplasm forms the foundation of this book. Biologists acknowledge the cytoplasm's gel-like character, but textbooks nevertheless build on aqueous solution behavior. A gel is quite different from an aqueous solution—it is a matrix of polymers to which water and ions cling. That's why gelatin desserts retain water, and why a cracked egg feels gooey.

The concept of a gel-like cytoplasm turns out to be replete with power. It accounts for the characteristic partitioning of ions between the inside and outside of the cell (Chapter 6). It also explains the cell's electrical potential: potentials of substantial magnitude can be measured in gels as well as in demembranated cells (Chapter 7). Thus, the gel-like character of the cytoplasm accounts for the basic features of cell biophysics.

The focus then shifts from statics to dynamics. Here again the question is whether adequate explanations can emerge from the cell's gel-like character. Contrary to the common perception, gels are not inert. With modest prompting, polymer gels undergo structural transitions that can be as profound as the change from ice to water, which is why they are classified as phase-transitions. Polymer-gel phase-transitions are commonly exploited in everyday products ranging from time-release pills to disposable diapers. They have immense functional potential.

Whether the cell exploits such functional potential is the focus of the book's second half. I explore the possibility that the phase-transition could be a common denominator of cell function. By cell function I mean material transport, motility, division, secretion, communication, contraction—among other essential functions.

So there we stand. Driven by the notion of simplicity, I suggest that much of cell biological function may be governed by a single unifying mechanism—the phase-transition. This is a heady notion, which should be met with due skepticism. Nevertheless, it is a course that follows naturally if the cytoplasm and its working organelles are gel-like, for the phase-transition is the gel's central functional agent. Whether ordinary gels and cytoplasmic gels operate by the same working principle is the theme of this book.

***

The book is designed for those with minimal background in biology. It could have been written for experts, rich with theory and dense with detail. My experience, however, is that too-deep immersion often obscures the underlying logic—or lack of it. The strategy is to expose the material to scrutiny by focusing on the underlying principles. As such, the material should pose little difficulty for students in any of the physical, biological, or engineering sciences—graduate or undergraduate. In fact, the response to drafts convinces me that the material should be accessible to any intelligent person with open-minded curiosity.

To keep the text streamlined, I have taken a middle ground between the extensive referencing characteristic of scholarly works and the absence of any referencing characteristic of lay books. References seemed necessary in controversial areas, but I felt free to omit them in areas I felt were more standard. I hope this omission will be forgiven. I also hope to be forgiven for having glossed over work that may have consumed someone's lifetime, or for having failed to cite some work that may seem conflicting. I can assure you that any such omissions were not purposeful.

I may have conveyed the impression that everything contained in this book is original, but that is not the case. The physical features of the protein-water-ion complex have been elaborated early on by a cadre of scientific pioneers including Albert Szent-Györgyi, A. S. Troshin, and Gilbert Ling. Ling in particular has spent his long career courageously advancing scientific frontiers in the face of frequent derision for the apparent extremity of his views. In my estimate the evidence powerfully supports his constructs. Indeed, the threads of his ideas weave the very fabric of this book, particularly the early chapters. The material on phase-transitions rests on the seminal contributions of the late Toyo Tanaka of MIT, who elaborated the phase-transitions' underlying mechanisms and forecast their biological relevance. Largely, the book ties together natural principles advanced by others, adding a few thoughts here and there to integrate these principles and broaden their relevance.

The end result is a fresh view of how cells function. Fresh views do not sit well with scientific establishments and I am not naïve enough to think that the ideas offered here will be embraced with enthusiasm, even if some features may be deemed attractive. In the interest of progress, however, maintaining a competing paradigm alongside the orthodox one can be advantageous, as shortcomings in either can be more readily exposed. But the challenge of dual maintenance is not easy—no easier perhaps than the challenge of having two lovers and remaining faithful to both at the same time. It is nevertheless in this constructive spirit that this fresh paradigm is offered.

GHP
Seattle, November, 2000

Science is built on the premise that Nature answers intelligent questions intelligently; so if no answer exists, there must be something wrong with the question.

Albert Szent-Györgyi, *The Living State: With Observations on Cancer,* Academic Press, 1972.

SECTION I

TOWARD GROUND TRUTH

This section deals with generally accepted views of cell biology. By peeling back layers of assumption, it attempts to get at the core of truth. In the end, an unorthodox conclusion is drawn about the nature of the cell.

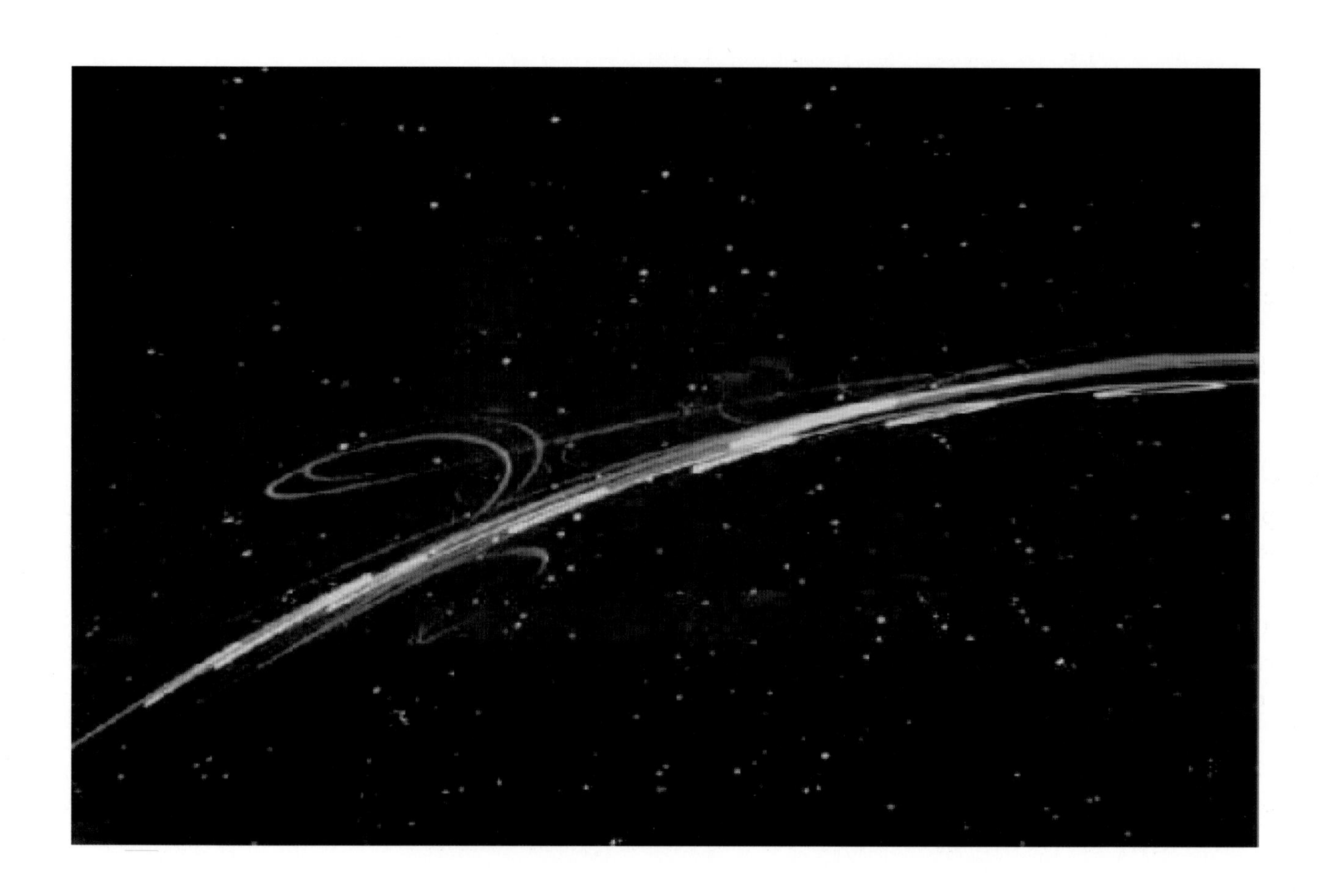

Epicycles

CHAPTER 1

DEBUNKING MYTHS

Long ago, scientists believed that the center of the universe was the earth: The sun could be seen to traverse the heavens, so it was logical to conclude that the earth must lie at the center point.

But this view eventually encountered difficulties. As the growth of mathematics increased the power of astronomy, it became possible to compute orbital pathways. The planets' paths around the earth turned out to be less simple than anticipated; each planet followed an orbit called an epicycle (Figure, opposite), which was sufficiently intricate to imply that something was surely amiss.

What was amiss is no longer a mystery. Although the persistent notion of an earth-centered universe may gratify our collective egos, Galileo showed that it was the sun that held this honor. With the sun at the solar system's center, orbital paths no longer required complex epicycles; they became a lot simpler. What had earlier seemed a reasonable hypothesis supported by seemingly indisputable visual observation, turned out to be dead wrong. A complicated paradigm was replaced by a simpler one.

IS LIFE REALLY ANY DIFFERENT NOW?

In the field of cell biology at least, complicated paradigms raise similar concern. On the surface everything seems to be in order. Virtually all known cellular processes are by now accounted for by well-described mechanisms: ions flow through channels; solutes are transported by pumps; vesicles are moved by motors; *etc.* For every problem there is a solution. But as we shall see as we probe beneath the surface of these solutions, a bewildering level of complexity hints at a situation that could parallel the epicycles.

I propose to step back and regroup. For genuine progress, foundational concepts must be unquestionably sound; otherwise an edifice of understanding may rise over a crevasse of uncertainty—no apparent problem until the edifice grows weighty enough to crack the foundation and tumble into the abyss. Firm ground needs to be identified.

I begin by considering two elements thought to be fundamental to cell function: membrane pumps and channels. Pumps transport solutes across the cell boundary against their respective concentration gradients. Channels permit the solutes to trickle back in the opposite direction. Through a balance between pump-based transport and channel-based leakage, the characteristic partitioning of solutes and ions is thought to be established.

Table 1.1. *Concentration of principal ions inside and outside of a typical mammalian cell*

Ion	Molecular Weight	Intracellular Concentration (mM)	Extracellular Concentration (mM)
Na	23	5-15	145
K	39	140	5
Cl	35.5	5-15	110

Thus, potassium concentration is relatively higher inside the cell, and sodium is relatively higher outside (Table 1.1).

That pumps and channels exist seems beyond doubt—or to put it more precisely, the existence of proteins with pump-like or channel-like features cannot be doubted. Genes coding for these proteins have been cloned, and the proteins themselves have been exhaustively studied. There can be no reason why their existence might be challenged.

Where some question could remain is in the functional role of these proteins. What I will be considering in this chapter is whether these proteins really mediate ion partitioning. Because a "pump" protein inserted into an artificial membrane can translocate an ion from one side of the membrane to the other, can we be certain that ion partitioning in the living cell necessarily occurs by pumping?

This task of checking this presumption may in this case be approached through the portal of historical perspective. Scientists on the frontier often dismiss history as irrelevant but in this particular instance a brief look into the trail of discovery is especially revealing.

ORIGINS

The emergence of pumps and channels was preceded by the concept of the cell membrane. The latter arose during the era of light microscopy, prior to the time any such membrane could actually be visualized. Biologists of the early nineteenth century observed that a lump of cytoplasm, described as a "pulpy, homogeneous, gelatinous substance" (Dujardin, 1835) did not mix with the surrounding solution.

To explain why this gelatinous substance did not dissolve, the idea arose that it must be enveloped by a water-impermeant film. This film could prevent the surrounding solution from permeating into the cytoplasm and dissolving it. The nature of the membranous film had two suggested variants. Kühne (1864) envisioned it as a layer of coagulated protein, while Schültze (1863) imagined it as a layer of condensed cytoplasm. Given the experimental limitations of the era, the nature of the putative film, still not visualizable, remained uncertain.

The idea of an invisible film was nevertheless attractive to many of the era's scientists, and was increasingly conferred with special attributes. Thus, Theodore Schwann (1839) viewed this film as "prior in importance to its contents." The membrane grew in significance to become the presumed seat of much of the cell's activity. Yet this view was not accepted by all. Max Schültze, often referred to as the father of modern biology, discounted the evidence for a cytoplasmic film altogether, and instead regarded cells as "membraneless little lumps of protoplasm with a nucleus" (Schültze, 1861). In spite of Schültze's prominence, the concept of an enveloping membrane held firm.

The modern idea that the membrane barrier might be semi-permeable came from the plant physiologist Wilhelm Pfeffer. Pfeffer was aware of the ongoing work of Thomas Graham (1861) who had been studying colloids, which are large molecules suspended indefinitely in a liquid medium—*e.g.*, milk. According to Graham's observations, colloids could not pass through dialysis membranes although water could. To Pfeffer, colloids seemed to resemble the cytoplasm. If the dialysis membrane were like the cell membrane, Pfeffer reasoned, the cell interior would not dissipate into the surrounding fluid even though the membrane might still be water-permeable. Thus arose the idea of the semi-permeable membrane.

Pfeffer took up the semi-permeable membrane idea and pursued it. He carried out experiments on membrane models made of copper ferrocyanide, which acted much like dialysis membranes in that they could pass water easily but solutes with great difficulty. It was on these experiments that Pfeffer based the modern cell-membrane theory (Pfeffer, 1877). The membrane at this stage was presumed permeable to water, but little else.

Although Pfeffer's theory held for some time, it suffered serious setbacks when substances presumed unable to cross the membrane turned out to cross. The first and perhaps most significant of these substances was potassium. The recognition, in the early twentieth century, that potassium could flow into and out of the cell (Mond and Amson, 1928; Fenn and Cobb, 1934), prompted a fundamental rethinking of the theory.

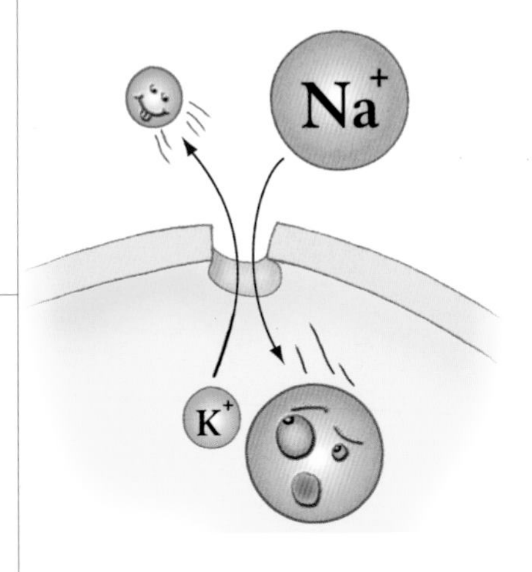

Figure 1.1. Atomic sieve theory. The size of the channel was postulated to be critical for passing potassium and blocking sodium. Observed passage of sodium compromised the theory.

ORIGIN OF THE CHANNEL

Faced with the need to explain the potassium-permeability issue, Boyle and Conway (1941) proposed an elegant solution: the potassium channel. Since the hydrated potassium ion was known to be smaller than the hydrated sodium ion, 3.8 Å vs. 5 Å, Boyle and Conway proposed transmembrane channels of critical size—large enough to pass potassium and its shell of associated water, but small enough to exclude sodium with its shell. The membrane was effectively a sieve that passed small ions, but excluded larger ones.

The Boyle-Conway atomic sieve theory was attractive in that it could also account for several known features of cell behavior without too much difficulty. It explained the accumulation of potassium inside the cell as an attraction to the cell's negatively charged proteins (a so-called Donnan effect). It explained the cell potential as arising from a charge separation across the membrane (a capacitive effect). And it accounted for the changes of cell volume that could be induced by changes of external potassium concentration (an osmotic effect). The sieve theory seemed to explain so much in a coherent manner that it was immediately granted an exalted status.

But another problem cropped up, perhaps even more serious than the first. The membrane turned out to be permeable also to sodium (Fig. 1.1). The advent of radioactive sodium made it possible to trace the path of sodium ions, and a cadre of investigators promptly found that sodium did in fact cross the cell boundary (Cohn and Cohn, 1939; Heppel, 1939, 1940; Brooks, 1940; Steinbach, 1940). This finding created a problem because the hydrated sodium ion was larger than the channels postulated to accommodate potassium; sodium ions should have been excluded, but they were not. Thus, the atomic-sieve theory collapsed.

Collapse notwithstanding, the transmembrane-channel concept remained appealing. One had to begin somewhere. A channel-based framework could circumvent the sodium problem with separate channels for sodium and potassium: if selectivity were based on some criterion other than size, then distinct channels could suffice. A separate channel for sodium could then account for the observed leakage of sodium ions into the cell.

But leakage of sodium introduced yet another dilemma, of a different nature. Sodium could now pass through the channel, flowing down its concentration gradient and accumulating inside the cell. How then could intracellular sodium remain as low as it is?

Origin of the Pump

The solution was to pump it out. In more-or-less the same manner as a sump-pump removes water that has leaked into your basement, a membrane pump was postulated to rid the cell of the sodium that would otherwise have accumulated inside.

The idea of a membrane pump actually originated before the sodium problem. It began at the turn of the last century with Overton, a prominent physiologist who had advanced the idea that the membrane was made of lipid. Realizing that some solutes could cross an otherwise impermeable lipid membrane, Overton postulated a kind of secretory activity to handle these solutes. Through metabolic energy, the membrane could thus secrete, or pump, certain solutes into or out of the cell.

The pump concept resurfaced some forty years later (Dean, 1941), to respond specifically to the sodium-permeability problem. Dean did not have a particular pumping mechanism in mind; in fact, the sodium-pump was put forth as the least objectionable of alternatives. Thus, Dean remarked, "It is safer to assume that there is a pump of unknown mechanism which is doing work at a constant rate excreting sodium as fast as it diffuses into the cell." With this, the sodium pump (later, the Na/K exchange pump) came decidedly into existence.

By the mid-twentieth century, then, the cell had acquired both channels and pumps. With channels for potassium and sodium, along with pumps to restore ion gradients lost through leakage, the cell's electrophysiology seemed firmly grounded. Figure 1.2 says it all.

Figure 1.2. *The sodium pump, adapted from a drawing by Wallace Fenn (1953), one of the field's pioneers.*

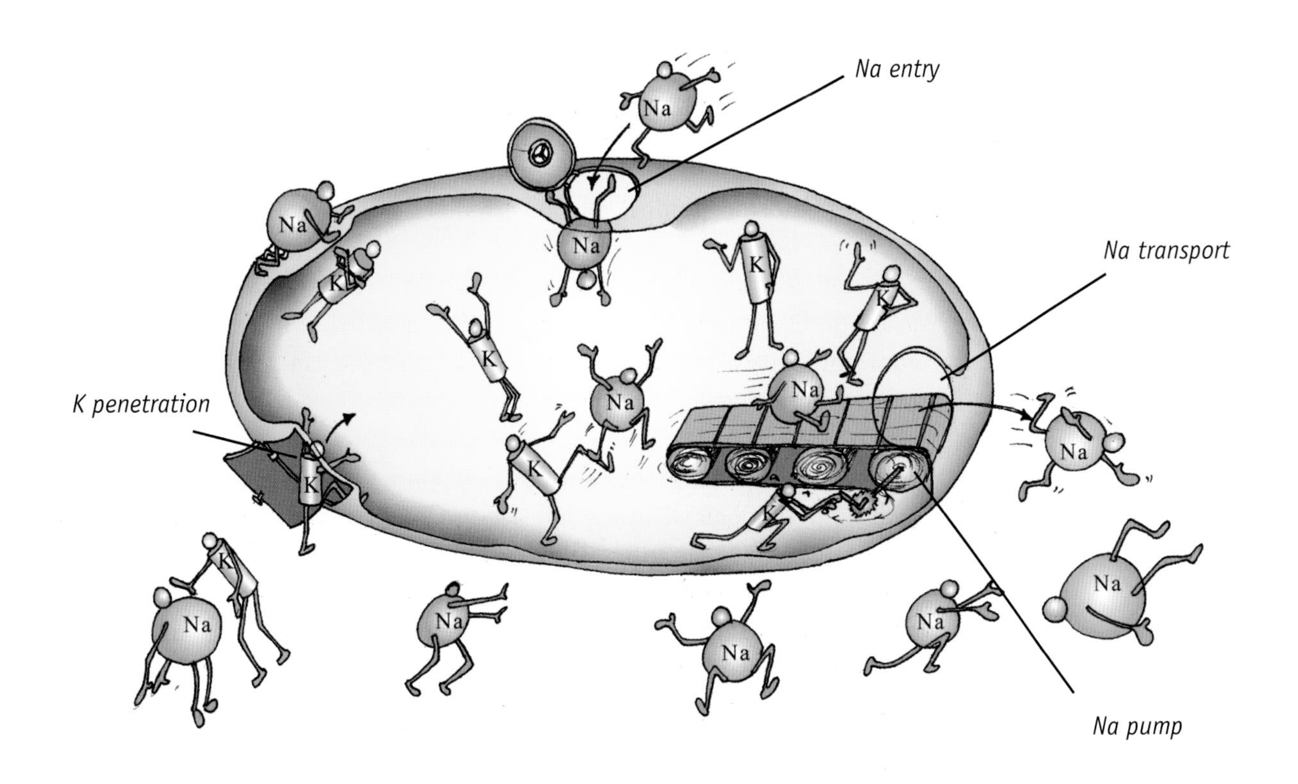

REFLECTIONS

Why have I dragged you through so lengthy a review? My purpose was to demonstrate how channels and pumps arose. They came into being not because some alert scientist stumbled upon them during a groundbreaking session at the electron microscope, but as *ad hoc* hypotheses needed to patch otherwise flagging theories. The channel arose when a putatively ion-impermeant membrane was found to pass potassium; the channel could pass potassium while properly excluding sodium and other larger hydrated ions. Then, sodium was found to enter the cell and instead of reconsidering the channel concept, a second channel specific for sodium was postulated. Sodium permeability also implied a persistent leak into the cell. To keep gradients from collapsing, a sodium pump was postulated.

Once this hypothetical framework gained a foothold, it expanded boundlessly. For the same reason that a sodium pump was needed, it became evident that pumps for other solutes were needed as well. Virtually all of the cell's solutes partition far out of electrochemical equilibrium (Stein, 1990) and therefore need to be pumped. Hydrogen-ion pumps, calcium pumps, chloride pumps, and bicarbonate pumps to name a few, soon came into being over and above the postulated sodium/potassium pumps. Yet, even at the height of this intense activity, Glynn and Karlish (1975) in their classic review had to reluctantly admit that notwithstanding an enormous thrust of experimental work on the subject, still no hypothesis existed to explain how pumps pump.

The channel field exploded similarly. With the advent of the patch-clamp technique (see below) in the late 1970s, investigators had gained the capacity to study what appeared to be single ion channels. It seemed for a time that new channels were being identified practically monthly, many of them apparently selective for a particular ion or solute. The number of channels has risen to well over 100. Even water channels have come into being (Dempster *et al.*, 1992). Elegant work was carried out to try to understand how channels could achieve their vaunted selectivity (Hille,

1984). At least some channels, it appeared, could pass one ion or solute selectively, while excluding most others.

Given the astonishing expansion of activity in these fields, could there be any conceivable basis for doubt? Mustn't the concepts of pumping and channeling be as firmly grounded as any biological principle?

It is tempting to answer by placing side-by-side the key experiments originally adduced to confirm pumping and channeling together with the published challenges of those experiments (Troshin, 1966; Hazlewood, 1979; Ling, 1984, 1992). I hesitate to recapitulate that debate because the challenges are largely technical. Readers willing to invest time in acquiring familiarity with technical details are invited to consult these sources and consider whether the published concerns are valid or not.

Another approach is to consider evidence that could potentially lie in conflict with these concepts. Unlike mathematical theorems, scientific theories cannot be proven. No matter how much evidence can be marshaled in support of a theory, it is always possible that some new piece of evidence will be uncovered that does not fit, and if such evidence is both sound and fundamental, the theory may require reconsideration. As we shall see in the next sections, certain basic questions concerning pumps and channels have not yet been adequately dealt with.

Channels Revisited

The existence of single ion channels appeared to be confirmed by groundbreaking experiments using the patch-clamp technique. In this technique the tip of a micropipette is positioned on the cell surface. Through suction, a patch of membrane is plucked from the cell and remains stuck onto the micropipette orifice (Fig. 1.3A). A steady bias voltage is placed across the patch, and the resulting current flow through the patch is measured. This current is not continuous; it occurs as a train of discrete pulses. Because the pulses appear to be quantal in size, each pulse is

A

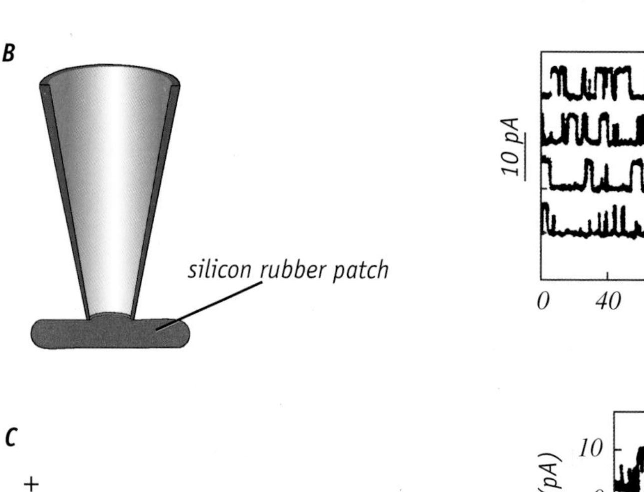

B

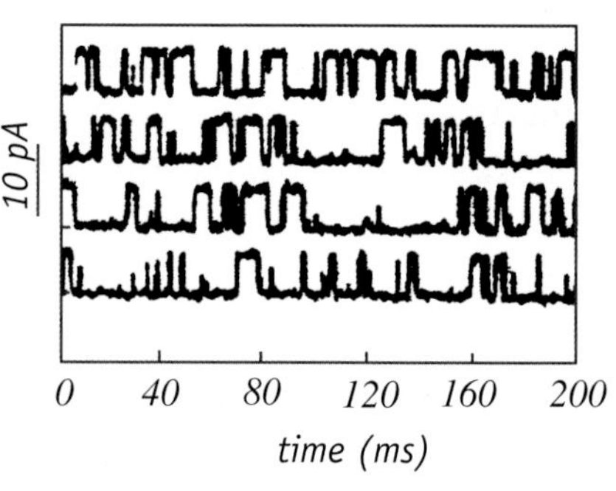

C

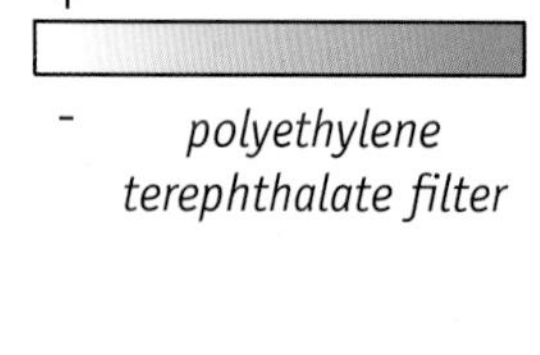

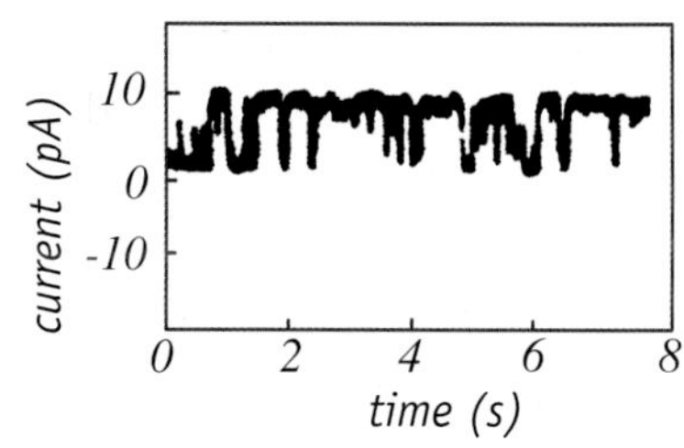

D

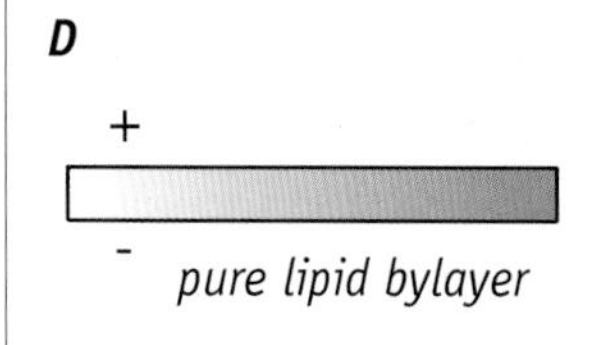

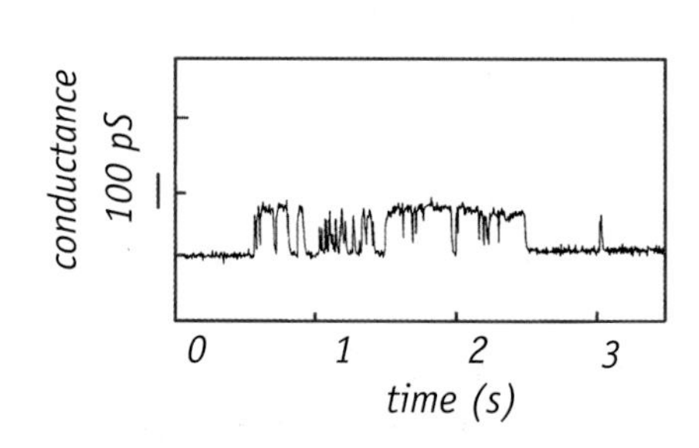

Figure 1.3. *"Single channel" currents recorded in situations depicted at left of each panel: (A) after Tabcharani* et al. *(1989); (B) after Sachs and Qin (1993); (C) after Lev* et al. *(1993); (D) after Woodbury (1989). Note the similarity of experimental records, implying that the discrete currents are not necessarily related to features specific to biological channels.*

assumed to correspond to the opening of a single ion channel.

This dazzling result has so revolutionized the field of membrane electrophysiology that the originators of the technique, Erwin Neher and Bert Sakmann, were awarded the Nobel Prize. The observation of discrete events would seem to confirm beyond doubt that the ions flow through discrete channels.

Results from the laboratory of Fred Sachs, on the other hand, make one wonder. Sachs found that when the patch of membrane was replaced by a patch of silicon rubber, the discrete currents did not disappear (Sachs and Qin, 1993); they remained essentially indistinguishable from those measured when the membrane was present (Fig. 1.3B). Even more surprisingly, the silicon rubber sample showed ion-selectivity features essentially the same as the putative membrane channel.

A similarly troubling observation was made on polymer samples (Lev *et al.*, 1993). Current flow through synthetic polymer filters was found to be discrete, just as in silicon rubber (Fig. 1.3C). The filters also showed features commonly ascribed to biological channels such as ion selectivity, reversal potential, and gating. Yet, the sample was devoid of any protein or lipid.

In yet another set of experiments, channel-like behavior was observed in pure lipid-bilayer membranes (Woodbury, 1989). Following brief exposure to large concentrations of lipid vesicles ejected from a pipette tip approximately 0.5 mm distant, these membranes showed typical channel-like fluctuations (Fig. 1.3D). Conductance changed in ways usually considered to be indicative of reconstituted protein channels—including step conductance changes, flickering, ion selectivity, and inactivation. But no channels were present; the membranes contained only lipid.

What are we to do with such observations? It is clear from these three studies that the discrete currents previously taken to confirm the existence of single biological channels seem to be general features of current flow through small samples. The currents presumably arise from some common feature of these specimens that is yet to be determined, but evidently not from single channels since they are absent. The channels may exist—but the prime evidence on which their existence is based is less than conclusive.

Ironically, the silicon-rubber test had actually been carried out as a control in the original patch clamp studies (Neher *et al.*, 1978). The authors did sometimes note "behavior contrary" to what was expected (p. 223); but such behavior was dismissed as having arisen from irregularities of the pipette tip. The possibility that small samples in general might give rise to channel-like behavior was apparently not considered.

Setting aside the above-mentioned concern, a second point to consider is the manner in which the channel achieves its specificity. Channels exist for each one of the cell's ions; additional channels exist for amino acids, peptides, toxins, and sugars, most of these being otherwise unable to cross the lipid bilayer; and as I mentioned, there are also channels for water. Thus, a plethora of channels exists, most engineered to be solute-specific. How is such specificity achieved?

To explain such exquisite specificity, models of some complexity have evolved (Hille, 1984; 1992). One model contains 16 different transition states, plus additional sub-states. Another contains 27 states. Most models are sufficiently complex that the solution requires numerical methods. Indeed, calculating the trajectory of a molecule diffusing through a channel during a 100-picosecond time window is the work of a supercomputer. Thus, models of daunting complexity are required for understanding the channels' apparent selectivity. It is not a simple process.

The naive question nevertheless lingers: How is it that small solutes do not pass through large channels? To imagine how one or two small ions might be excluded from a large channel as a result of a distinctive electric field distribution is not too difficult to envision. But the theory implies that smaller solutes should be excluded as a class—otherwise, independent channels for these solutes would not be required. This enigma harks of the dog-door analogy (Fig. 1.4). Why bother adding a cat door, a ferret door, a hamster door and a gerbil door when these smaller animals could slip readily through the dog door? Must some kind of repellent be added for each smaller species?

The seriousness of the size problem is illustrated by considering the hydrogen ion. The hydrogen ion is only about 0.5 Å in diameter (the hydrated ion somewhat larger). The sodium-channel orifice is at least 3 - 5 Å across, and channels specific for some of the larger solutes must have a

Figure 1.4. *Dog-door analogy. Door is large enough to pass all smaller species.*

minimum orifice on the order of 10 - 20 Å. How is it possible that a 20 Å channel could exclude a diminutive hydrogen ion? It is as though a two-foot sewer pipe that easily passes a beach ball could at the same time exclude golf balls, as well as tennis balls, billiard balls, *etc.*

Arguably, the situation is not so black and white. Textbook depictions of the channel as a hollow tube oversimplify the contemporary view of the channel as a convoluted pathway; and the process of selectivity is thought to rest not on size *per se* but on some complex interaction between the solute's electric field and structural features of the channel's filter (*e.g.*, Doyle *et al.*, 1998). Also, channel selectivity is not absolute (Hille, 1972). Nevertheless, the issue of passing only one or a few among a field of numerous possible solutes including many smaller ones remains to be dealt with in a systematic manner. And the issue of non-biological samples producing single-channel currents certainly needs to be evaluated as well. What could all this imply?

Pumps Revisited

Like channels, pumps come in many varieties and most are solute-specific. The number easily exceeds 50. The need for multiple pumps has already been dealt with: unless partitioning between the inside and outside of the cell is in electrochemical equilibrium, pumping is required. Because so few solutes are in equilibrium, one or more pumps are necessary for each solute.

A question that arises is how the cell might pump a solute it has never seen. Antibiotics, for example, remain in high concentration outside the bacterial cell but in low concentration inside. Maintaining the low intracellular concentration implies the need for a pump, and in fact, a tetracycline pump for *E. coli* has been formally proposed (Hutchings, 1969). A similar situation applies for curare, the exotic arrow poison used experimentally by biophysicists. Because curare partitioning in the muscle cell does not conform to the Donnan equilibrium, a curare pump has been proposed (Ehrenpries, 1967). To cope with substances it has never seen, the cell appears to require pumps over and above those used on a regular basis—on reserve.

How is this possible? One option is for existing pumps to adapt themselves to these new substances. But this seems illogical, for if they could adapt so easily why would they have been selective to begin with? An alternative is for the cell to synthesize a new pump each time it encounters a foreign substance. But this option faces the problem of limited space: Like the university parking lot, the membrane has just so many spaces available for new pumps (and channels). Given chemists' ability to synthesize an endless variety of substances—10 million new chemical substances have been added to the American Chemical Society's list of molecules during the last quarter century alone (*N. Y. Times*, Feb 22, 2000)—how could a membrane already crowded with pumps and channels accommodate all that might eventually be required? Could a membrane of finite dimension accommodate a potentially infinite number of pumps?

A second question is how the cell musters the energy required to power all of its pumps. Where might all the ATP come from? Since ions and other solutes cross the membrane continually even in the resting state (in theory because of sporadic channel openings), pumps must run continuously to counteract these leaks. Pumping does not come free. The sodium pump alone has been estimated, on the basis of oxygen-consumption measurements, to consume 45 - 50% of all the cell's energy supply (Whittam, 1961). Current textbooks estimate a range of 30 - 35%.

To test whether sufficient energy is available to power pumping, a well-known experiment was carried out long ago by Ling (1962). Ling focussed on the sodium pump. The idea was to expose the cell to a cocktail of metabolic poisons including iodoacetate and cyanide, and to deprive it of oxygen—all of which would deplete the cell of its energy supply and effectively pull the pump's plug. If these pumps had been responsible for maintaining sodium and potassium gradients, the gradients should soon have collapsed. But they did not. After some eight hours of poison exposure and oxygen deprivation, little or no change in cellular potassium or sodium was measurable.

Ling went on to quantitate the problem. He computed the residual energy—the maximum that could conceivably have been available to the cell following poisoning. This residual was compared to the energy required to sustain the ion gradient, the latter calculable from the known sodium-leak rate. Using the most generous of assumptions, a conservative estimate gave an energy shortfall of 15 to 30 times (Ling, 1962). The pump energy needed to sustain the observed gradient, in other words, was concluded to be at least 15 to 30 times larger than the available energy supply.

This conclusion stirred a good deal of debate. The debate was highlighted in a *Science* piece written by the now well-known science writer Gina Kolata (1976), a seemingly balanced treatment that gave credence to the arguments on both sides. Kolata cited the work of Jeffrey Freedman and Christopher Miller who had challenged Ling's conclusion about the magnitude of the energy shortfall. Ling's late-coming rebuttal (1997) is a compelling "must-read" that considers not only this specific issue but also the process of science. The energy-shortfall claim was nevertheless left to gather dust with the advent of pump-protein isolation—forgotten by all but a modest cadre of researchers who have remained steadfastly impressed by the arguments (*cf.* Tigyi *et al.,* 1991).

In retrospect, any such niggling debate about the magnitude of the sodium-pump-energy shortfall seems academic, for it is now known that numerous other pumps also require power. Over and above sodium and potassium, the cell membrane contains pumps for calcium, chloride, magnesium, hydrogen, bicarbonate, as well as for amino acids, sugars, and other solutes. Still more pumps are contained in organelle membranes inside the cell: In order to sustain intra-organelle ion partitioning, organelles such as the mitochondrion and endoplasmic reticulum contain pumps similar to those contained in the surface membrane. Given leak rates that are characteristically proportional to surface area, we are not speaking here of trivial numbers of pumps: Liver cell mitochondria contain 20 times the surface area of the liver cell membrane (Lehninger, 1964), and the area of the muscle's sarcoplasmic reticulum is roughly 50 times that of the muscle cell membrane (Peachey, 1965). Membranes of such organelles must therefore contain pumps in numbers far higher than those of the cell membrane—all requiring energy.

In sum, pumping faces obstacles of space and energy. The membrane's size is fixed but the number of pumps will inevitably continue to grow. At some stage the demand for space could exceed the supply, and what then? Pumping also requires energy. The Na/K pump alone is estimated to consume an appreciable fraction of the cell's energy supply, and that pump is one of very many, including those in internal membranes. How is the cell to cope with the associated energy requirement?

COULD CHANNEL AND PUMP PROTEINS PLAY ANOTHER ROLE?

The sections above have outlined certain obstacles faced by the pump – channel paradigm. But proteins exhibiting pump-like or channel-like behavior have been isolated and their existence needs to be explained. If not specifically for pumping and channeling, why might they be present?

One plausible hypothesis is that they exist for some different purpose. Given their superficial location, "pump" and "channel" proteins could trigger a chain of events leading ultimately to action in an intracellular target. Conformational changes are known to occur not only in pump and receptor proteins but in channel proteins as well (Kolberg, 1994). If such changes were to propagate inward, the pump or channel protein would effectively play the role of a receptor.

A scenario of this sort need not contradict the proteins' classical "pumping" or "channeling" action: As the conformational change proceeds, any bound ion will shift in space along with the protein (Fig. 1.5). If the charge shifts against the voltage gradient, the result will be interpreted as pumping; if it shifts with the gradient the ion will be presumed to have passed through a channel. Thus, pumping or channeling would be a natural inference, even though the protein's functional role is as neither a pump nor a channel.

Figure 1.5. *Translocation of charges across the membrane will be registered as current pulses.*

A good example to illustrate this kind of behavior is colicin Ia, a toxin molecule that insinuates into bacterial membranes. As it does, it is thought to create a channel that allows ions to pass, thereby collapsing ion gradients and killing the cell. The protein's action is associated with substantial conformational change (Fig. 1.6); some 50 amino acids flip from one side of the membrane to the other—along with their bound ions. Such charge shift would constitute a pulse of current similar to the currents depicted in Figure 1.3; hence, channeling is implicit. Whether such "channel" current mediates the protein's toxic function, however, is less certain. Toxicity could as well lie in the protein's altered configuration, inhibiting some vital process through interaction with cell proteins.

Another example is rhodopsin. Rhodopsin is a retinal receptor molecule that undergoes conformational change in order to signal the presence of light. Rhodopsin exists in another form called bacteriorhodopsin. Also driven by absorbed light energy, bacteriorhodopsin can translocate protons across the bacterial cell membrane. Thus, rhodopsin is a light-driven receptor while bacteriorhodopsin is presumed to be a light-driven pump. Again, the charge movement observed in bacteriorhodopsin may not necessarily be the main event—the protein could function as a receptor of

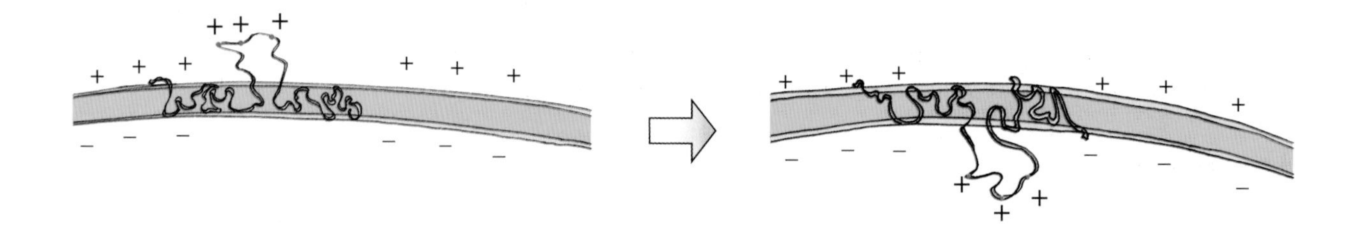

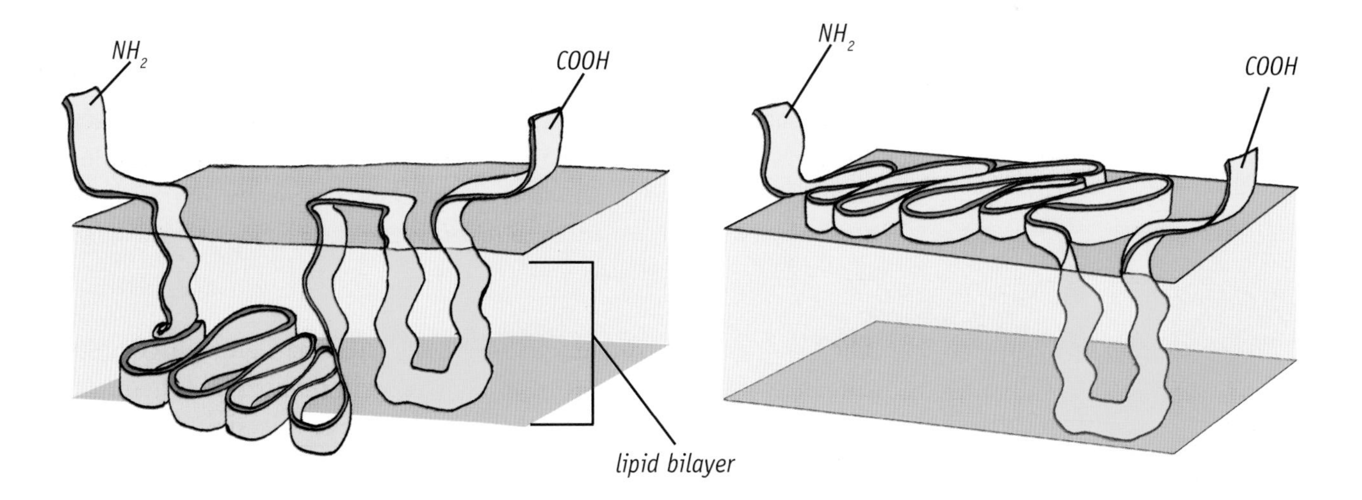

light just as rhodopsin does in the retina, triggering a response through conformational change (Lewis *et al.,* 1996). As with the channeling action of colicin Ia, the pumping action of bacteriorhodopsin might then be an incidental byproduct, and not necessarily the primary event.

Figure 1.6. *Channel protein that opens by flipping from one side of the membrane to the other. After Slatin* et al. *(1994).*

Whether the observed pumping or channeling seen in such molecules could be of functional significance would depend on their magnitude: If the "pump" molecule translocates relatively few ions, its contribution to cellular ion separation would be small. And if the "channel" fails to carry the lion's share of ion traffic through the membrane, it too would play little or no role in partitioning. Even though experimentally observable, pumping and channeling by these proteins would then remain functionally insignificant—much like the heat generated by a light-bulb.

It seems, then, that this section's conundrum may be resolvable. Although tests imply that the proteins under consideration can "pump" or "channel," such processes may be incidental to the proteins' main functional role as receptors. Receptor proteins are often closely linked to pump and channel proteins in order to "modulate" their activity; here they merge into a single unit whose contributions to ion partitioning may be entirely secondary.

Conclusion

Some bold leaps have been taken in this chapter. We began by granting ourselves license to explore two basic elements of modern cell biology—channels and pumps. As we reviewed their origin, we found that they arose as postulates, put forth to rescue attractive theories that would otherwise have collapsed.

The framework surrounding those postulates then grew in complexity. Channels and pumps multiplied rampantly in number and their features grew devilishly intricate. In order to achieve selectivity, the channel needed many states and sub-states; and the pump was required to handle substances it had never before seen. These complexities hinted that something could well be amiss.

Some things were indeed amiss, or at least questionable. For the channels it was the lead provided by the patch-clamp experiments. Those experiments had been taken as proof of the existence of discrete biological channels, but that evidence has been thrown into doubt by the demonstration that similar results could be obtained when channels were absent. Also considered was the selectivity issue. It was not clear how the channel could pass one solute primarily, while systematically excluding others of the set—particularly its smaller members (the dog-door problem).

Questions were also raised about pumps. One issue is the means by which a membrane of finite surface area could accommodate a continually growing number of pumps (and channels)—what happens when space runs out? A second issue is the nettlesome one of energy-balance: if the cell's energy supply is marginally adequate to handle sodium pumping, what resources are available to power all of the rest of the many pumps?

Although this chapter's goal was to begin constructing a functional edifice, the challenge of finding solid foundational ground has not yet been met. The foundation remains uncertain. Nor can the soundness of sub-structural layers be presumed, for the pump and channel questions seem profound enough to hint that the problems could originate more deeply.

The Death of Marat, 1793, by Jacques Louis David.

CHAPTER 2

THE CROAK OF THE DYING CELL

Given the turn of events in Chapter 1, we need to ask whether the uncertainties end there. Pumps and channels did not emerge in a vacuum. They arose out of an established conceptual framework and it may be that the original sin lies within the framework—much like the epicycles.

The framework in question is the cell membrane. Pumps came into being when a membrane that was presumed impermeant to solutes was found to be permeant. At first it was a single solute, then another, and then additional ones. Instead of abandoning the notion of impermeance, the passage of each solute was accommodated by presupposing another channel (and pump). Serious consideration was not given to the alternative possibility—that the continuous barrier framework itself might be erroneous.

When dealing with membranes, I recognize we are treading on hallowed ground. The continuous phospholipid-barrier concept has become so deeply ingrained in modern thinking that merely putting it on the table for discussion seems akin to reconsidering the virtues of motherhood. Yet, as a major limb of the logic tree, the issue cannot be ignored.

To approach this issue, we consider what happens when the membrane is disrupted. If ion partitioning (Table 1.1) requires a continuous barrier, violating the barrier should collapse these critical gradients. The violated cell should also lose enzymes and fuel, metabolic processes should grind to a halt, and the cell should be brought quickly to the edge of death. On the other hand, if the barrier were not continuous to begin with then membrane disruption could prove relatively inconsequential.

Wanton Acts

To disrupt the membrane experimentally, scientists have developed means uncannily similar to those developed to inflict wounds on humans, namely cellular swords and guns. Such instruments are not concocted by enraged scientists bent on revenge; they are designed to probe. Consider the three implements shown in Figure 2.1:

• The *microelectrode* (or micropipette) is a tapered glass cylinder filled with an electrolyte solution. The tip is plunged into the cell in order to probe its electrical properties. The microelectrode tip may seem diminutive by conventional standards, but to the 10-μm cell, invasion by a 1-μm probe is not entirely dissimilar to a human torso being impaled by a fence post.

• *Electroporation* is a widely used method of punching multiple holes in the cell membrane. The holes are created by shotgunning the cell with a barrage of high-voltage pulses, leaving the membrane riddled with orifices large enough to pass genes, proteins and other macromolecules.

• The *patch-clamp* method involves the plucking of a 1-μm patch of membrane from the cell for electrophysiological investigation.

Although these insults may cause fatal injuries in some cells, the surprise, as we see below, is that they are not consistently consequential.

Consider the microelectrode plunge. Not only does impalement violate the membrane, but it also mangles organelles unfortunate enough to lie along the track of penetration. The anticipated surge of ions, proteins, and metabolites might be thwarted if the hole were kept plugged by the microelectrode shank, but this cannot always be the case. For example,

Stuart Taylor and colleagues have routinely used micropipettes to microinject calcium-sensitive dyes at multiple sites along the isolated muscle cell. Achieving this with a single micropipette requires repeated withdrawals that leave micron-sized holes. Yet for up to several days after injection, the cell continues to function normally (Taylor *et al.,* 1975).

The results of patch removal are similar. Here again, the hole in question is roughly a million times the size of the potassium ion. My colleagues, Guy Vassort and Leslie Tung, who routinely do these kinds of experiments, tell me that following removal of the 1-µm patch, the 10 µm isolated heart cell will frequently live on and continue beating.

A possible explanation for survival in all of these situations is that the wound rapidly reseals. With rapid healing, the deadly surge of ions could be stemmed. Residual membrane could be imagined to spread over the damaged zone as soap film spreads over the surface of water, covering the wound and thereby resealing the cell. By this "band-aid" mechanism, the wounded cell could be rescued.

The difficulty with this otherwise attractive argument is that the membrane is not at all like a soap film. The soap film may be many layers

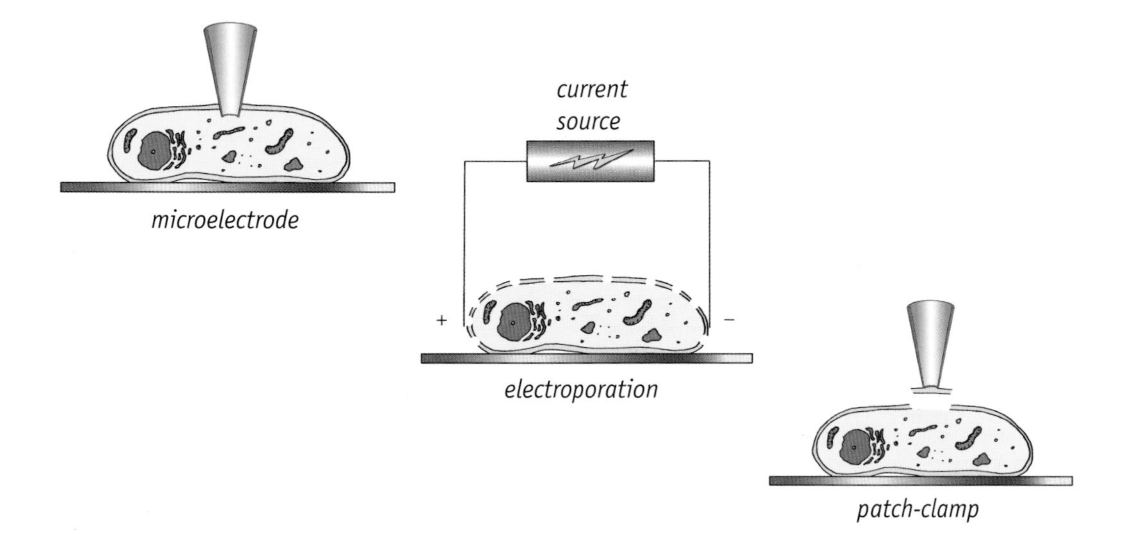

Figure 2.1. *Three methods of poking holes in the cell membrane.*

thick, and can spread by thinning. The membrane is a single lipid bilayer—it can spread only if constituent molecules spread. But X-ray diffraction (Korn, 1966) and fluorescence studies (Discher *et al.*, 1994) reveal a constant molecular packing density. Although molecules may diffuse within the bilayer, they cannot separate appreciably without causing membrane disruption. And if they could, it would still be necessary to find a reason why they should: what might induce the bilayer to spread?

In fact, the absence of resealing is confirmed by direct evidence. In the case of the withdrawn microelectrode, the hole can be seen either by direct microscopic observation (Nickels, 1970) or with even greater clarity if the withdrawing microelectrode leaves behind a deposit of electron-dense sediment (Nickels, 1970; 1971). The potential for resealing has also been examined in cut cells. Sectioned cells survive for use in electrophysiological studies, but electron micrographs show no evidence of membrane resealing, either in muscle cells (Cameron, 1988) or in giant axons (Krause *et al.*, 1994). Radical changes certainly ensue—exocytotic vesicle fusion may lead to some membrane replenishment (Bi *et al.*, 1995); and near the wound there may be a combination of vesicle accumulation (Krause *et al.*, 1994), cytoskeletal/vesicular coagulation (Fishman *et al.*, 1990; Spira *et al.*, 1993) and vacuole accumulation (Casademont *et al.*, 1988), but electron micrographic evidence confirms that the wound is not covered by a new membrane.

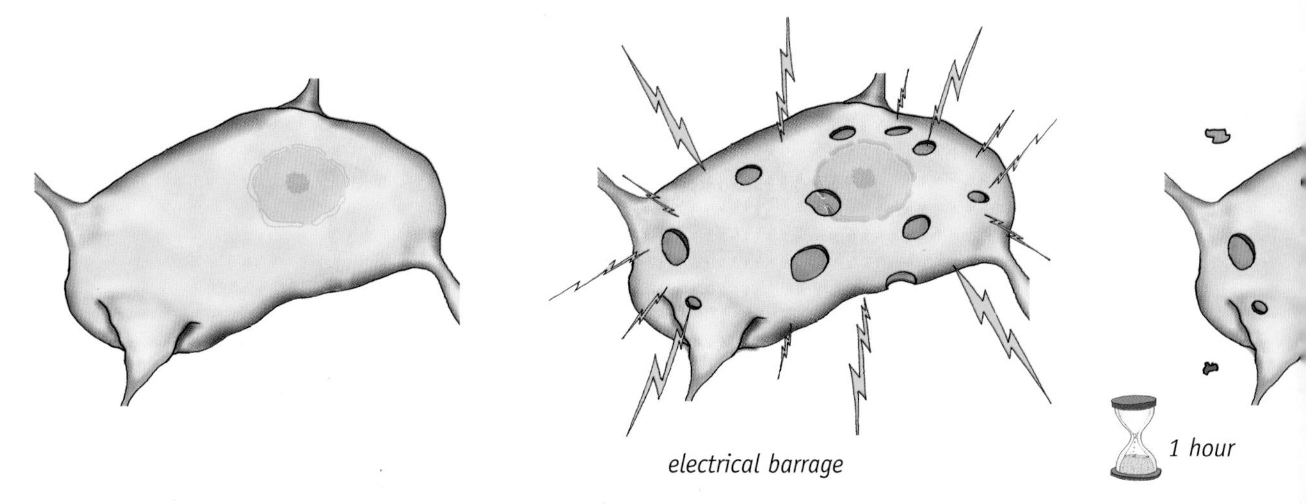

Long-term orifices are also confirmed during electroporation. Electroporation is a technique developed principally to transfer foreign molecules into cells (Fig. 2.2). Although permeability can sometimes be short-lived, this is not necessarily the rule. When molecules are introduced into the bath well after the end of the electrical barrage, entry into the cell is still possible. The window of opportunity depends on molecular size. In the case of huge molecules such as DNA, substantial penetration is observed when the molecules are added from 20 minutes to an hour after the shock terminates (Xie *et al.*, 1990; Klenchin *et al.*, 1991). Albumin, a common extracellular protein, can also penetrate for approximately one hour (Prausnitz *et al.*, 1994), and for smaller substances with molecular mass on the order of 1,000 Daltons, post-electroporation flow can occur for hours to days (Schwister and Deuticke, 1985; Serpersu *et al.*, 1985). Thus, the pores can remain open for long periods during which substances thousands of times the size of an ion can pass.

Figure 2.2. *Substances delivered substantially after the period of electroporation are able to enter the cell.*

Clearly, then, wanton acts such as punching holes in the membrane do not necessarily wreak havoc within the cell. The holes are huge relative to the ion. Yet, there is no evidence that the cell really cares. In spite of stringent ion-concentration requirements, the cell can sail through such insults with little sign of any abnormal behavior.

If the examples above seem too technical, consider the behavior of the common alga *Caulerpa,* a single cell whose length can grow to several

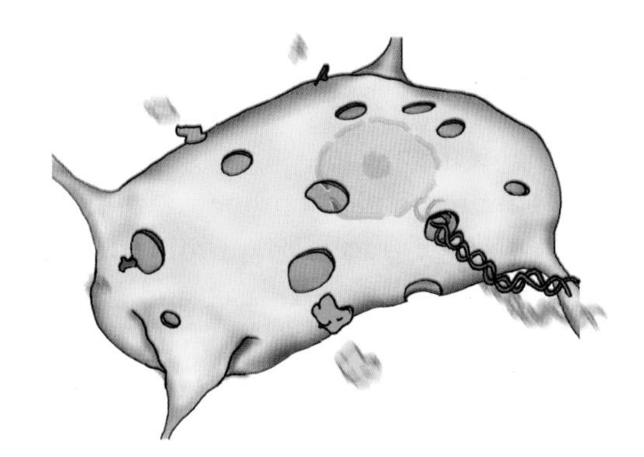

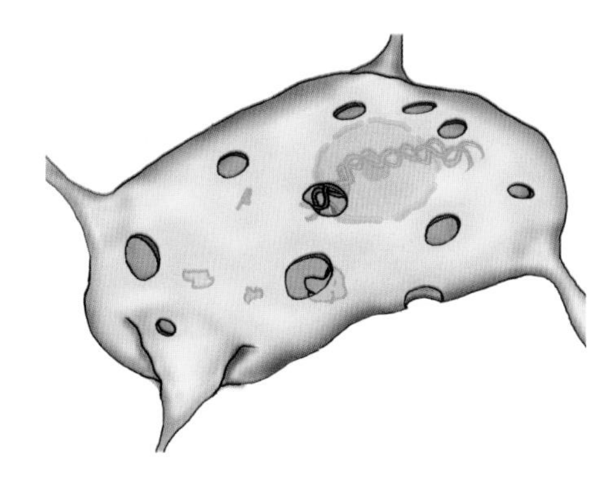

meters. This giant cell contains stem, roots, and leaves in one cellular unit undivided by internal walls or membranes (Jacobs, 1994). Although battered mercilessly by pounding waves and gnawed on relentlessly by hungry fish, this single cell cares not a whit; such breaches of integrity do not threaten its survival. In fact, deliberately cut sections of stem or leaf will grow back into entire cells. Severing the membrane is irrelevant.

Yet another example lies within the domain of experimental genetics, where otherwise genteel scientists will brutally slice innocent cells in two in order to monitor the fates of the respective fractions. When cultured epithelial cells are sectioned by a sharp micropipette, the non-nucleated fraction survives for 1-2 days while the nucleated, centrosome-containing fraction survives indefinitely and can go on to produce progeny (Maniotis and Schliwa, 1991). The cell does not seem to mind that it had just been sliced like a tomato. Sectioned muscle and nerve cells survive similarly (Yawo and Kuno, 1985; Casademont *et al.*, 1988; Krause *et al.*, 1984).

Then, there is the crawling cell that tears itself apart as it journeys onward (*see* Preface). Leaving behind a trail of cellular fragments is the rule rather than the exception (Chen, 1981). In the case of fibroblast cells, fragments down to 2% of initial volume continue to show standard behaviors such as ruffling, blebbing, filopodia production, and contact avoidance (Albrecht-Buehler, 1980). The fragments remain "alive" for up to eight hours.

Finally, and perhaps not surprisingly in light of all that has been said, ordinary mammalian cells are continually in a "wounded" state. Cells that suffer mechanical abrasion such as skin cells, gut endothelial cells and muscle cells are particularly prone to wounds—as confirmed by entry into the cell of large tracers that ordinarily fail to enter, such as serum albumin, horseradish peroxidase, and dextran (mol. wt. 10,000). Such wounded cells appear otherwise structurally and functionally normal (McNeil and Ito, 1990; McNeil and Steinhardt, 1998). The fraction of wounded cells in different tissues is variable (Table 2.1). In cardiac muscle cells it is ~20%, but under isoproterenol stimulation the fraction rises to 60% (Clarke *et al.*, 1995). Thus, tears in the cell membrane are neither exotic nor arcane; they occur as a common event in normal, physiologically functioning tissue.

It appears we are stuck on the horns of a dilemma. If we insist that a continuous barrier envelops the cell, we need to reconcile the aforementioned series of observations and we need to explain why breaching the barrier is not more consequential than it seems to be. The anticipated collapse of ion gradients should quickly destroy cellular proteins (Choi, 1988; Berridge, 1994). On the other hand, if we concede that the barrier may not be continuous so that creating yet another opening makes

Table 2.1. *Cell wounding under physiological conditions (from McNeil and Steinhardt, 1997).*

Organ	Cell types	Percentage wounded
skeletal muscle	skeletal muscle cells	5-30
skin	epidermal cells	3-6
G.I. tract	epithelial cells	–
cardiac muscle	cardiac myocytes	20
aorta	endothelial cells	6.5
inner ear	auditory hair cells	–

little difference, we then face an obstacle of a different nature. We challenge the long-held dogma of the continuous barrier, as well as the evidence on which the continuous barrier concept is presumably based.

Just how strong is such evidence?

MEMBRANE CONTINUITY?

The continuous phospholipid-membrane concept arose during the era of light microscopy. Because ordinary light microscopes cannot resolve structures on the order of 10 nm, what was identified as a membrane was probably the dense underlying cytoskeletal layer, which is 100 times thicker. The cytoskeleton is a subject of significance, to which considerable attention will be devoted (Chapter 10). The fact that it may appear continuous in the light microscope implies nothing about molecular scale continuity—or indeed whether the unseen lipid-bilayer might be continuous.

With time, the continuous phospholipid-membrane concept nevertheless drew broad support. Lipid arrays formed spherical vesicles that seemed analogous to cell membranes. Electron microscopy revealed the cell periphery to have a characteristically trilaminar structure that was inferred to correspond to the lipid bilayer. And freeze-fracture images showed striking images of the membrane's faces. What had been inferred early on from the light microscope seemed all but confirmed by these methods.

At the same time, it was becoming clear that the cell membrane was by no means phospholipid alone. With the isolation of membrane-based proteins, it became clear that the protein content of the membrane was appreciable. The current textbook estimate for typical membranes is on the order of 50%. In the inner membranes of mitochondria and chloroplasts, the ratio by weight of protein to lipid rises to 3:1, and in certain bacterial membranes the ratio is as high as 4:1 or 5:1 (Korn, 1966, Table

1). Thus, the popular idea of a lipid bilayer studded with occasional proteins is misleading. The generic membrane is as much a protein structure with lipid inclusions as a lipid structure with protein inclusions. See Figure 2.3.

For some, this revelation may come as a surprise. The idea of a continuous phospholipid membrane is so deeply ingrained in current thinking that it is difficult to assimilate the fact that membranes can comprise 50% - 80% protein. I venture to say that if the protein had been identified before the lipid, the rubric of a protein membrane might have been adopted in place of the lipid membrane.

But the "protein" membrane begs the issue of continuity. Membrane proteins are typically large molecules that fold back upon themselves many times as they course through the membrane. The resulting crevices create natural flow pathways like cracks in concrete, through which water and ions can seep. These proteins impact nearby membrane-lipid molecules as well, altering their natural arrangement and creating additional

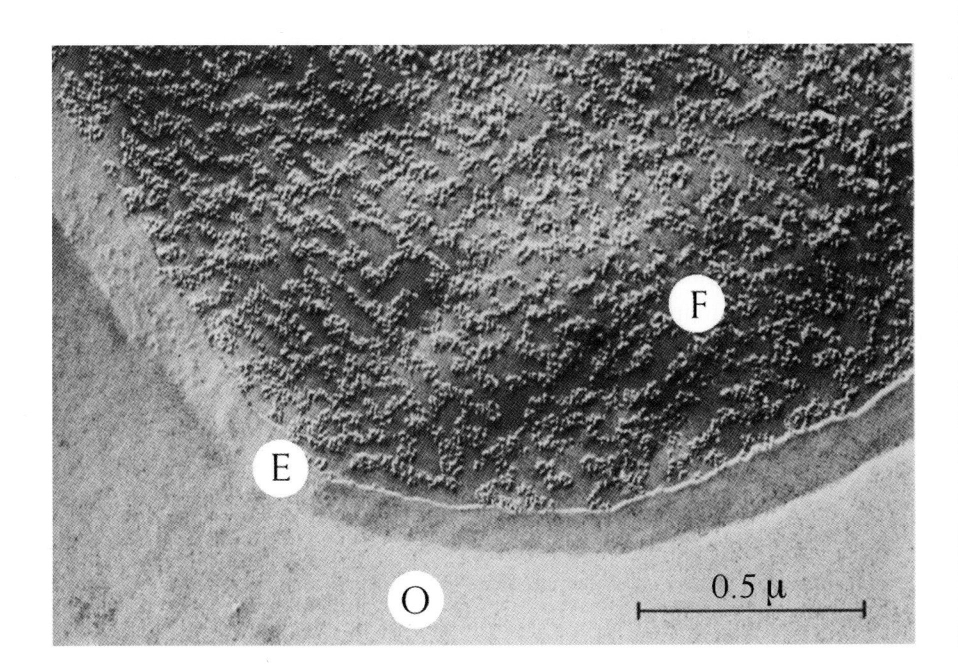

Figure 2.3. *Disposition of proteins in the cell membrane. This red blood cell was quick-frozen, and then fractured in preparation for electron microscopy. The fractured surface (F) reveals many particles thought to be membrane proteins. The thin region (E) reveals the cell's exterior surface, while the region (O) is outside the fractured zone. After Pinto da Silva and Branton (1970).*

ion-flow pathways (Heimburg and Biltonen, 1996). Indeed, it is difficult to find any report of a protein stuck into a model membrane that does not increase permeability.

Given the leakiness of the protein-dominated membrane, it is worth reflecting on how some well-known observations that have seemed anomalous might now reconcile.

The first is an old experiment by a then-prominent physiologist who studied transmembrane potassium flux (Solomon, 1960). Flux was measured under control conditions and under conditions in which membrane lipids were partially removed. Potassium flux was the same—it did not depend on whether the lipids remained intact. The absence of physiological impact led the author to question whether the membrane is "superfluous to the maintenance of cellular integrity and intracellular function."

Another well-known observation is on the transmembrane flux of water. Exchange of water through a lipid bilayer is extremely slow. Exchange through the cell boundary is an order of magnitude faster at least, and this excess has been explained by invoking specialized water channels that penetrate the bilayer (Dempster *et al.*, 1992). But water channels are superflous within the discontinuous barrier framework; water can exchange naturally in the context of protein hydration (Chapter 4).

Another observation easier to accommodate with the protein-dominant membrane is a change of cell-surface area. In response to acute osmotic challenge, a doubling of membrane-surface area is not unusual in some cells. How such a change could be accommodated by a membrane has remained unclear, given that molecular packing undergoes no measurable change in the lipid bilayer even when the membrane is strained almost to the point of fracture (Discher *et al.*, 1994). In a membrane replete with convoluted, accordion-like proteins, on the other hand, accommodating such surface-area change is less of an issue; the proteins simply unfold (Fig. 2.4).

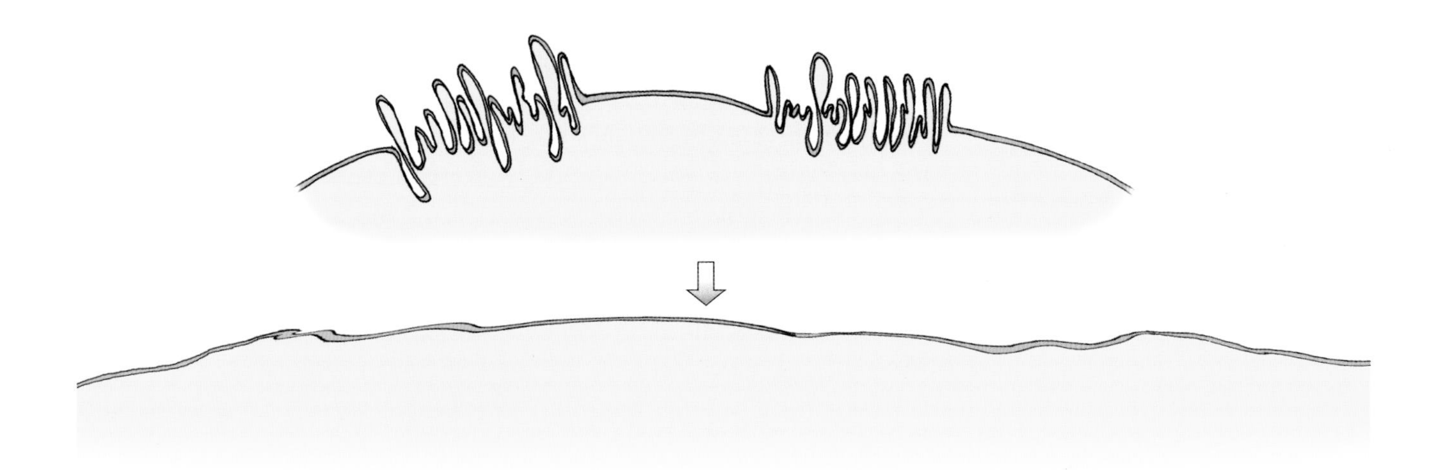

While these explanations may make sense, one is left with an uneasy feeling of having seen electron micrographs showing a continuous membrane surrounding the cell. Can such evidence be ignored? We are conditioned to interpret the trilaminar border that bounds the cell as phospholipid. Yet, when all lipid in mitochondrial membranes is dissolved by acetone treatment, the border persists (Fleischer *et al.,* 1965). And membranes of *E. Coli B* have a trilaminar structure even though they contain essentially no unsaturated fatty acids—ordinarily a major membrane-lipid component (van Iterson, 1965). Given these facts, one must wonder whether the trilaminar structure seen in the electron microscope can really represent the membrane phospholipid, or whether instead it represents some cell-interfacial feature highlighted by the staining procedure.

Figure 2.4. *Protein unfolding can increase membrane-surface area.*

This issue is given much attention in a review by Hillman (1994). One of several questions raised by Hillman is why the trilaminar cell borders seen in electron micrographs are not replete with channel, pump, and receptor proteins. Nominally 20 to 30 nm in size, such proteins are larger than the 10-nm trilaminar border, and if they are as abundant as anticipated, they ought to show up either as stained blobs, or if they do not take up stain, as gaps. Yet the tri-laminar border is almost always uninterrupted. What can this mean?

Finally, if the membrane does not function as a continuous ion barrier, one may rightly ask why it should be there to begin with. I can think of at least three possible roles (Chapter 16): a scaffold into which membrane proteins are inserted; a partial barrier to retard the loss of soluble proteins and metabolites from the cell; and a deflector of transmembrane ion flow to those (protein) regions where useful triggering action can take place.

In sum, the view that the cell is surrounded by a continuous ion barrier does not seem well substantiated. The membrane is dominated by proteins and glycoproteins, which are able to exchange water, salts, and other small solutes. The barrier is leaky. Leakiness may explain the documented passive diffusion into the cell of peptides < 100 amino acids in size (Lindgren *et al.*, 2000). I hope this conclusion is not misinterpreted as implying that the membrane is unimportant, or that there is no membrane. The cell evidently does have a membrane—it is just that the role of this organelle may be different from the one that is ordinarily envisioned.

CONCLUSION

Continuing to move boldly, we took it upon ourselves in this chapter to reconsider the notion of the continuous ion barrier. If the barrier were continuous, we reasoned, violating its continuity by tearing large holes should allow ions to surge across the cell boundary and solutes to leak out, dramatically altering the cell's makeup, shutting down cell function, and eventually killing the cell.

But that did not happen. Whether created by shoving a micropipette into the cell, plucking a patch from the membrane, riddling the membrane with an electrical barrage, or slicing the cell into two, the wounds seemed to matter little; the cell could often continue to function as though there had been no violation. It was as though function could be sustained by the cytoplasm alone.

To gain perspective on this unsettling result we considered the evidence thought to underlie the barrier's presumptive continuity and found it to be less than certain. We noted that the lipid bilayer contained an abundance of proteins, which confer leakiness on the membrane. With a leaky nature, one can rationalize why poking another hole might not matter much.

The leaky membrane paradigm also lends understanding to the difficulties with channels and pumps. These elements arose to provide needed pathways for ions to flow through a membrane that was presumed impermeant. With permeance, such elements become redundant. Why their imputed function might seem out of accord with evidence is understandable.

Perhaps it is becoming clear that excavating toward the core of truth requires the peeling back of multiple layers of assumption. We are not yet through. The next chapter considers a layer of assumption even more fundamental than those considered—the assumption that led to the continuous barrier notion in the first place. Why was it necessary to postulate a continuous barrier?

Woman IV, 1952-1953, by Willem de Kooning

CHAPTER 3

CYTOPLASMIC DISCOMFORT

Up to now our attention has been focused on the cell membrane and its components. Ion-separation issues have been dealt with in terms of membrane pumps and channels, and pump and channel issues have been dealt with in terms of the continuous barrier concept. Here we extend this chain. We ask whether continuous barrier issues may arise from something even more fundamental—the nature of the enclosed cytoplasm.

How was it that the membrane-barrier concept came into being in the first place? Recall that the barrier arose as a 19th century postulate advanced to explain why substances within the cell would not mix freely with the surrounding medium. Other than by the presence of a membranous barrier, how could the cell be kept distinct? The logic of this argument is difficult to fault, but the underlying premise needs to be recognized. The premise is that most substances inside the cell are free to diffuse; otherwise, why bother to postulate a barrier? The barrier concept, in other words, arose from the presumption that cytoplasmic constituents are effectively dissolved in aqueous salt solution and free to diffuse.

How could it be otherwise? At this stage a rational alternative may be difficult to envision but by the end of a few chapters I would hope that an alternative will be clear, natural—and possibly even correct. The alternative places protein surfaces on center stage, ordering nearby water molecules into a "structured" state and providing adsorptive sites for charged solutes. The resulting protein-ion-water matrix has a gel-like character very different from the liquidity of an aqueous solution. Al-

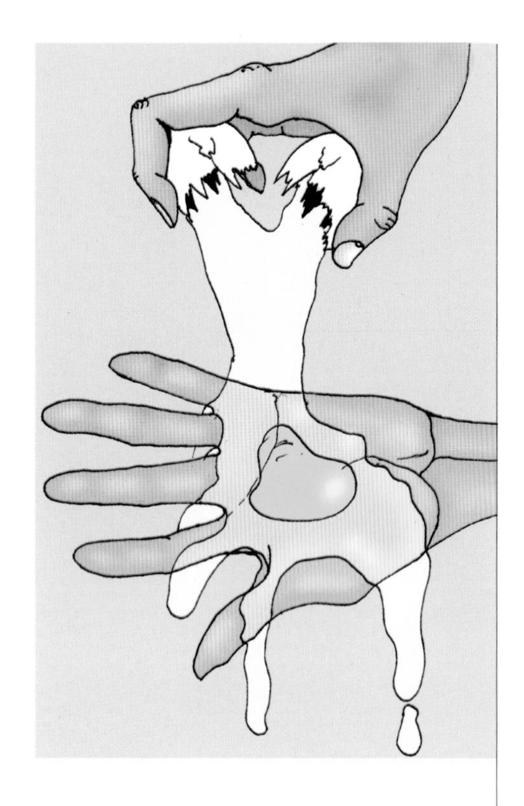

Figure 3.1. *Cytoplasm resembles a gel.*

though such attributes may be unfamiliar to classically trained biologists, they are exploited regularly by chemists and engineers for practical device design (Chapter 8).

In considering the nature of the cytoplasm, think of the familiar egg (Figure 3.1). Certainly you have cracked open a raw egg—would you characterize it as an aqueous solution? Or, is fresh egg-white more gel-like?

To get a rigorous handle on the cytoplasm's nature, we consider its physical features. If the cytoplasm were an aqueous solution, then water and solutes should behave in a predictably conventional manner. If the cytoplasm is gel-like, these consituents will behave differently. The distinction is significant, for if it turned out that solutes were not free to diffuse, this would obviate the need for a continuous barrier.

Thus, we examine the behavior of the cell's water, and then its ions.

THE BIG CHILL

In Seattle, where I am fortunate to live, temperatures rarely dip below freezing but on the occasional winter night they may descend to as low as minus 15°C. At such temperatures a rhododendron here or there will die and a few others may exhibit leaf damage, but mainly, such hardy plants survive even the coldest of winters without a hitch.

In order to survive, plants (or animals) cannot freeze. That is, the water inside the cell cannot turn to ice, for ice crystals rip mercilessly through cells and wreak havoc on any structure that happens to lie in their growth path. The force of ice crystallization is astonishingly high. My back yard pond for example will occasionally freeze over, and when cold weather is anticipated, the common practice is to float a thick wooden board on the water surface so as to absorb the impact of the crystallizing ice. Otherwise, the pond's concrete wall may be cracked by the force of the growing ice.

The ravages of ice-crystal formation are also well known to those involved with the freezing of cells for electron-microscopic analysis. When freezing is achievable on the microsecond time scale, as is often the case at the specimen surface, local protein structure is largely unaffected because ordinary crystalline ice does not form. Deeper into the sample where freeze time slows to milliseconds, even the most sophisticated of ultra-rapid freeze devices cannot prevent the formation of ice crystals. Where such crystals do form, the protein matrix is destroyed, sometimes beyond recognition. For survival, freezing of the cell's water (albeit not necessarily the water outside the cell) is to be avoided at all costs.

Nature's defense against the big chill is to lower the temperature at which the cell freezes. Since the cell is filled with solutes, cell water will not freeze at 0° but at some lower temperature. The principle is the same as the melting of ice provoked by the sprinkling of salt: the salt lowers the freezing temperature by virtue of the solution's colligative properties. For the cell, however, the maximum depression that can be achieved by this route is only about 2°C (de Vries, 1982). Thus, at -15°C the rhodendron should easily freeze even if some heat is generated by slowed metabolic processes. And at -30°C, a temperature not so unusual in extreme climes, plants and cold-blooded animals should expire without question. How remarkable it is that in the high Arctic, insects such as the wooly bear caterpillar survive temperatures that consistently dip below -50°C.

Something is evidently amiss. If cell water is anything like water in my pond, even with dissolved solutes in high concentration the water ought to freeze at temperatures not far below 0° — but it does not. As the temperature descends, the cell's water molecules are restrained from entering the characteristic ice-crystal configuration; they are somehow held captive. How they are held captive will be considered in Chapter 4. For the present, the bottom line is that something about cell water seems different from ordinary bulk water.

Given this difference, the question arises how solutes might be impacted. Are ions in such "unfreezable" water as diffusible as they are in ordinary aqueous solution? Or are they restricted?

Ion Diffusibility

Propelled by thermal forces, ions in ordinary aqueous solution will suffer continual random motions, eventually distributing themselves uniformly over space. Like the pinch of salt thrown into chicken soup, the molecules will spread so as to eradicate any concentration gradient. The distribution of molecules will as a rule be uniform.

Most rules have exceptions and two relevant ones should be noted. First, the cell's internal compartments such as mitochondria and endoplasmic reticula tend to accumulate ions; within such organelles ions may be highly concentrated. Second, the proteins' charged surfaces may similarly attract and concentrate ions. Remote from such structures, however, the expectation of free diffusibility implies uniform cytoplasmic ion distribution.

To test for uniformity, several methods are commonly employed. In one method, ions are radio-labeled and their distribution is detected on film. A second method uses the electron microscope to map the distribution of detectable ions or ion-surrogates. A third method involves the use of X-ray microanalysis of quick-frozen samples to ascertain elemental distribution. With these methods it is possible to analyze whether or not the cell's ions are distributed uniformly.

The type of cell most used for this kind of study is skeletal muscle (Fig. 3.2). Because this cell's internal organization is predictably regular, it is easy to correlate ion distribution with structure. The contractile machinery of skeletal muscle consists of a parallel array of strands called myofibrils, each girdled by a mesh-like reticulum. The regularly repeating array of proteins along the myofibril gives rise to the characteristic banding pattern, the (dark) A-bands and (light) I-bands.

The band-to-band distribution of ions has been mapped, with consistent results among all the above-mentioned methods (Edelmann, 1988). Namely, the number of potassium and sodium ions is higher in the A-band than the I-band. What is paradoxical is that the A-band volume contains more protein and therefore less space for solvent. Thus, the larger number of ions in the smaller volume translates into an even larger concentration disparity. With the micro-analytical method, the A/I concentration difference is on the order of two times, and some of the other methods report larger values. These gradients persist indefinitely.

A second level of organization at which the presence of gradients can be checked is between the inside and the outside of the myofibril—again on a ~1-μm scale. Under physiological conditions, average potassium concentration is approximately 50% higher inside the myofibril than in the space surrounding the myofibril (Stephenson, 1981).

A third level of distinction is in the zone just inside the cell membrane. Ion distribution within this zone was studied in cardiac muscle cells using X-ray microanalysis (Wendt-Gallitelli *et al.*, 1993). Sodium concentration was always highest nearest the membrane and decreased toward the interior of the cell. The gradient was steepest when the cell was in its activated state with sodium concentration just inside the membrane at 40 mM *vs.* 17 mM some 200 nm away. Evidence of a gradient remained for up to three minutes after electrical stimulation had ceased—far longer

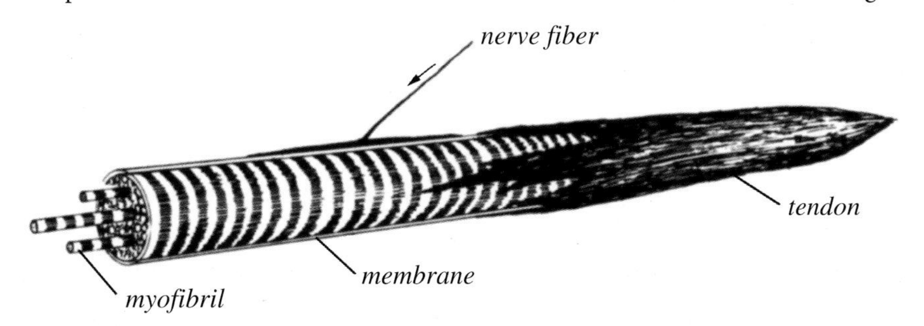

Figure 3.2. *Essential structure of a skeletal muscle cell.*

than the milliseconds predicted from ordinary diffusion theory.

In addition to gradients of sodium and potassium there is evidence for sustained gradients of calcium. With stimulus pulses of long duration applied to skeletal muscle cells, the cellular region nearest the stimulator's anode becomes calcium deficient and the deficiency persists for as long as the stimulus train persists (Trube *et al.*, 1981). Gradients are also seen in cross-sectional views of activated fibers, with "hot-spots" containing up to 100 - 200 times more calcium than several micrometers away (Rozycka *et al.*, 1993). These hot spots do not fade with time although they may gradually shift in space. Thus, calcium is distributed non-uniformly, just like sodium and potassium. Sustained ion gradients are also characteristic in heart cells, where a comprehensive review concluded that spatial non-uniformity is the rule rather than the exception (Carmeliet, 1992).

Figure 3.3. *Membrane disruption does not lead to immediate release of potassium.*

Local gradients make sense because they allow for functional compartmentalization. Consider calcium. This ion activates virtually every one of the cell's numerous processes (Alberts *et al.*, 1994). If calcium were to flood the cell uniformly, all processes would be activated simultaneously; each time the cell was activated to secrete it would also be signaled to divide. A scenario of this sort evidently makes little sense—processes need to be activated selectively, and this is possible only because the cell is able to sustain the steep calcium gradients. A similar theme applies to potassium. High potassium is required for function of certain enzymes, low potassium for others (Stryer, 1988). With uniform potassium concentration, only a fraction of the cell's enzymes could operate at the same time. Local concentration gradients allow for selective local control of enzymatic function.

But I digress. The main point is not so much that the gradients keep the cell from going berserk but merely that they exist. Ions in contiguous zones are maintained at disparate concentrations. Such ions are evidently not freely diffusible, for if they were, the gradients would quickly disappear. Something restrains the ions.

GLOBAL DIFFUSIVITY

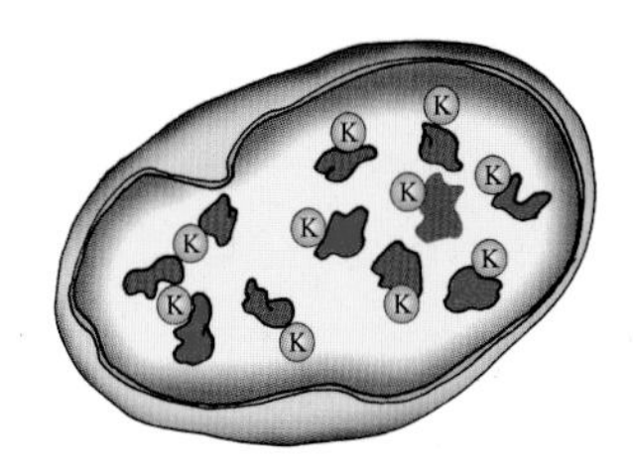

exposure to detergent

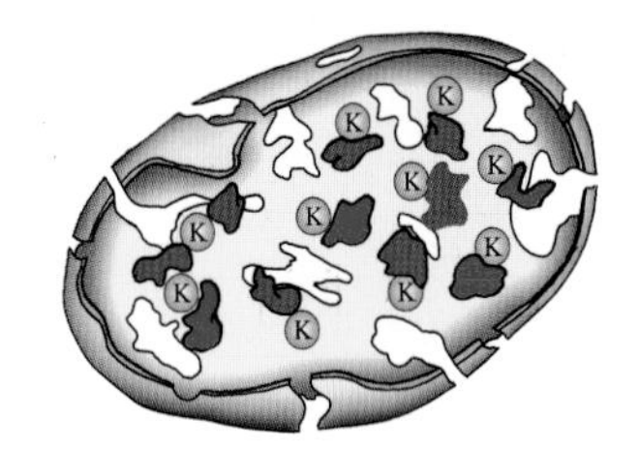

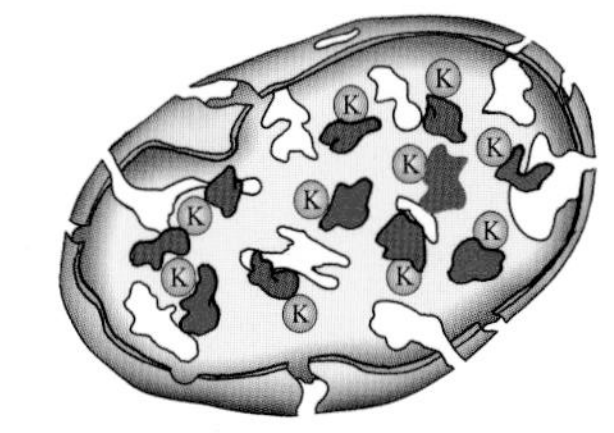

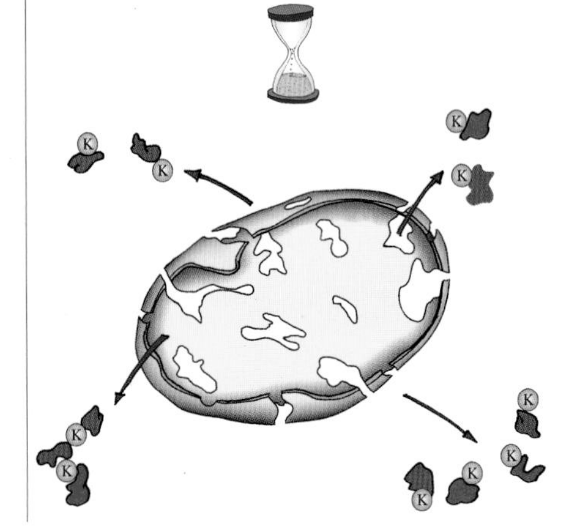

Diffusibility can also be checked on a global scale. In the previous section we considered diffusion among contiguous micro-regions within the cell. Here we consider whether ions have the capacity to diffuse into or out of the cell.

In testing for trans-boundary diffusion, any lingering uncertainty about the effect of a membranous barrier can be eradicated by simply removing the membrane; ions should then be perfectly free to pass across the boundary if they are diffusible. For example, if the membrane-free cell is placed in a bath of low potassium, the concentrated potassium ions inside the cell should immediately flow outward to eradicate the concentration gradient.

This very experiment was deftly executed by Kellermayer *et al.* (1986). Kellermayer pre-loaded cells with radioactively labelled potassium and examined how rapidly the labelled ions left the cell once the membrane had been disrupted (Fig. 3.3). Holes in the membrane were created by exposure to a detergent. With the detergent Brij 58, electron microscopy confirmed that the membrane contained gaping holes within two minutes; but potassium did not begin flowing appreciably out of the cell until five to six minutes. And when it did finally flow in earnest, the potassium came in a large bolus, together with the cell's proteins.

A more recent repeat of this experiment with another cell type yielded even more dramatic results (Cameron *et al.*, 1996). With eye-lens cells, full loss of potassium required more than eight hours following membrane disruption. The time scale for protein loss was similar, and other ions (sodium, calcium, magnesium) flowed about as slowly. Yet, large holes in the membrane could be documented only minutes after detergent exposure.

These results say a lot. The experimental framework is analogous to a dam in which a large hole has been blasted. Once the dam's integrity is violated, water should flow immediately. If flow does not begin at once we are forced to reconsider what might be going on. The high water could for example be frozen, or in some other way restrained.

Similarly for the cell, because potassium flow does not commence immediately after the holes appear, we infer that diffusion is restrained by some feature of the cytoplasm. The feature might be related to the cytoplasm's water, which was deduced to be different from ordinary water; or, it might be related to the proteins, which leaked together with the ions. Protein-based restriction is implicit in the sustained concentration gradients seen between A- and I-bands: A-band proteins may somehow constrain ions and restrict their diffusion into the I-band (*cf.* Chapter 6).

Any such restrictiveness should not be surprising: if it is not a membranous barrier that is responsible for ion partitioning it must be something else. Ion adsorption to protein surfaces is a phenomenon that will be considered in detail in the next several chapters.

CONCLUSION

The raw egg did not mislead—the cytoplasm is not the aqueous solution it is cracked up to be. Although ordinary water freezes at 0° C, cell water does not. Even after accounting for freeze-temperature depression arising from the presence of dissolved particles, cell water should still freeze at just below 0°C, but far lower temperatures are required. The cytoplasm's water behaves anomalously—molecules are somehow kept from entering the ice phase. Something restrains them.

A similar restraint appears to hold a grip on ions. Although ions in activated cells are known to diffuse readily, in the unactivated cell, restriction is observed both on a microscopic scale and on a global scale where ions stubbornly refuse to diffuse across the opened cell boundary. Proteins also do not easily diffuse out (Clegg and Jackson, 1988). Sustained gradients are not compatible with the idea of free diffusibility; ions are somehow restrained, just like water.

The charm of these observations is that they help resolve the chapter's conundrum. If ions are restricted, the specter of ions and other solutes diffusing through the portals of the leaky cell membrane and obliterating critical concentration gradients is no longer a concern. With restriction, the ions will stay put.

PERSPECTIVE

As this first section of the book draws to a close it seems appropriate to reflect on where we have come. The goal was to approach a core of certainty by penetrating beneath layers of the "generally accepted." Successive layers of assumption have been peeled back. The process has amply confirmed what should have been obvious at the outset: assumptions are weeds that choke the flowers of truth.

The process began with the question of ion partitioning, a feature presumed to rest on pumping and channeling. Although "pump" and "channel" proteins clearly exist, the concept of physiological ion separation by pumping and channeling was found to be illogical and out of accord with certain basic evidence.

We then realized that the pump-channel paradigm itself arose from an underlying assumption—the cell is enveloped by an ion-impermeant barrier. We found however that any such barrier must be leaky, for the membrane contains 50% - 80% protein which provides ample pathways for leakage. This helped explain why slicing the cell could be so innocuous—slicing merely increased the number of flow pathways, and not anything more fundamental.

We then realized that the impermeable barrier paradigm likewise rested on an underlying assumption. The assumption here was that cytoplasmic ions were free to diffuse, but simple experiments showed that cytoplasmic ions were restrained.

What we have unearthed, then, is a nested set of assumptions. From the bottom up: Free diffusibility begat the continuous membranous barrier; the continuous barrier begat pumps and channels; and pumps and channels begat the accepted process of ion separation. A chain of illegitimacy, if you will, began with the original sin of free diffusibility (Fig. 3.4).

This chain of doubt underlies all of modern cell biology. I see no way around it. Cell biological paradigms rest on a foundation of flimsy assumptions. Why these paradigms have become bewilderingly complex is not difficult to fathom, for simple paradigms rarely emerge from unsound foundations—we learned that from the epicycles.

The conclusion drawn here is obviously unconventional, and before going any further I would urge you to stand back and reflect on whether I may have led you along the garden path. A good place to begin is with one of the field's standard texts such as the impressive volume by Alberts *et al.* (1994). Authoritative reviews of this sort offer a counterpoint to what I have been presenting and should allow you to make informed judgment on the merit of this book's message.

At this stage I would like to suggest that we have proceeded far enough along the current line of approach. Having penetrated through layers of assumptional "begats," we have acquired direction enough to sense that ground truth should not be far distant—we should not need to beget ourselves back to Adam and Eve.

The footing we seek to build is assembled in the book's next several chapters. There we set aside the standard building kit of the "generally accepted" and employ the bricks and mortar of elementary chemistry and physics to deduce how proteins, ions, and water interact. Once this milestone is reached we should be in a sound position to launch an exploration of cell dynamics—the manner in which the cell carries out its functional tasks. And that, I promise, is where the fun begins.

Figure 3.4. *Bewilderingly complex mechanisms for cell function rest on a series of questionable assumptions.*

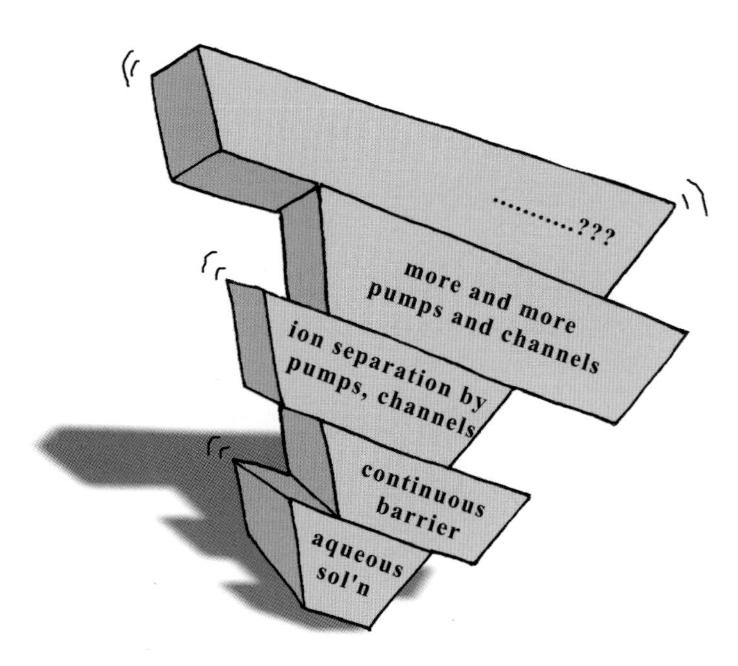

SECTION II

BUILDING FROM BASICS

This section pursues elementary cell function, with emphasis on water and protein surfaces. We see how the interaction of these two elements can give rise to ion partitioning, cell potential, and several other of the most basic features of cell physiology in ways that differ from current explanations.

The Great Wave off Kanagawa, 1823-1829, by Katsushika Hokusai

CHAPTER 4

WATER

To chemists, water is a substance whose simple structure belies formidably complex properties. To life-scientists water is a featureless sea that bathes the protagonists of life. This sea is ignored, as Albert Szent-Györgyi put it, "as the fish forgets the water in the ocean." Water is considered little more than a neutral carrier.

The previous chapter hinted, however, that water inside the cell might not be so featureless or ordinary: as the temperature drops well below the anticipated freezing point, ice does not form. Cell-water molecules have difficulty orienting into a characteristic ice-crystal configuration, implying that the motional freedom of those molecules is somehow constrained.

The current chapter explores whether cell-water molecules are indeed constrained beyond ordinary bulk water. If they are, then the consequences should not be trivial, for as we shall see, the properties of restricted water differ from those of ordinary water. We tread on a minefield of implication.

ORIGINS

The idea that cell water might be restricted, or "structured" is not new; it can be traced back a full century (Fischer and Moore, 1907). The proposal was strongly advocated by the Russian school headed by Aphanasij Troshin. Troshin's 1956 monograph, *Problems of Cell Perme-*

ability, challenged the traditional concept of the cell as an aqueous salt solution surrounded by a membranous barrier. In his view it was not the cell membrane that was responsible for the partitioning of solutes such as sodium and potassium, but the special physical properties of the cytoplasm including its water. Although no specific theory was put forth, the clarity of Troshin's arguments and the strength of his evidence inspired a strong following not only in Russia but among the eastern European scientists as well (Ernst, 1970; Szent-Györgyi, 1972; Tigyi *et al.*, 1991).

Cell water also formed the centerpiece of Szent-Györgyi's 1972 classic book on *The Living State*. According to Szent-Györgyi (who won the Nobel Prize for his discovery of vitamin C), the living state's dominant feature is macromolecule-water interaction: Macromolecules induce structure in surrounding water molecules, and the structured water protects, separates, and also links the macromolecules.

Quite independently, a similar albeit more extensive theory was developed by Gilbert Ling. In his first book *A Physical Theory of the Living State*, published in 1962, Ling briefly suggested that cell water may differ from bulk water, and then went on to present detailed evidence that the water inside the cell was largely structured (Ling, 1965). Structuring was argued to originate at the charged protein surfaces, to which water molecules could adsorb because of their polar nature. Once a single layer adhered, additional layers would adhere one upon the other, forming a multi-layered network. Because structured water was a poor solvent for most ions, Ling argued, many ions including sodium were naturally excluded from the cytoplasm. The characteristically high intracellular accumulation of potassium, on the other hand, was claimed to arise from that ion's unusually high affinity for the proteins' negatively charged carboxyl groups. Thus, essential ion partitioning could be explained by this theory—resting to a large extent on the physical properties of structured water.

Although Ling's thesis was warmly embraced by Russians and eastern

Europeans, it was received with skepticism by western scientists, most of whom were unaware of the complementary work of Troshin. Particularly difficult for these scientists was the implicit dismissal of membrane pumps, which had begun attracting considerable attention. In Ling's theory the pumps were unnecessary: partitioning of all solutes could be explained by a combination of association with proteins and reduced solvency of cell water. Ling's view of cell water implied rejection of the seemingly well-established channel - pump paradigm.

Any hopes the followers of Troshin and Ling may have had to advance their views were dashed by the polywater debacle. In the late sixties a Russian group (Derjaguin, 1966; Anisimova *et al.*, 1967) described special columns of water grown in quartz capillaries. Because this water had unusual properties it was deemed "anomalous" and was later renamed "polywater" by western scientists (Lippincott *et al.*, 1969, 1971), whose enthusiastic acceptance lent an aura of significance to this Russian-based phenomenon. Within a few years it became clear to most that these water columns were artifact—they were apparently not structured water *per se* but gels built around impurities leached from the quartz capillaries (Lippincott *et al.*, 1969, 1971; Brummer *et al.*, 1972).

But stunning breakthroughs by physical scientists have reopened the door of consideration. I am referring largely to the work of Israelachvili, Pashley, Granick, Parsegian and others, whose experimental studies of solvents in the vicinity of surfaces leave little doubt that surfaces profoundly impact nearby water. We consider some of these studies next.

SURFACE WATER

The principle underlying much of this modern work can be understood from the diagram of Figure 4.1. Take a drop of liquid and put it between a heavy ball and a table. Pressure exerted by the ball will force the liquid out, first rapidly then more slowly. The surprise is that the ball never quite reaches the table; a film of water remains indefinitely sandwiched

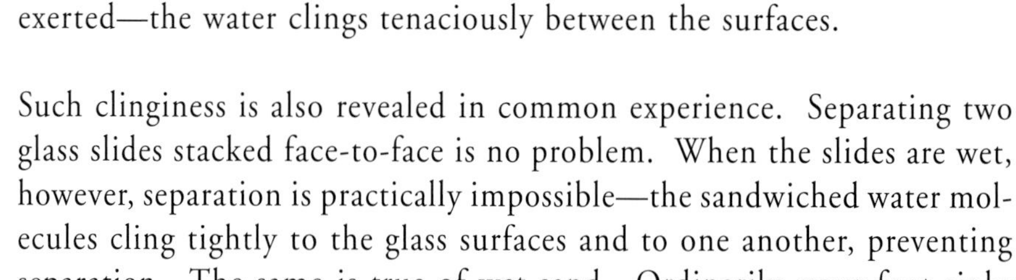

between the surfaces. This film supports the entire weight of the ball. To squeeze out the remaining film, an extraordinary pressure must be exerted—the water clings tenaciously between the surfaces.

Such clinginess is also revealed in common experience. Separating two glass slides stacked face-to-face is no problem. When the slides are wet, however, separation is practically impossible—the sandwiched water molecules cling tightly to the glass surfaces and to one another, preventing separation. The same is true of wet sand. Ordinarily, your foot sinks deeply into sand at the beach, leaving a large featureless imprint; but in wet sand your foot hardly sinks at all. Water clings to the sand particles, bonding them together with enough strength to support your full weight.

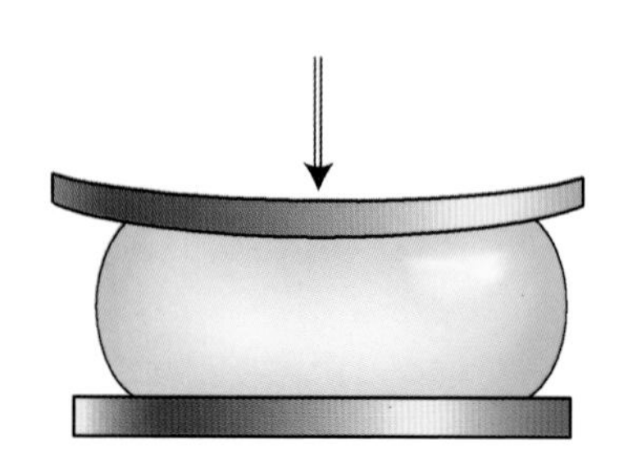

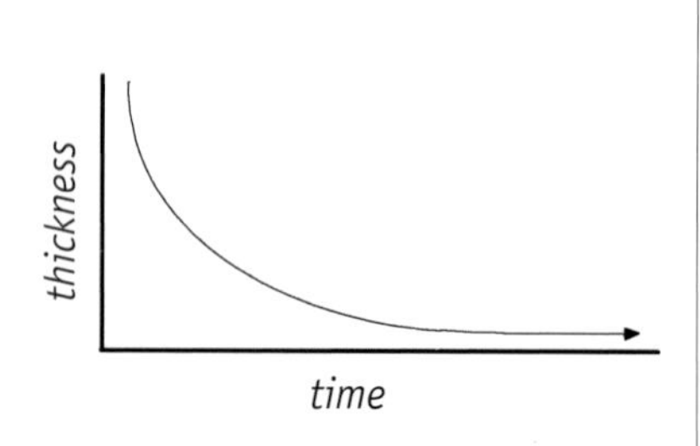

Figure 4.1. *Effect of pressure exerted on fluid sandwich between surfaces. After Granick (1991).*

To understand the nature of such clinginess, Israelachvili and colleagues measured the force required to displace solvents sandwiched between parallel mica surfaces (Horn and Israelachvili, 1981; Israelachvili and McGuiggan, 1988; Israelachvili and Wennerström, 1996). The closer the surfaces got, the higher the force required (although van der Waal's attractive force began to dominate once the two surfaces practically touched). The overall behavior was largely classical, but there was a surprise.

The force-separation relation was not purely monotonic. Superimposed on the anticipated response was a series of regularly spaced peaks and valleys of force (Fig. 4.2). The spacing between peaks was equal to the diameter of the sandwiched molecules, and this correspondence persisted no matter what the molecular diameter of the sandwiched fluid. Such astonishing correspondence implied that the oscillations must have arisen from a layering of the sandwiched molecules—each layer corresponding to each oscillation. Molecular layering should not have been a complete surprise, for the same had been implied many years ago by Sir Wm. Hardy (1931) in his studies of lubricants around metallic surfaces. If the first layer sticks to the surface, the second layer sticks to the first, and so on, this would explain why nearby surfaces cling to one another.

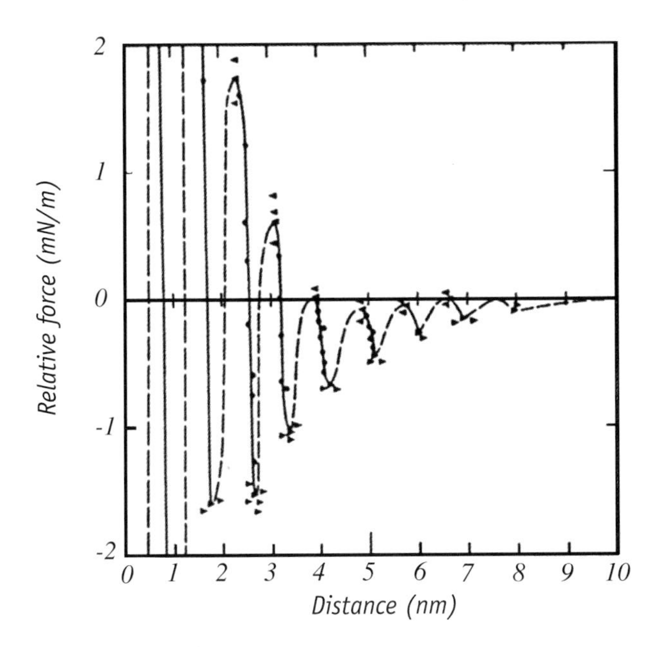

Figure 4.2. *Effect of separation on force between closely spaced mica plates. Only the oscillatory part of the response is shown. After Horn and Israelachvili (1981).*

Under the specific conditions of the Israelachvili experiment, some eight to ten peaks were detectable as the surfaces approached one another, implying the ordering of a like number of solvent layers. The experiment was repeated by Granick (1991), who could detect up to 12 peaks, implying 12 strata of layered solvent. Whether the number of peaks might increase with future instrumentation advances is unclear; what is clear is that surfaces have the capacity to organize solvent into at least 10 -12 layers.

A second indication of multi-layer structuring comes from measurements of the sandwiched fluid's viscosity. Ordinary solvents have a well-known bulk viscosity. But if solvent molecules are linked to one another, the viscosity will increase. Thus, the motivation arose to measure the viscosity of the fluid sandwiched between plates. This was achieved by shearing one plate past the other, measuring the force required to do so, and using this force to compute the viscosity (Granick, 1991). When plate separation was large, viscosity was no different from ordinary bulk viscosity. But as the plates got closer together the viscosity began to increase, steeply. The increase began to be detectable at a separation of 5 nm, or roughly 12 solvent diameters in that study. By 2.5 nm the viscosity had grown to ten times that at 3.5 nm.

Several studies imply that layering effects can extend out considerably beyond the few nanometers documented above. When the aforementioned viscosity measurements were carried out with polished glass surfaces and water, the viscosity increase was detectable at surface separations of 500 nm (Szent-Györgyi, 1972). When condensate films of water were grown from vapor onto crystalline quartz plates (Pashley and Kitchener, 1979), adherent water extended out to 150 nm; followup studies gave 200 nm (Fisher *et al.*, 1981). This implies hundreds of adherent water layers. And protein-trapping effects commonly induced by surfaces can extend out to 180 nm (Xu and Yeung, 1998).

These experiments extend those of Israelachvili and Granick and imply that surfaces have the capacity to stratify vicinal solvent molecules to distances up to 400 - 500 layers. Layering evidently rests on the nature of the surface, and the question whether layered water is similarly nucleated by biological surfaces is not addressed in these studies. The issue had been addressed in the many earlier studies of Troshin, Ling, and Szent-Györgyi which, together with the more modern non-biological studies, imply a kind of universality of the structuring phenomenon.

Addressing the issue of biological water structuring will consume the rest of this chapter. Some basics must be understood first. We need to consider the physical properties of water, as well as the physico-chemical properties of the surfaces that interact with water. If this material seems somewhat textbookish, I apologize.

Water Basics

Water has long been appreciated both as a liquid and as a solvent. Relative to other solvents, its properties are unusual because of the combina-

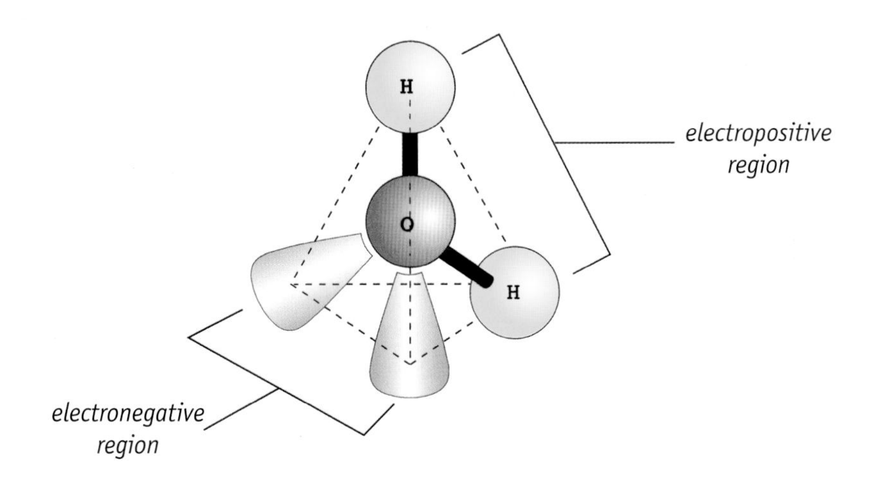

Figure 4.3. Dipole-like charge distribution of the water molecule. Adapted from *Alberts* et al. *(1989)*.

tion of small size and high charge. The molecule has two electropositive hydrogens at one end, and an electronegative oxygen at the other. It is generally modeled as four charges located along the four arms of a tetrahedron (Fig. 4.3). This confers a sense of directionality to the molecule—a dipole that allows strong polar interactions with a high degree of spatial orientation.

Each water molecule contains two donors, the protons, and two acceptors, the two unshared pairs of electrons of oxygen. This makes the molecule highly reactive. It is able to form hydrogen bonds with up to four nearest neighbors, or possibly six if hydrogen bonds bifurcate (Robinson *et al.*, 1996). The water-water interactions arising from hydrogen bonding are relatively weak, about 3 - 5 kcal/mole. Nevertheless, self-association confers on water most of its physicochemical properties.

Water exists in three commonly known phases—vapor, liquid and solid—which differ in hydrogen-bond association. In the vapor phase thermal energy is high and molecules undergo incessant random motion. In the liquid phase randomness is reduced and molecules form transient hydrogen bonds with one another on a time scale of 10^{-11} sec; they may aggregate transiently in chains originally called "flickering clusters" (Frank and Wen, 1957), although the manner in which this happens remains under debate. Modern molecular models suggest that at room temperature, 75-80% of water is associated with three or four neighboring water molecules, yielding an extensive three-dimensional network (*cf.* Robinson *et al.*, 1996). Because of the brief duration of these bonds, water remains a liquid.

In the solid phase, or ice, water molecules take on a distinct arrangement in which each molecule is bonded to four others (Figure 4.4). Many crystalline arrangements are possible, but in the ice crystal's two-dimensional projection each hydrogen atom lies on a straight line between two oxygen atoms. The straightness of the hydrogen bond confers stability, and accounts for ice's solidity. It also explains why the structure of ice is a rather open one, with density (at 0°C) of only 0.92. Ice cubes float.

Figure 4.4. *Molecular structure of ice. Hydrogen bonds interconnect molecules. Adapted from Stillinger (1980).*

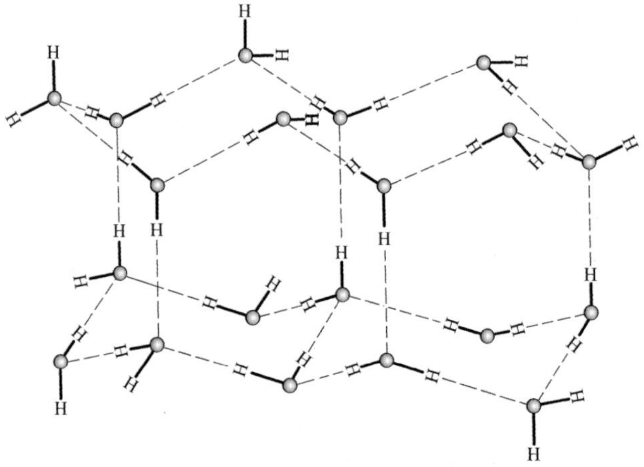

Lying somewhere between ordinary liquid water and ice is the so-called structured or layered water. Because it is found near interfaces it is also called interfacial water. Interfacial water is distinct enough from water's other three phases that Sir Wm. Hardy early in the 20th century suggested it to be a "fourth phase" of matter, beyond solid, liquid, or vapor. In structured water the molecules are linked, albeit less regularly than in ice. Because the hydrogen bonds are bent (*cf.* Pauling, 1959), structured water's density is higher than that of ice, allowing molecules to crowd together more closely (Garrigos *et al.,* 1983). The bent bonds may lend compliance to the structure, so that the array is less rigid than ice and is not necessarily of fixed dimension.

In fact, there are at least two types of interfacial water, depending on whether the interfacial surface is hydrophobic or hydrophilic. Hydrophobic surfaces induce bonding between neighboring water molecules. Bonding gives rise to 5Å pentagonal elements, which grow into vast cage-like networks called clathrates (Wiggins, 1990; Vogler, 1998). Such induced water bonding explains why hydrophobic surfaces are sparingly soluble in water: it is not because hydrophobic surfaces associate so strongly with one another, although they ultimately interact by short-range van der Waals force, but because the surrounding water is preoccupied with extensive self-association, and therefore unavailable for solubilization.

The second type of interfacial water is the one near hydrophilic surfaces. Charged and polar elements of hydrophilic surfaces react strongly with dipolar water, which defeats water's tendency to self-associate. The reaction creates the layers of adherent water described above. Thus, hydrophilic surfaces will induce stratification, whereas hydrophobic surfaces will induce clathrate formation.

Which type of water can be expected in the cell? The picture is evidently complex because the cell contains surfaces of every conceivable ilk. Hence, some water of each type is anticipated—clathrate, layered, and bulk. On the other hand, a single macromolecular species dominates the cytoplasm: protein. Whatever water type ordinarily surrounds protein should therefore make up the lion's share of water in the cell.

Proteins contain both hydrophilic and hydrophobic elements. Typically, hydrophobic elements lie buried in the protein's core, leaving the hydrophilic elements superficial and exposed to the water. Because they contain many charges and polar elements, such surfaces should induce water layering. If layering is substantial, it could account for cell water's known resistance to freezing—for layered water molecules are constrained and thereby unable to rearrange easily into the ice-like configuration.

The possibility of water layering is considered next, as we examine the nature of protein surfaces in some detail.

Figure 4.5. *Organization of water near a protein surface. (A) Water dipole adsorbs onto protein-charge site. (B) Adsorbed dipole induces additional dipoles to adsorb. (C) Additional protein-charge sites reinforce and extend dipole network.*

PROTEIN SURFACES

The characteristic feature of protein surfaces is that they are studded with charge. Proteins are built of a long string of amino acids, giving rise to a backbone with side-chains. Charges arise both from backbone carbonyl (-) and amino (+) groups and from the side-chains (+ or -). Thus, each amino acid will present several charges. Since these charges are situated mainly on the protein's surface rather than in its interior, the surface will be densely coated with charge.

These charges should have a profound effect on vicinal water. Because of its dipolar nature, the water molecule will be attracted; to minimize potential energy it will adsorb (Figure 4.5A). Water adsorption to proteins is not a new idea. It was put forth by Linus Pauling (1945), who argued that the seat of adsorption was the amino-acid side chains, while still earlier views (Jordan-Lloyd and Shore, 1938) maintained that responsibility lay both in the side chains and the backbone amino and carbonyl groups. Either way, vicinal water molecules are expected to be oriented by surface charge, and this expectation is now confirmed by modern experimentation (Toney *et al.,* 1994). The issue is whether it is merely a layer or two, or the multiplicity of layers implied from the studies described above.

Before tackling this question it is helpful to address a potentially confounding issue for the biological situation: competition. The cell contains not only water but ions as well. These ions will compete with water for the proteins' charged sites. Bear in mind, however, water's overwhelming dominance. Computed the standard way, the concentration of water in the cell is 55 molar. The concentration of potassium, easily the cell's most abundant ion, is only 0.1 molar. With 500 waters for every potassium, it is clear that the presence of ions does not seriously alter the thesis—surface charges should orient water molecules.

Why, then, should the oriented water molecules form strata? This feature follows directly from the molecule's dipolar nature (Figure 4.5B). Once a single molecule is adsorbed, others will align by simple attraction. They align laterally as one dipole orients alongside another, and they align vertically, in multilayers. Layering is the inevitable consequence of the molecule's strongly dipolar nature.

Just how many layers may form depends principally on two competing factors: the strength of the organizing center and the strength of thermal forces that tend to disrupt the induced organization. The thermal force is significant but not overwhelming: at room temperature the thermal energy is some ten times weaker than the cohesive energy of the hydrogen bond (Szasz *et al.*, 1994). Thus, thermal forces will tend to be only modestly disruptive. As for the organizing center, a major role should be played by surface-charge distribution. If charge spacing is sparse, there will be little opportunity for lateral reinforcement; organization will dissipate with distance from the charge, and one may expect only clusters of organization around each charge (Fig. 4.5B).

For serious reinforcement, the charges on the protein surface must be densely packed. Perhaps it is intuitively obvious that the most effective surface distribution should be a checkerboard-like pattern whose unit spacing is an integer multiple of the water-molecule diameter (Ling, 1965; Szent-Györgyi, 1972); structuring induced by one charge could then reinforce the structuring induced by the next (Fig. 4.5C).

The picture presented in Figure 4.5 is obviously oversimplified. First, in representing the molecule as a simple bean, it subverts water's well-known

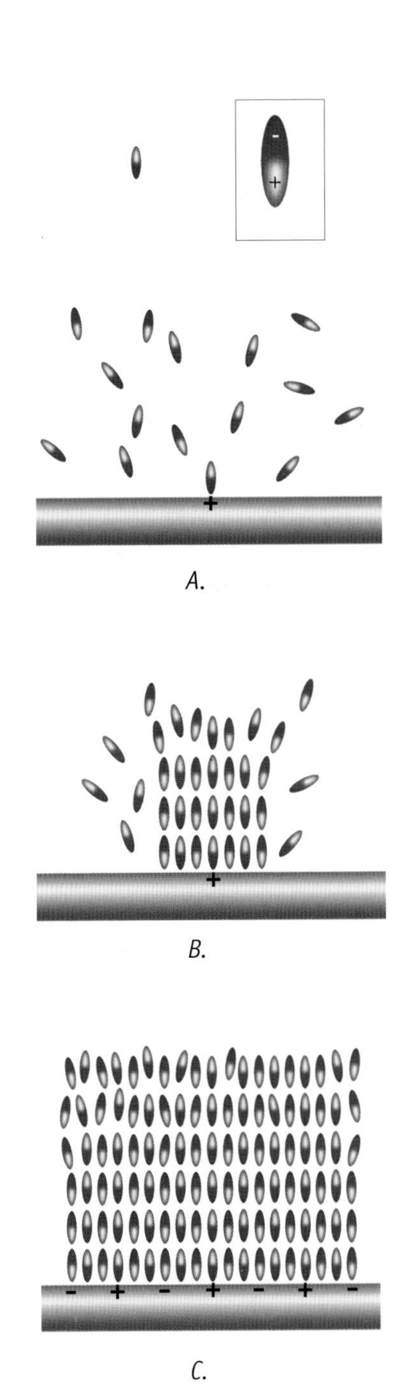

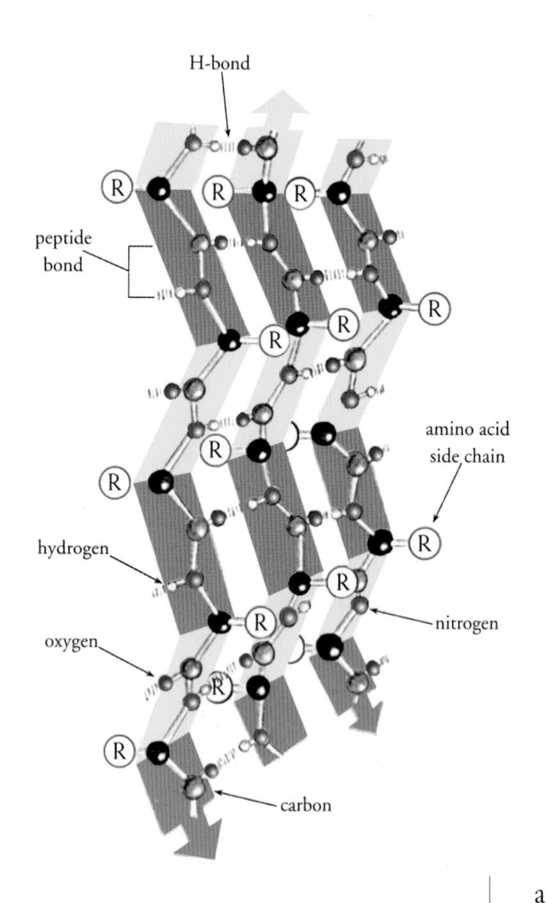

Figure 4.6. *Structure of the beta-sheet. After Alberts* et al. *(1994).*

tetrahedral structure. Second, it implies that water molecules are static; water molecules should be in perpetual kinetic motions similar to those of more distant layers, the extent of motion decreasing with proximity to the surface. The molecular positions shown represent a statistical average. Third, the figure implies that surfaces are infinitely far from one another, whereas a nearby surface should actually reinforce the ordering. And fourth, protein surfaces are not flat. The figure is nevertheless useful for providing an intuitive picture of the anticipated layering.

To determine whether the protein-surface organization implied by Figure 4.5C is actually realized, consider the proteins' two principal building blocks, the beta-sheet and the alpha-helix.

The beta-sheet is a dominant surface feature in many globular proteins. An ideal sheet is shown in Figure 4.6. It is built of a long polypeptide chain that folds back upon itself multiple times, each turn running in a direction opposite to its neighbor. The turns are held together by hydrogen bonds. Backbone carbonyl and amino groups, which carry negative and positive charge respectively, repeat approximately every 7 Å.

The alpha-helix is a structure included in many globular proteins as well, and is also the principal element used to build fibrous, rod-like proteins. Successive turns of the helix are held together by hydrogen bonds (Fig. 4.7). Frequently, two such helices twist around one another to form a coiled-coil. Along the length of the coil's surface is a regular alternation of charge. In the myosin rod for example, negative and positive charges alternate axially with a periodicity of 9.3 amino-acid residues (McLachlan and Karn, 1982). This translates to a linear repeat of approximately 14 Å, or 7 Å between unlike charges. Again, regularity prevails.

Given that the building block charges are regularly disposed, charge regularity should exist in the protein as well. A good example to look at is actin, a ubiquitous protein found in all eukaryotic cells. Crystallization of actin has made possible the construction of surface-charge maps. These maps show regions of stacked negative and positive charge on both the molecule's front and back surfaces (Hennessey *et al.*, 1993).

The question you may be asking is whether such charge repeats could properly reinforce the dipole ensemble. Would the water dipoles fit integrally into the surface-charge array? This question is akin to asking whether your feet can be positioned on two arbitrarily spaced floor targets. The answer in both cases is yes, within a range—for like your legs, the water molecule's two hydrogens are somewhat separable. Thus, the water molecule has no exact "size," and consequently, the required value of surface-charge repeat is unlikely to be critical although some optimum is inevitable.

To gauge the reasonableness of such a water-structuring scheme, one may

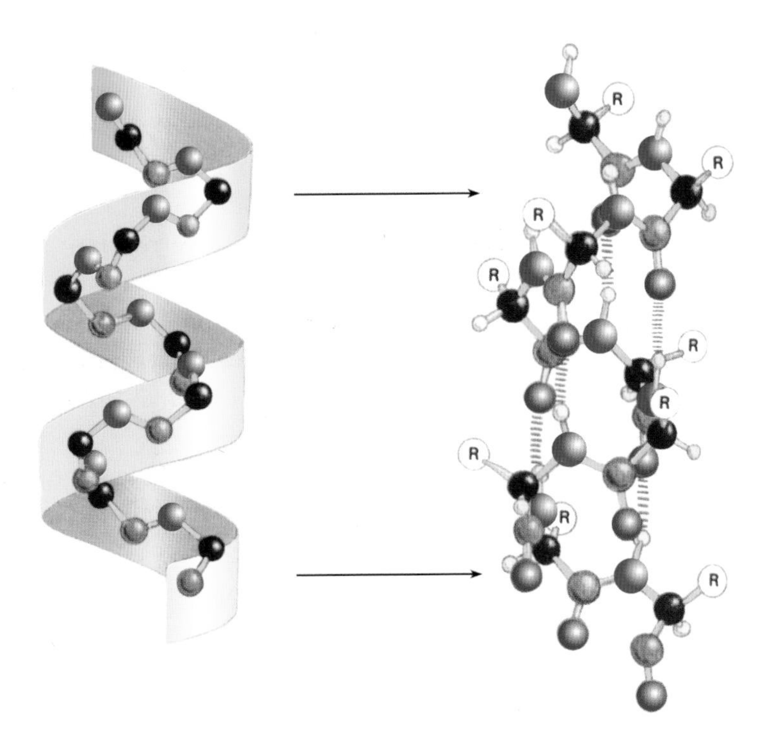

***Figure 4.7.** Structure of the alpha helix. After Alberts* et al. *(1994).*

compare water's size with the size of the surface-charge repeat. The surface area of the water molecule is recently estimated at ~9 $Å^2$ (Robinson and Cho, 1999) with a linear dimension that can range at the extremes from 2 to 4 Å. Building block repeats of 7 Å and 14 Å, as observed, will thus accommodate some two to seven water molecules. For strong reinforcement, charge spacing every few water molecules would seem reasonable.

A prediction of this paradigm is that all else equal, surfaces with the most regular charge spacing should structure the most water. To check this, we identify a situation with extreme water structuring and see whether the charge repeat is correspondingly regular. The paradigm that comes to mind is nature's "anti-freeze" proteins, which presumably mediate their freeze resistance by structuring many layers of vicinal water molecules, thereby inhibiting water's transition into the ice state. As excellent structurers, anti-freeze proteins are anticipated to have surface charges that repeat extremely regularly.

Indeed, the hallmark of such proteins and glycoproteins is their charge regularity. The protein backbone is an alpha-helix with a characteristic alanine-alanine-threonine motif that repeats again and again. The charge repeat along the helix occurs with an axial periodicity of 11 amino-acid residues, or every 16 Å (Chou, 1992). The extreme regularity of the repeat confirms our expectation, and implies that the 16 Å value may be one that is exceptionally effective in structuring water. This 16 Å repeat is perhaps more effective than, say, the 14 Å repeat characteristic of the myosin-rod.

In sum, the features required for effective water structuring appear to be present on protein surfaces. Surfaces are dense with charge, and the charges occur in regularly repeating patterns. It may not be a stretch to speculate that such charge regularity exists by design—principally to organize vicinal water. Indeed, water structuring as a central feature of protein design has been suggested earlier (Watterson, 1991, 1997).

SURFACE INTIMACY

How much of the cell's water could its proteins organize? If cytoplasmic proteins were dilute, the structured fraction should remain small. If the cytoplasm is crowded with proteins, then the structured fraction will grow higher and with sufficient crowding water structuring could approach global scale.

Thanks largely to the pioneering work of Keith Porter, it is now clear that the cytoplasm is relatively crowded. Porter has been able to visualize a fine meshwork of delicate protein strands termed the microtrabecular lattice that invests the cytoplasm (Porter *et al.*, 1983). For some years this latticework could be observed only by high-voltage electron microscopy and was therefore controversial. It is now visualizable also by standard electron microscopy (Tromibitás and Pollack, 1995). This matrix pervades the cytoplasm and endows it with numerous closely spaced surfaces (Figure 4.8). Analysis of high-voltage electron micrographs reveals that within this trabecular lattice more than half the water lies within 5 nm of some obvious surface (Clegg, 1988).

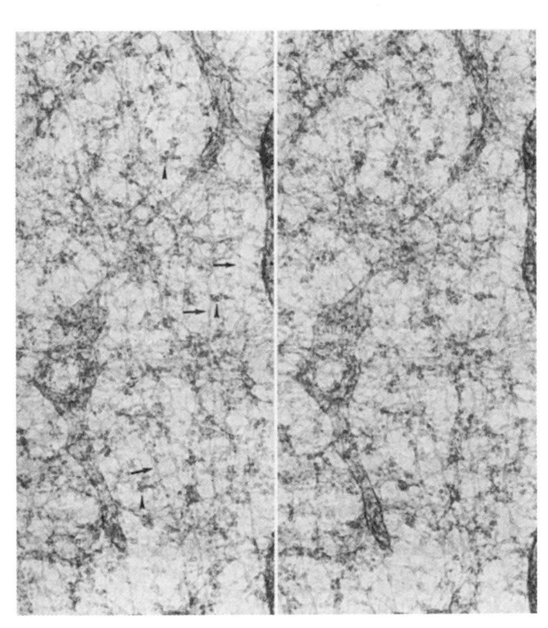

Figure 4.8. *Stereo view of cytoplasm of a BSC-1 cell, illustrating fine stranded microtrabeculae and cytoplasmic matrix. Arrowheads indicate ribosomes; arrows, microtubules. Micrographs from Schliwa (1986).*

If you are given to numbers, you may be interested in the fraction of the cell occupied by the latticework. The volume fraction taken by the microtrabecular network, along with the microtubules and intermediate filaments that compose the cytoskeleton, is measured to be just under 20% (Gershon *et al.*, 1985).

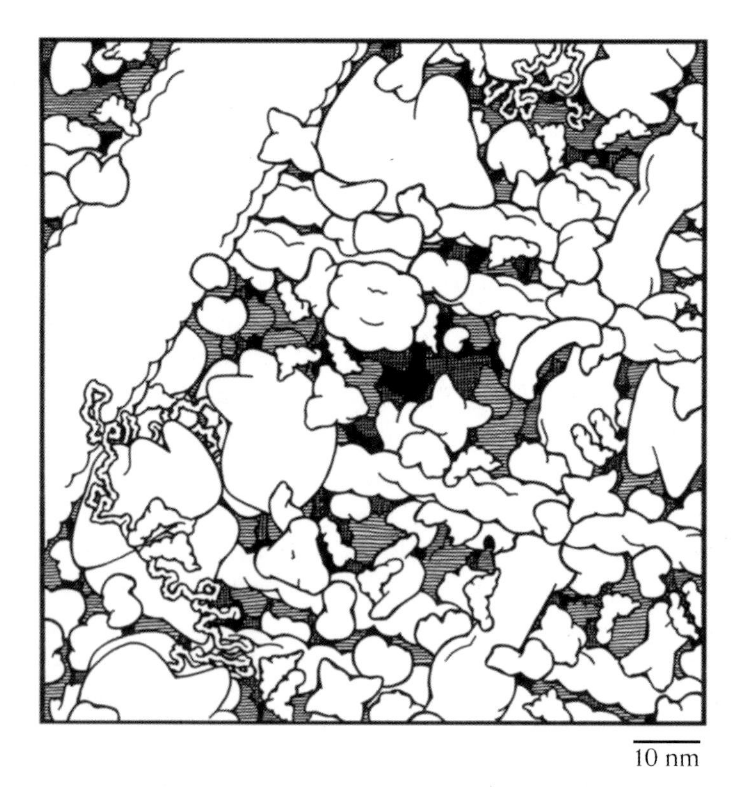

Figure 4.9. *Scale rendering of yeast-cell cytoplasm. Drawing includes cytoskeletal elements and other proteins. After Goodsell (1991).*

The surface area presented by these structures is estimated to be 80,000 μm^2 for a 16-μm cell, or about 100 times the cell-membrane-surface area. Assuming each protein to be a sphere for simplicity, surface-to-surface distance can be computed to be ~6 nm, not much different from the 5-nm value above.

Similar values can be obtained from scale renderings of the structure-filled meshwork such as the one shown in Figure 4.9. Rarely (in two dimensions) are the gaps more than a few nanometers.

While these visual methods all reveal surface-to-surface distances on the order of 5 nm, the electron micrographs on which they are based identify only the molecules with high electron density. Regions that appear empty on the micrographs may nevertheless contain globular proteins, enzymes, and other intracellular macromolecules that do not show up because of lower electron density or smaller size. These entities fill the spaces like balls in cages and must thereby diminish surface-to-surface gaps by a considerable extent. Hence, the 5-nm value, corresponding to roughly 14 layers of water, is likely to be an upper bound.

Crowding is also characteristic of the space within organelles. Mitochondria, myofibrils, stress fibers, endoplasmic reticulum, *etc.*, contain densely packed structural proteins with surface-to-surface gaps also on the nanometer scale, diminished further by interposed enzymes and other molecules. Thus, both inside and outside the organelles, typical surface-to-surface gaps are unlikely to contain more than some 10 - 15 water layers, and probably fewer.

An independent and perhaps more quantitative approach to estimating gap size is based on stoichiometric considerations. This is a matter of comparing the number of water molecules with the number of amino-acid charges. The calculation gives a ratio of 31:1 (Ling *et al.*, 1993, p. 201). Thus, each amino-acid charge is, on average, contiguous with 31 water molecules. If all protein charges are assumed to lie on the surface, and water molecules are assumed to form a cube adjacent to each charge, then the cube's linear dimension will be $31^{1/3}$ or approximately three water molecules. The surface-to-surface gap will then be two times three molecules, or six water molecules wide.

A similar value is obtained using a different kind of stoichiometric approach. A single layer of water covering the protein surface amounts to ~28% of total weight, virtually independent of the protein's shape (Fisher, 1964). Two water layers will amount to 56%, and three to 84%. Thus, the typical cellular water fraction of 70% is accounted for by two to three water layers, which gives about five molecules between surfaces.

In sum, electron microscopic approaches give an upper bound of surface-to-surface gap width of 10 - 15 water molecules, while stoichiometric approaches pinpoint the value at around five to six molecules. Gap size will vary from place to place, but the rather surprising conclusion is that the surfaces are astonishingly close. It is as though cellular macromolecules are like loosely fitting jigsaw-puzzle pieces.

Given such proximity, what can we expect of the state of cell water? In non-biological systems we saw clear evidence for 10 to 12 structured water layers adjacent to surfaces, with some experiments implying up to many hundreds of layers. Surface-to-surface gaps in the cell are perhaps as narrow as five or six water layers. Considering the gaps' narrowness, and macromolecular surfaces seemingly designed to confer structure, it could well be that a major fraction of cell water is multilayered.

Such expectation is largely theoretical, however; the bottom-line question is whether indeed the implied structuring is confirmed by direct physical evidence.

Evidence for Cell-Water Structuring

Apart from viscosity considerations to be dealt with later, experimental approaches to the question of water structuring fall into three principal categories: nuclear magnetic resonance, ultrahigh-frequency-dielectric dispersion, and quasi-elastic neutron scattering—tongue-twisting labels to which simple acronyms have thoughtfully been assigned. All of these methods test whether cell water molecules are more restricted than bulk water molecules, which are relatively free to dance.

In the nuclear magnetic resonance (NMR) method, the sample is placed in a static magnetic field. In the direction perpendicular to this static field, an oscillating magnetic field is imposed. The oscillating field brings water's two hydrogen nuclei (protons) into an excited state. Then, the oscillating field is suddenly turned off and the hydrogen nuclei "relax" as they dissipate their extra magnetic energy. Relaxation to the ground state requires a certain amount of time, the chief determinant of which is the degree of interaction with neighboring molecules. Thus, hydrogen nuclei in restricted water will relax more quickly than in unrestricted water.

For cell water the relaxation time is indeed shorter than for ordinary water (Odeblad *et al.*, 1956; Bratten *et al.*, 1965; Cope, 1969; Hazlewood *et al.*, 1969; Cameron *et al.*, 1988; López-Beltrán *et al.*, 1996). These results imply that cell water is relatively more structured, although they do not reveal whether the restriction applies to cell water in its entirety or only to a fraction of its water. Restriction is confirmed also by related measurements of water's self-diffusion coefficient (Rorschach *et al.*, 1991), and particularly by measurements of a large chemical shift relative to ordinary salt solution (Kasturi *et al.*, 1987), which implies global structuring.

More revealing perhaps are those NMR studies that examine cell water over a period during which the cell is functioning. Hamster-ovarian and HeLa cells were examined as they passed through their mitotic cycle (Beall, 1980; Rao *et al.*, 1982). Water was restricted during all phases except active mitosis, when a decrease of structuring seemed to correlate with the decrease of chromatin surfaces available to confer structure. A similar temporal correlation was found in fertilized sea-urchin egg cells (Cameron *et al.*, 1987). Here again, the extent of water structuring during the cell cycle was found to correlate with the amount of filamentous protein, in this case actin. These studies are significant in that they monitor a phenomenon in the same tissue as a function of time.

The ultrahigh-frequency-dielectric dispersion (UHFD) method provides a second opportunity to probe water's restrictiveness. Water molecules tend to line up in an electric field because of their dipolar nature. If the field direction is reversed, the molecules will about-face, like soldiers responding to command. As long as the frequency of the imposed field is not too high, molecules will continue to flip on command; but when the frequency is raised beyond a critical value the molecules will no longer be able to respond in timely fashion. The effective dipole is then seemingly weaker, and the "dielectric constant" is said to be lower.

For ordinary water, the critical frequency for this weakening is 20 GHz.

For restricted water, where flipping is more difficult, the critical frequency will be lower. In ice, for example, whose molecules are extremely restricted, the critical frequency drops precipitously to 10 KHz. Measurements in cells give values intermediate between ice and water. In brine-shrimp cysts, which are particularly useful for such studies because of their high protein content, measurements give values of 5 - 7 GHz (Clegg *et al.*, 1982; 1984). Measurements on rabbit tissues show a similar degree of restriction (Gabriel *et al.*, 1983). A study in plant cells (Pissis *et al.*, 1987) showed 30% of water molecules tightly restricted, 70% restricted to a lesser extent, and essentially no bulk water.

A third method to study water structure is quasi-elastic neutron scattering (QENS). Slow neutrons impinging on water will be scattered in all directions. The energy spectrum of the scattered neutrons will be narrow if the water is restricted, and increasingly broad depending on translational and rotational freedom of the bombarded molecules. From QENS data, one can extract both the translational and the rotational diffusion coefficients. The method is advantageous over the previous two in that it measures the mean behavior of all sampled molecules. Whereas some of the NMR and UHFD methods can be unduly influenced by a small fraction of water, the QENS method is a reflection of the average.

In frog muscle cells, the rotational diffusion coefficient is reduced by two times relative to ordinary KCl solution (Heidorn, 1985; Rohrschach *et al.*, 1987). In brine-shrimp cysts, translational and rotational diffusion coefficients are reduced respectively by three and 13 times (Trantham *et al.*, 1984). Thus, substantial restriction of cell water is indicated.

In sum, all three physical methods show that a substantial fraction of intracellular water is structured. Whether this fraction approaches 100%, as implied by the narrowness of the surface-to-surface gaps, or is less, is not yet established with certainty.

If water is adsorbed onto macromolecular surfaces, many of the "anomalous" physical features of cell behavior that we have been alluding to fall into place. These features include: cell water's elevated viscosity up to several orders of magnitude above bulk water's viscosity (Sato *et al.*, 1984; Luby-Phelps *et al.*, 1986; Wang *et al.*, 1993; Bausch *et al.*, 1999)—although fluorescence methods frequently yield values not very much different from bulk water (Fushime and Verkman, 1991; Luby-Phelps *et al.*, 1993); predictable cell size which depends on the volume of adsorbed water; resistance to water loss in desert plants facing months without rainfall; immunity to freezing even in arctic regions; and the cell's ability to remain intact even when demembranated. These features arise directly from water's proclivity to adhere to and structure around cellular macromolecules.

In fact, virtually all of the cell's physical features should differ from those anticipated from bulk water because structured water's physical properties are at variance with those of bulk water. Table 4.1 summarizes many of these differences. The table shows differences of viscosity, compress-

Table 4.1. *Comparison of some properties of pure and vicinal water, (from Clegg and Drost-Hansen, 1991).*

Property	Bulk	Vicinal
Density (g/cm^3)	1.00	0.97
Specific heat (cal/kg)	1.00	1.25 +/- 0.05
Thermal expansion coefficient ($^{\circ}C^{-1}$)	$250 \cdot 10^{-6}$ (25°C)	$300\text{-}700 \cdot 10^{-6}$
(adiab.) Compressibility coefficient (Atm^{-1})	$45 \cdot 10^{-6}$	$60\text{-}100 \cdot 10^{-6}$
Excess sound absorption ($cm^{-1} \cdot s^2$)	$7 \cdot 10^{-17}$	$\sim 35 \cdot 10^{-17}$
Heat conductivity ((cal/sec)/cm^2/°C/cm)	0.0014	~ 0.01-0.05
Viscosity (cP)	.089	2-10
Energy of activation ionic conduction (kcal/mol)	~ 4	5-8
Dielectric relaxation frequency (Hz)	$19 \cdot 10^9$	$2 \cdot 10^9$

ibility, thermal expansion, *etc.* Water adsorbed to macromolecular surfaces is distinct, and cells containing such water will behave differently from predictions based on bulk water.

An exception to the rule of substantial water structuring around protein surfaces is the case of isolated proteins suspended in solution, where structuring is commonly inferred to be restricted to a few layers only. Such inferences are not necessarily relevant to the situation inside the cell because the extensiveness of structuring rests on preservation of the protein's natural configuration, which may not be the case when the proteins are isolated and suspended in dilute solution. Within the cell, the conclusion that a large fraction of water is organized differently from bulk water is practically universal: it has been drawn by all scientists I know of who have recently taken the time to review the biological water field (Wiggins, 1990; Watterson, 1991; Clegg and Drost-Hansen, 1991; Mentré, 1995; Cameron *et al.*, 1997; Vogler, 1998). It is neither new nor radical.

Conclusion

This chapter began with the question whether cell-water molecules might be constrained because of their resistance to freezing. A basis for constraint was identified in the proteins' hydrophilic surfaces. These surfaces contain abundant, regularly disposed surface charges that attract and orient vicinal water into layers. Because the gaps between surfaces are unexpectedly narrow—perhaps five to six water molecules on average—much of the cell's water is potentially structurable, and physical evidence confirms appreciable structuring.

The question that naturally arises is: how might the presence of structured water impact cell function? For functional impact, the most relevant consideration may be the handling of solutes, and this subject is addressed in the next several chapters. If solvency of structured water differs from that of bulk water, the difference could provide a basis for the partitioning of solutes, which could, in turn, provide a potentially straightforward answer to the question of ion separation across the cell boundary.

We deal first with uncharged solutes, and then charged solutes.

5

CHAPTER 5

SOLUTES

In the previous chapter we concluded that cell water was different from bulk water. Water dipoles adsorb onto the charged hydrophilic surfaces that pervade the cytoplasm; the adsorbed dipoles induce additional dipoles to adsorb, creating layers of organized water. The presence of such water in the cell is broadly confirmed (Clegg, 1984; Ling, 1992; Mentré, 1995; Cameron *et al.*, 1997). We now begin assessing the implications.

SOLUTE EXCLUSION

If there is a universal solvent, certainly it is water. Water's broad solvency arises from the molecule's strongly dipolar nature. Dipoles aggregate around the solute's charges, effectively stabilizing the reactant and permitting it to remain independent. That is why sodium and chloride do not aggregate in solution in spite of their opposite charges.

In water that is structured, however, the dipoles are preoccupied. As a result, most solutes are relatively insoluble in structured water, and partition into ordinary bulk water where hydrogen bonds are more freely available for solute hydration. A similar albeit more extreme case is that of ice, where rigidly locked hydrogen bonds tend to exclude solutes altogether. In terms of solvency, then, structured water lies somewhere between ice and ordinary bulk water.

To gain more rigorous insight into how solute exclusion works, we need to consider energetics. For a system at equilibrium, free energy is at a minimum. Thus, the partitioning of a solute between structured and non-structured water can be determined by looking at the standard-free-energy difference between the two phases and establishing the partitioning ratio that minimizes the free energy.

The standard-free-energy difference consists of two terms: an enthalpy (energy) term and an entropy term. The energy term is related to the solute molecule's volume: To move from bulk water to structured water, the solute needs to fill the hole left in the bulk water and excavate a new hole in the lattice of structured water. Digging a hole in the structured lattice is challenging, and larger holes require more work than smaller holes. Thus, more energy will be required to dissolve larger solutes than smaller ones.

An additional factor can sometimes enter into the energy term and modify this simple volume dependence. This factor has to do with the arrangement of surface charges on the solute (Ling, 1992). If the charges are arranged such that they fit particularly well into the structured water lattice, less energy will be required to excavate the hole and solubility will be enhanced. For solutes that fall into this category (see below), solubility will be higher than predicted on the basis of volume alone.

The second term, entropy, is related to the solute's motional restriction. A solute residing in structured water cannot translate or rotate very freely because the environment imposes restriction. Since nature tends to maximize randomness, the imposed restriction is unfavorable; it acts to keep solutes out.

Like the energy term, the entropy term depends on the solute's molecular volume. To visualize this, consider the analogy of the spider web (Fig. 5.1). The butterfly caught up in the web and stuck at multiple sites will be essentially immobile whereas the smaller mosquito will remain free to squirm. The larger species suffers more motional restriction than the smaller. Analogously for solutes, the larger the molecular size the greater the motional restriction and hence the more profound the exclusion.

In sum, hole excavation (energy) and motional restriction (entropy) are both unfavorable for solvency in structured water; solutes should be excluded. Because both terms are proportional to solute volume, larger solutes should suffer more exclusion than smaller ones. The only deviation from this rule is for those solutes whose surface-charge groups fit snugly into the water lattice. Those special solutes may experience less exclusion than the rest. Otherwise, the principle is straightforward.

Exclusion from Cells

We next consider whether solutes are excluded by cell water. If water structuring is a cytoplasmic attribute (Chapter 4), size-based exclusion should be evident.

The most comprehensive approach to this question was taken by Ling *et al.* (1993), who studied intracellular-extracellular partitioning of 21 un-

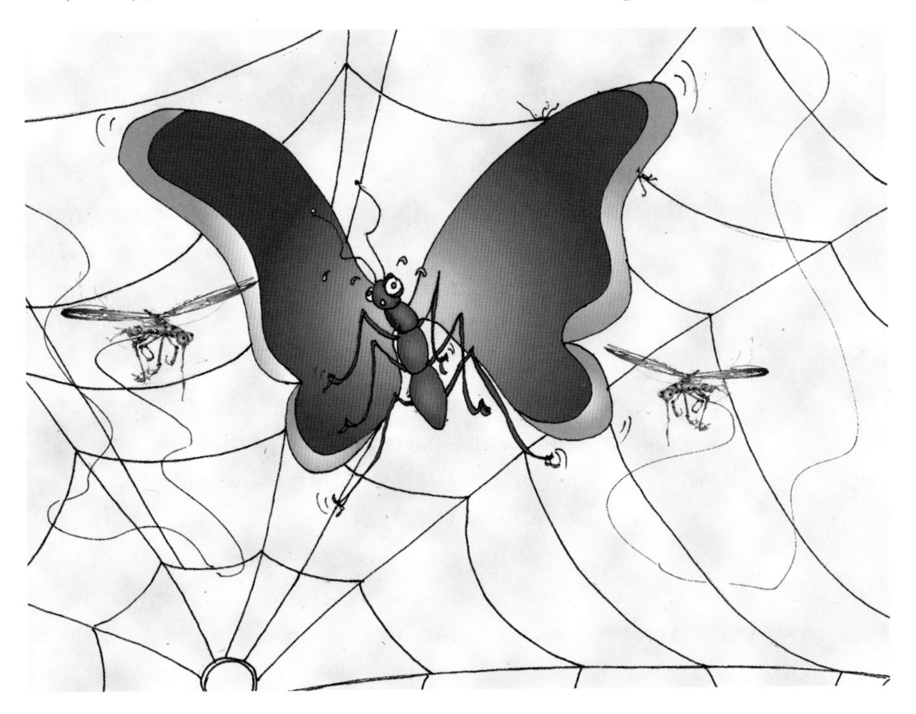

Figure 5.1. *Motional restriction is more pronounced for larger species than for smaller.*

charged solutes including sugars, alcohols, and other non-electrolytes that ranged in mass between 32 and 900 Daltons. Each solute was radio-labelled. The labelled solute was placed in an incubating solution containing a muscle, and the concentration inside the muscle was measured after a time period extending up to six days. Temperature was kept at 0° C to ensure that the cell did not appreciably metabolize any of those solutes that were metabolizable (some were not), which could confuse the results. With these precautions in place, the authors determined how solutes of different size partitioned themselves.

The results were straightforward. First, outside and inside concentrations were consistently proportional even after days of exposure: the higher the concentration of a solute outside the cell, the higher the concentration inside. This simple proportionality made it easy to determine each partition coefficient accurately. The proportionality also implied that the partitioning mechanism was unlikely to involve anything very complicated—no more complicated perhaps than a simple physical difference between the respective fluids.

The second and main finding was that the inside to outside partition ratio diminished as the molecular size increased; *i.e.*, exclusion from the cell was based on size. Smaller solutes entered the cell rather easily with steady-state partition coefficent near 1.0. Larger ones encountered more difficulty, partition coefficients dropping progressively down to 0.08. Of the 21 solutes studied, 14 fell on a single curve (Fig. 5.2). For those solutes, exclusion depended uniquely on molecular weight or, in turn, on computed solute volume, as expected.

For the remaining seven solutes the partition coefficient also increased with molecular size, but the curve was shifted slightly to the right (Fig. 5.2). The rightward shift implies that this class of solutes enters the cell relatively more easily. At least five of the seven solutes in question (*e.g.*, ethylene glycol, glycerol) are recognized cryoprotectants, which lower the cell's freezing point much like anti-freeze proteins (Chapter 4). If these cryoprotectants act by inducing vicinal water to become more tightly structured, then their snug fit into the lattice is implied by their function. The rightward shift is understandable.

An implication of the solute-exclusion principle is that solutes within the cell should not diffuse very rapidly. After all, if solutes are excluded from the water lattice why would they have much proclivity to diffuse through it? If such rationale is not immediately obvious, think of ice—the most extreme example of a structured water lattice. Imagine how rapidly a solute like sugar diffuses from one side of an ice cube to the other.

The expectation of slow intracellular diffusion is confirmed by experimental evidence, although the results are somewhat mixed. At one end of the spectrum are studies indicating only a modestly reduced diffusion rate relative to the bulk: high energy phosphates for example are reported to diffuse at half the rate as in bulk water (Yoshizaki *et al.*, 1982; Hubley

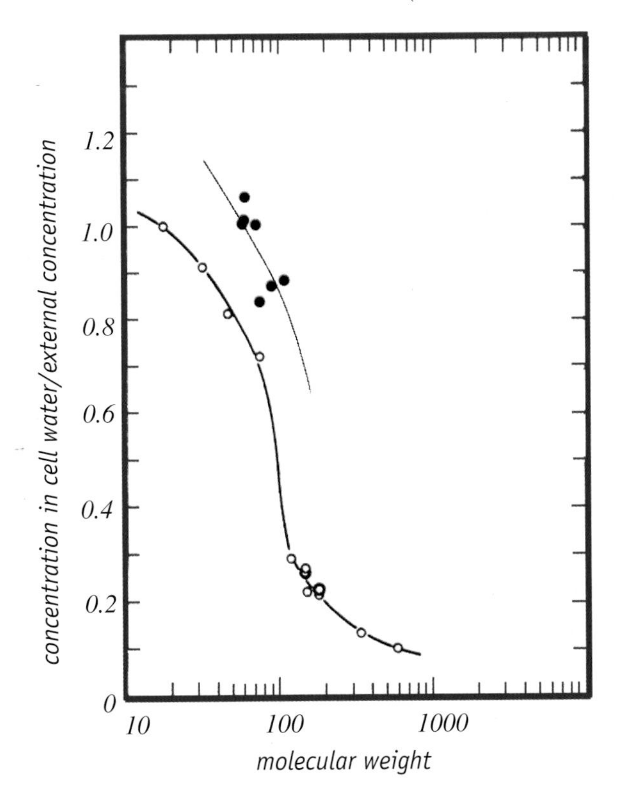

Figure 5.2. *Experimental partitioning of solutes of different size between inside and outside of cell. After Ling* et al. *(1993).*

et al., 1995). Other solutes show diffusion rates much reduced. For various dextrans, diffusion is reported to be 5 - 15% of that in free solution (Peters, 1984; Luby-Phelps *et al.*, 1986), with micron-scale diffusion sometimes requiring days (Kraft *et al.*, 1995). For large solutes including soluble proteins, the diffusion rate is typically on the order of 10% to 1% of that in free aqueous solution (Wojcieszyn *et al.*, 1985; Maughan and Lord, 1988). Some diffusion impairment is anticipated from the presence of intracellular obstacles, but the magnitude is not quantitatively explainable (Gershon *et al.*, 1985); some other feature of the environment seems responsible, and structured water is certainly a candidate.

On the other hand, virtually all intracellular processes require rapid solute movement—which implies diffusional rates qualitatively higher than much of the evidence suggests. Solutes do move rapidly when cells are activated; such movement will be treated in detail in later chapters, and appears to be associated with transient water destructuring. Here we deal with the unactivated state, where impaired diffusion is consistent with water structuring.

* * *

The chapter's objective has been met. We set out to determine how solutes behave in structured water. Theoretical arguments predicted size-dependent exclusion, and this prediction was confirmed—solutes are excluded from the cytoplasm according to size. The solute-exclusion paradigm is simple, and offers the beginnings of a framework for function. All is well—but one issue remains dangling.

Remember the raw egg? From the egg's consistency (particularly when fresh), we likened the cytoplasm to a gel, and it was on this basis that we first speculated that the cytoplasm might behave differently from an ordinary aqueous solution. Cytoplasmic water has been amply addressed, but the issue of whether the cytoplasm is gel-like has not yet been dealt with.

Armed with some understanding of water and solutes, we can now address this question. The question is significant, for gel science and technology are rapidly expanding disciplines which have found application in fields ranging from timed drug-release capsules to compact-disc lenses, and have potentially much to teach students of the cell. Thus, we consider whether the cytoplasm has a gel-like character.

CELLS AND GELS

What exactly is a gel? You may be surprised to learn that an agreed-upon definition is lacking. It is much like the U.S. Supreme Court's definition of pornography—it's hard to define but you know it when you see it. And so it is with the gel. Perhaps the most useful definition is one that came from a paper by Almdal *et al.* (1993) entitled *Towards a Phenomenological Definition of the Term 'Gel.'* In this view, gels are soft, solid or solid-like materials that consist of two or more components, one of which is a liquid of some abundance. Thus, common gelatin gel ("Jello" in the U.S.) is built of a meshwork of cross-linked collagen strands plus ~95% water.

Why does the water stay put? In the cell we surmised that the water stays put because it is adsorbed onto protein surfaces. This explains why cells don't dry out and why cell volume is predictable. A similar retentive force must exist in the gel, for its water is also held. The blob of Jello does not immediately shrivel up and collapse; its 5% collagen retains the 95% water. Other (hydro)gels may retain their 99.9% water (Osada and Gong, 1993).

The explanation for water retention commonly given by gel scientists is that the surfaces of the matrix are hydrophilic—they adsorb water. Matrix polymers that are slightly hydrophilic will adsorb modest amounts of water, while polymers that are strongly hydrophilic can adsorb water up to thousands of times their volume. Thus, the term hydrophilic implies not just one but many layers of adsorbed water. Although gel scientists refer to such water as "adherent" water (Jhon and Andrade, 1973), there is little reason to presume it is any different from the "structured" water that adheres to proteins in the cell; the semantics may differ but the principle is the same.

The extent of water structuring in the cell depends on the proximity of apposing surfaces, and the same is true in the gel. A study of common gels showed that up to a saturating limit, the degree of water structuring varied with the degree of polymer cross-linking. When polymer strands were loosely arranged, NMR evidence confirmed little water restriction; but as more extensive cross-linking brought filament surfaces closer together the extent of restriction likewise increased (Yasunaga and Ando, 1993).

Water structuring also implies resistance to freezing, and such resistance is as much a feature of gels as of cells. To confirm this, purchase a package of Knox gelatin, make a 1% solution in a glass container, cover it with plastic wrap and put it in the freezer without jarring it to nucleate ice formation. Even at –15° C it will remain gel-like for months. The constrained water does not enter the ice-crystalline state—yet another indication of the gel water's structured state.

Cells and gels are similar also in their physical consistency—although it goes without saying that not all cells have the consistency of Jello. Consistency depends on the concentration of protein as well as the nature of the respective protein surfaces. Thus, the muscle cell, densely packed with highly charged protein filaments, will be Jello-like, whereas the embryologically undeveloped egg cytoplasm, relatively free of organelles and structure, will be more liquid-like (and will freeze more easily).

Another general basis for comparing gel and cell is in the respective behavior of solutes. If the cytoplasm is like a gel, then the size-based solute exclusion that characterizes the cell should be a feature of the gel as well. Extensive studies of solute permeation have been carried out in gelatin gels. On the basis of exclusionary behavior, water seemed divided into two fractions. In one fraction water was fully available to any size solutes, whereas in the other fraction, sugars, alcohols and other solutes were excluded on the basis of size (Gary-Bobo and Lindenberg, 1969).

Other types of gel also show solute exclusion. Sheet-like gels made of cellulose-acetate polymer contain 3-nm pores filled with water. A large fraction of this water will not freeze at temperatures as low as -60° C (Frommer and Lancet, 1972); nor does it readily admit solutes (Taniguchi and Horigome, 1975). In polyvinylmethyl ether, exclusion of sucrose was reported with a partition ratio between 0.1 and 0.4 (Ling *et al.*, 1980). In ion-exchange-resin gels, solutes were found to be excluded according to their molecular volume (Ginzburg and Cohen, 1964). Thus, the solute-exclusion paradigm applies as much to the gel as it does in the cell: to the extent that the gel's water is structured, size-based exclusion is confirmed.

Gels and cells are thus similar on many counts: water structuring, resistance to freezing, exclusion of solutes, and physical consistency. Polymer-gel science will have much to contribute to cell biology. Stay tuned.

Conclusion

With this chapter we began to consider the impact of structured water on cell function. Thermodynamic principles implied that solutes should be excluded from the cell and that the extent of exclusion would depend principally on the solute's volume. This expectation was confirmed. And it was likewise confirmed in the gel, which bears many physical chemical similarities to the cell.

Solute behavior is explored further in the next chapter. We see that exclusion applies not only to the non-electrolytes we have been considering, but also to ions. Solute exclusion will be a principal factor in the partitioning of ions between the inside and outside of the cell.

NERNST'S
IONIZED SALT
THIS SALT SUPPLIES CHARGE, A NECESSARY NUTRIENT
NET WT. 16 OZ. (1LB., 0 OZ.)

CHAPTER 6

IONS

Ions experience forces arising from their charge. Positively charged ions are drawn toward the proteins' negative surface charges, while negative ions are drawn toward the proteins' positive charges. Ions could thus adsorb as flies stick to fly paper. If so, partitioning between inside and outside of the cell would rest not only on solubility considerations, but also on the extent of adsorption to cytoplasmic proteins. Both of these factors need to be considered to understand ion distribution (Table 1.1)—a feature central to function.

SOLUBILITY

In the previous chapter we established that the major factor underlying solubility in the cell was size: the larger the solute, the lower the solubility in structured water. If this paradigm holds true for charged solutes as well, then the larger ions should be preferentially excluded from the cell. We see in a moment that this expectation holds in the general sense.

Establishing ion size, however, is not merely a matter of determining the dimensions of the bare ion. Those values are known from X-ray crystallography: sodium, for example has an ionic radius close to 0.95 Å, and potassium, 1.33 Å. But bare-ion dimensions have little meaning in aqueous solution because the ion attracts a shell of structured water, which confers solubility. The ion is said to have an "hydrated" diameter, which may exceed bare ion diameter by a considerable amount.

The standard way of inferring hydrated size is from diffusibility. In aqueous solution smaller solutes diffuse more rapidly than larger ones. This fact was recognized by Einstein, who reasoned that high solute mobility must be based on low friction. Einstein applied Stoke's law of friction and came up with the now-famous Stokes-Einstein relation. This relation predicts that a solute's diffusion coefficient should vary inversely with its hydrodynamic radius. Thus, size could be determined from the measured diffusion coefficient.

Applying this formulation yielded unexpected conclusions. For the smaller set of cations, hydrated diameter and bare ion diameter turned out to be inversely related. Sodium's hydrated diameter was inferred to be around 5 Å, while potassium's was 3.6 to 4.0 Å. This counter-intuitive result arises because the more compact sodium ion is known to set up a stronger electric field, which attracts more layers of dipolar water. The hydrated diameter difference is only about 25 - 30%, but the resulting volume difference calculates to approximately two times. Thus, hydrated sodium is twice as bulky as hydrated potassium, and should therefore suffer more profound exclusion from structured water.

Relative exclusion applies not only to sodium and potassium but to all ions in the so-called Hofmeister or lyotropic series. Biologically relevant members include: $Mg^{2+} > Ca^{2+} > Na^{+} > K^{+} > Cl^{-} > NO_3^{-}$. Ions at the left have the highest field intensity—they attract the largest shell of structured water and are thus classified as "structure-promoting." Those ions at the right, conversely, are "structure-breaking." Ions will partition into water whose hydrogen bonds are most readily available to satisfy their craving for solvation, bulk water preferred over structured because it is less self-associated. Partitioning toward the bulk will be most pronounced for ions that build the largest hydration shells (Wiggins, 1990; Wiggins and van Ryn, 1990; Vogler, 1998). Thus, Mg^{2+} and Ca^{2+} will shun structured water, whereas Cl^{-} and NO_3^{-} will enter more easily.

Lying somewhere near the middle of this continuum are Na^{+} and K^{+}.

Their contiguous positions, however, do not necessarily imply comparable solubilities. Extent of hydration changes markedly at a critical ion size where the strengths of ion-water and water-water interactions are equal. For monovalent cations this occurs at an ionic radius of 1.06 Å (Washabaugh and Collins, 1986; Collins, 1995)—roughly midway between Na^+ and K^+. Sodium's hydration-shell size will therefore be considerably larger than potassium's—just as Einstein concluded—and its solubility will be substantially lower.

It is tempting to suggest, therefore, that the reason sodium concentration is so low inside the cell is because of low solubility in cell water. Sodium may simply partition out of cell water and toward the bulk.

A direct test of this thesis is to offer sodium to the cytoplasm and see how much of it enters. The experiment has been carried out (Ling, 1978).

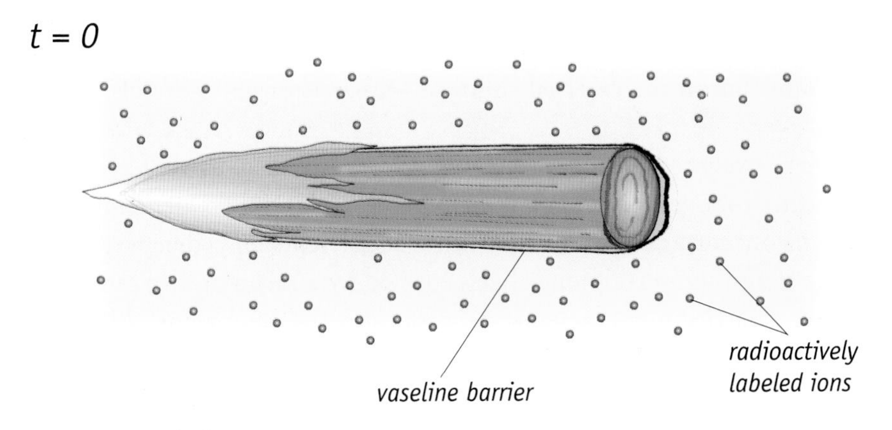

Figure 6.1. *Cut end of muscle is exposed to a bath containing the relevant ion. Diffusion rate into the muscle is measured.*

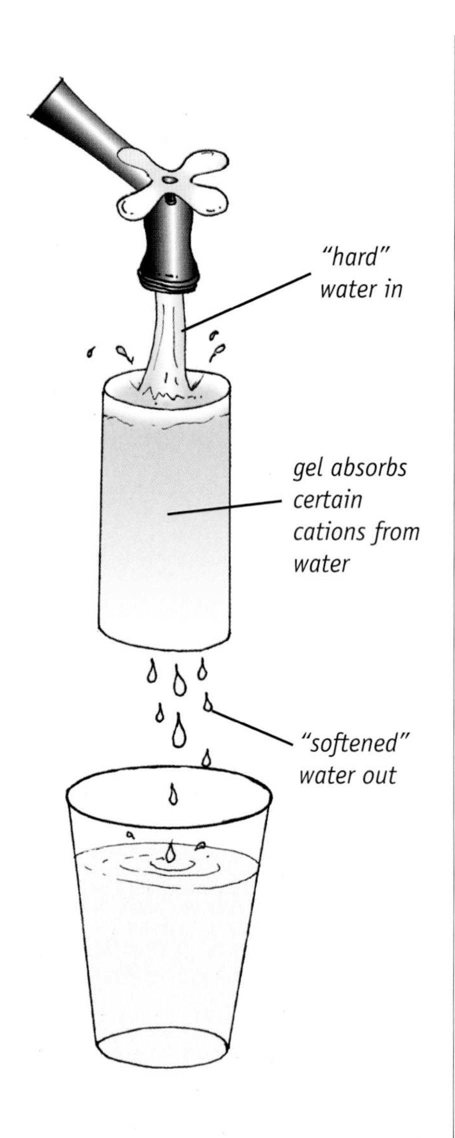

Figure 6.2. *Principle of ion-exchange gel.*

A frog muscle was sliced transversely—cut like a sausage (Fig. 6.1). The cut end of the muscle was exposed to a salt solution that contained radioactively labeled sodium ions. To preclude any ion exchange through the membranous boundary, the cylindrical surface of the muscle was suspended in air or coated with vaseline. By measuring the radioactive label accumulated inside the muscle after substantial exposure, partitioning between bulk water and cell water could be evaluated.

Despite its high concentration in the surrounding salt solution, sodium remained appreciably excluded from the cell interior. Exclusion could not have been due to resealing because no membrane is regenerated (Cameron, 1988). In the zone just inside the cut-end where the knife-induced injury disrupted the cells' proteins, sodium was not excluded (*cf.* Kushmerick and Podolsky, 1969). But several centimeters beyond, where the resting potential remained normal and the cytoplasm's pristine nature was uncompromised, sodium was excluded. Even after many days' exposure, sodium concentration continued to remain at a low value of 15% of that in the external solution—a partition ratio of 0.15.

Sodium is excluded not only from cytoplasm but also, as we have come to expect, from gels (Negendank, 1982). In gelatin gels, depending on collagen concentration, sodium partition coefficients ranged between 0.89 and 0.54; and in non-biological gels formed by a range of different sized polymers, partition coefficients were found to be as low as 0.22 (Ling, 1992). Thus, structured water excludes sodium both in the cell and in gels.

It is instructive to compare the degree of sodium exclusion achieved through partitioning (the cut-end experiment) with that characteristic of the intact cell. In the cut-end experiment the inside-to-outside partition ratio was 0.15, while in the cell it is typically 0.10. This similarity lends weight to the thesis that sodium's low concentration in the intact cell may be explainable principally on the basis of size-dependent exclusion.

ION ADSORPTION

Apart from solubility, a second factor that could affect ion accumulation inside the cell is protein affinity. In much the same way that water dipoles adsorb to protein surfaces, charged solutes may also adsorb. With sufficient affinity, clinging ions could concentrate themselves in some abundance.

The ion-affinity principle is neither mysterious nor arcane; it is exploited in common every-day devices such as water softeners (Fig. 6.2). The softener is a so-called ion-exchange resin, a polymer gel that "softens" the passing water because of its high affinity for the heavy metal ions that otherwise "harden" the water. A typical cation exchanger will be built of a negatively charged resin initially loaded with sodium. Since the resin's affinity for sodium is less than for the heavier potassium and calcium ions found in the wa-

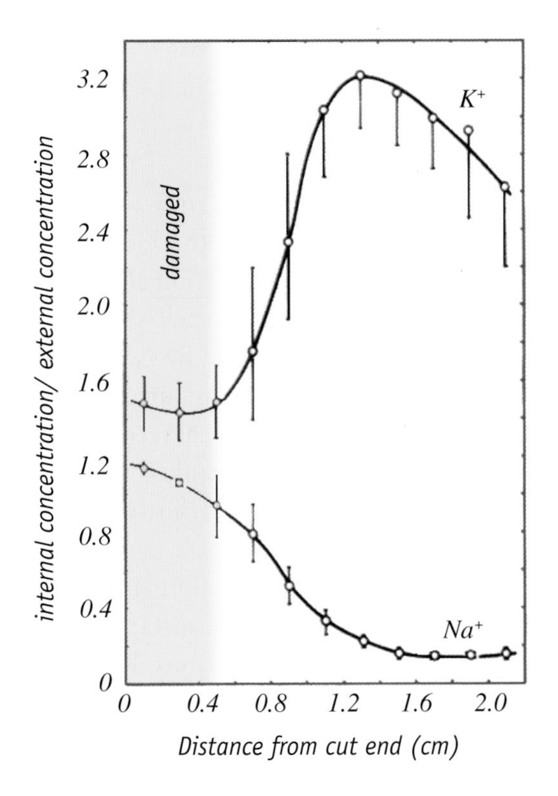

Figure 6.3. *Experimental results obtained using the "cut-end" muscle preparation. After Ling (1992). See text for description. Sodium ions in the bath are excluded from the cytoplasm, whereas potassium ions accumulate in high concentration.*

ter, the gel will lose its sodium in exchange for these cations. Adsorption of heavy cations softens the passing water.

This capacity for selective ion adsorption could well be shared by proteins—for proteins, after all, are biological polymers. Consider the cut-end experiment. Figure 6.3 shows that the same region of the cell that excludes sodium actually accumulates radioactively labeled potassium. The potassium concentration deep into the muscle cell is ultimately about three times the value in the immersion bath. Something in the cytoplasm evidently accumulates potassium while excluding sodium—much the same as the ion-exchange gel.

To understand the basis for selective adsorption we return to the cell's proteins, whose surfaces are densely coated with charge. Ions and water dipoles compete for these charges (Chapter 4). Positively charged sites will attract negative ions such as chloride, and negatively charged sites will attract sodium and potassium. For adsorption to take place, the ion's hydration shell must be removed. Shedding sodium's larger shell requires more energy than removing potassium's smaller one. Thus, potassium adsorption will be favored over sodium adsorption (*cf.* Ling, 1952; Joseph *et al.,* 1961).

Adsorption-affinity differences apply not only to potassium and sodium but to all ions. Once again they follow the Hofmeister series. Ions at the left adsorb poorly because their hydration shells are the largest; ions at the right adsorb more readily because their shells are smallest. Thus, Mg^{2+} and Ca^{2+} will ordinarily be displaced by Na^{+}, which is displaced by K^{+}. The situation differs, however, when surfaces to which these ions bind are closely juxtaposed. Negatively charged surfaces in close proximity of one another can be easily cross-linked by divalent cations such as Ca^{2+}, conferring an anomalously high affinity on these ions. The interplay between divalents and monovalents at closely apposed anionic surfaces will play a central role in intracellular dynamics (Chapters 9 and 10). For ordinary charged polymeric surfaces, however, the bottom line is that among biologically relevant cations, K^{+} adsorbs most strongly.

To determine how much potassium the cell's proteins can adsorb, consider the numbers. Negative surface-charge concentration inside the cell is calculated to be about 1.6 M (Wiggins, 1990). The concentration of potassium is typically 0.15 M. Thus, the number of negative surface charges outweighs the number of potassium ions by ten times. Even if 90% of the fixed charge sites were occupied by water dipoles, ample sites would remain available to accommodate every one of the cell's potassium ions (as well as other, less abundant ions). The capacity for ion adsorption is astonishingly high.

Evidence for Potassium Adsorption

To see whether such theoretical capacity translates into actual adsorption requires a look at the evidence. Although appreciable binding of potassium is implied by the cut-end experiment, and a very low potassium-diffusion rate is reported (Ling and Ochsenfeld, 1973), one needs to determine whether such results are representative of the broader spectrum of evidence.

The first evidence that comes to mind is the potassium-efflux experiment (Chapter 3). Following membrane dissolution, recall that potassium did not leak out of the cell until after a significant lag period, and when finally it did leak it went with the cell's proteins (Kellermayer *et al.*, 1986). This result implied a linkage between potassium and proteins. Experiments of similar nature repeated on red blood cells with different methodologies gave similar results (Kellermayer *et al.*, 1994): over a spectrum of experimental conditions that led to a twenty-fold range of potassium leakage rates, protein and potassium efflux rates were consistently the same. This confirmed the protein-cation linkage.

A second experiment that addresses the potassium-adsorption issue uses a clever "aqueous droplet" technique (Maughan and Recchia, 1985). In this approach a small droplet of pure water is deposited onto the surface of a demembranated muscle cell. The droplet forms a reservoir into which diffusible intracellular ions can partition (Fig. 6.4). From the observed partitioning, the investigators concluded that 52% of the cell's potassium was bound. The real bound fraction could be even higher than 52% because some of the potassium was likely drawn from the large free-diffusion spaces that typically open between the cell's myofibrils during demembranation.

A third class of experiment focuses on the way potassium is distributed within the cell. If protein surfaces adsorb potassium, regions high in protein should be rich in potassium, whereas protein-deficient regions ought to be potassium deficient. This prediction has been tested and confirmed again and again by the German biophysicist Ludwig Edelmann and others, using a variety of different approaches. Outlined below, these experiments were carried out on muscle cells, where alternating A- and I-bands (Fig. 3.2) provide natural zones of high and low protein density that are convenient for testing the prediction.

Figure 6.4. *Aqueous droplet technique. Free ions from the cell diffuse into the droplet until equilibrium is established.*

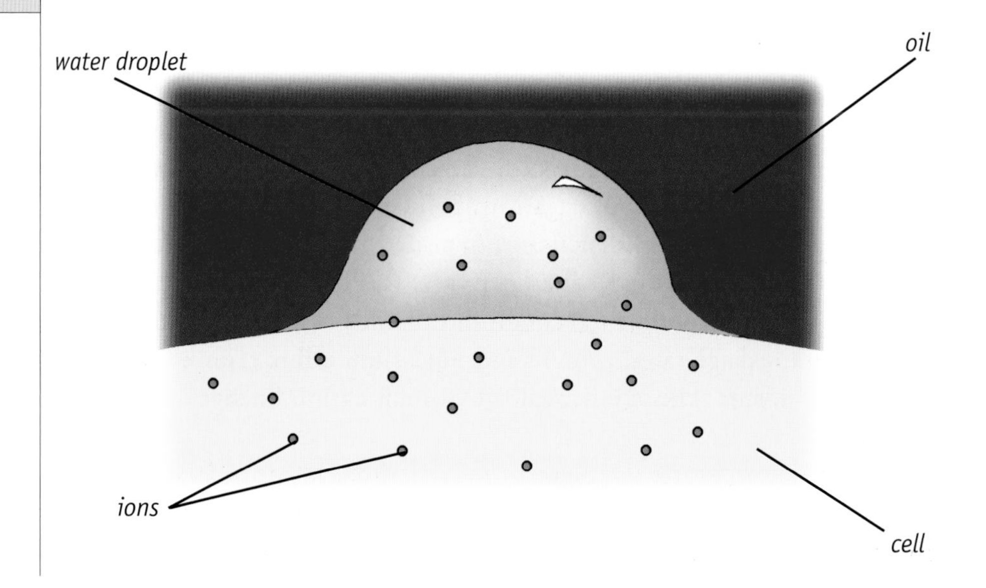

One approach involved X-ray microanalysis. This method localizes atoms relative to the cell's organelles by exploiting each atom's distinctive scattered energy profile. More potassium was contained in the A-band than in the I-band (Edelmann, 1978; Trombitás and Tigyi-Sebes, 1979). This distinction was later reconfirmed (Edelmann, 1983) in an experiment that also showed that a misleading result could be obtained if the potassium-rich Z-line, which bisects the I-band, was included in the sample (*cf.* Somlyo *et al.*, 1981). When the Z-line is properly excluded, microanalysis results, including followup experiments in another laboratory (von Zglinicki, 1988), uniformly confirm a higher potassium content in the A-band than in the I-band. Potassium concentrates where protein concentrates.

Another approach was based on mass spectrometry (Edelmann, 1980). The first step was to prepare an electron micrographic section of muscle. Then, a laser beam vaporized selected regions of the section and the vaporized ions were analyzed quantitatively by the mass spectrometer. This general approach is given the appropriate acronym LAMMA, or laser-microprobe-mass analysis. Like the X-ray microanalysis results, the LAMMA results showed more potassium accumulation in the A-band than in the I-band.

The LAMMA results confirmed a strong binding preference of potassium over sodium. Even when the bathing solution contained one tenth as much potassium as sodium, A-band potassium accumulation continued to exceed that of sodium, by a factor of two. These numbers imply a potassium-to-sodium preference on the order of 20:1, which is similar to the ratio in the living cell.

Yet another approach traps ions in place by quick-freezing, and localizes them by electron microscopy (Edelmann, 1978). Structures seen in the electron microscope are ordinarily visualizable because they are stained with heavy metals, which are good electron scatterers. When heavy metals are omitted, observable structures arise out of scattering from naturally available elements—here mainly potassium. Potassium "stained" the A-band but not the I-band. Selective accumulation in the A-band was even more clearly demonstrated when potassium was replaced by its surrogates such as cesium, thallium, or rubidium, which are superior scatterers by virtue of their increased weight; with these surrogates, the A-band preference over the I-band was dramatic (Edelmann, 1978).

This array of experiments confirms that potassium is concentrated where proteins are concentrated. They co-localize. The result is opposite that predicted if potassium were dissolved, because the protein-dense regions contain relatively less solvent space and should therefore have had fewer potassiums, not more potassiums. Linkage between protein and potassium is supported as well by the results cited earlier in this section: the aqueous droplet experiment; the membrane-disruption experiment; and the cut-end experiment. All imply that potassium is largely adsorbed onto protein surfaces.

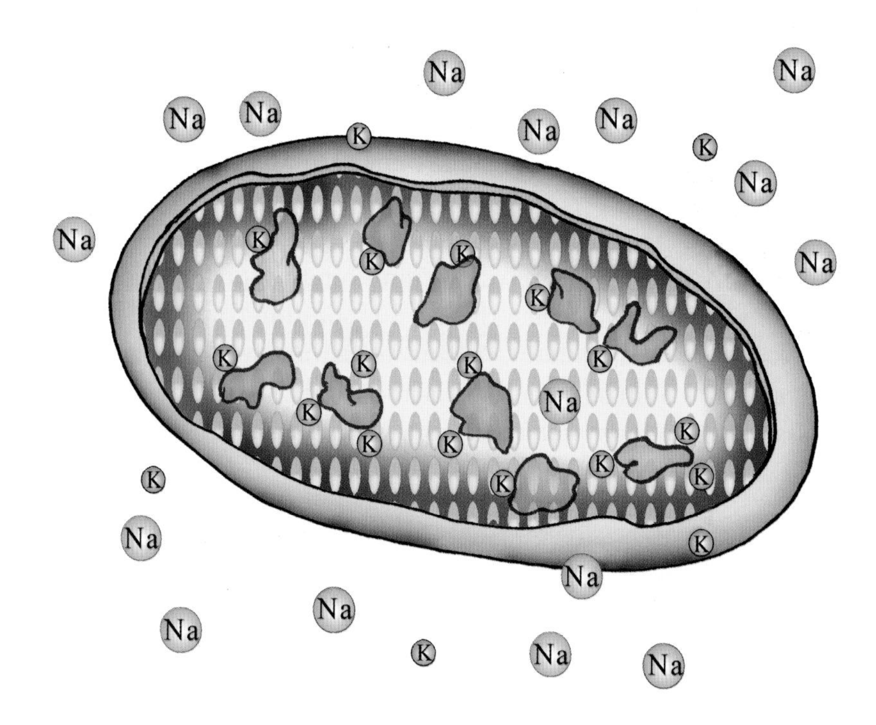

Figure 6.5. *Summary of ion partitioning mechanism. Cell proteins bind potassium preferentially, and cell water excludes sodium preferentially.*

CONCLUSION

In this chapter we considered why the cytoplasm contains high potassium and low sodium. We found that the difference arises directly out of the physical chemistry of these ions in relation to the cell's water and proteins (Fig. 6.5).

Cell water excludes ions because it is structured. Exclusion is more pronounced for sodium than for potassium because sodium's hydration shell is larger and hence more difficult to accommodate in the structured water lattice. Thus, intracellular sodium concentration remains low, whereas potassium can partition more easily into the cytoplasm.

A second discriminating site is the protein, whose surface charges provide adsorption sites for ions. These ions include Cl^- (Chapter 7) as well as cations such as $Na^+$and K^+. Potassium is adsorbed more readily than sodium because its smaller hydration shell can be removed with relatively less energy. It thereby concentrates itself inside the cell.

A salient feature of this partitioning mechanism is that the system is in equilibrium. No energy is required for maintenance. No special widgets are needed to keep potassium inside the cell or sodium outside the cell—all of this follows directly from the cytoplasm's basic physical chemical features.

Dr. Megavolt, 2000, by Dr. Austin Richards

7

CHAPTER 7

CELL POTENTIALS

Originally trained in electrical engineering, I was flabbergasted to learn in my first physiology class that the very same charges and potential gradients that drove induction motors, surrounded power lines, and gated semiconductor junctions also pervaded the living cell. Gauss's Law could hold immediate relevance after all. The charm of this revelation has yet to evaporate.

The existence of a potential difference between the inside and outside of a cell was first inferred more than 150 years ago when Matteucci placed a galvanometer between the muscle surface and a cut in the muscle's belly, thereby detecting an "injury" potential. But it was Gilbert Ling, the same Ling whose work I have been citing in these chapters, who was a key player in measuring the true intracellular potential. Ling had just arrived from China, a bright young scholar chosen from among many candidates to receive a Boxer Fellowship to study in the U.S. Landing in the laboratory of the prominent physiologist Ralph Gerard, Ling perfected the glass microelectrode, a solution-filled, tapered glass capillary that could be stuck into a cell to probe the inside-to-outside potential difference. Out of this simple invention grew virtually all modern concepts of cell electrophysiology, including the now widely used patch-clamp method to study membrane currents, which landed the developers a Nobel Prize.

Cell electrophysiology is the topic of this chapter, particularly the genesis of the cell potential. As we shall see, the cell potential arises inevitably and naturally from the physical features of the cytoplasm.

CHARGE

The proposal that the cytoplasm itself might hold relevance for creating the cell potential is not without experimental underpinning. Potentials of appreciable magnitude are seen in demembranated specimens (Naylor and Merrillees, 1964; Weiss *et al.*, 1967; Miller, 1979; Collins and Edwards, 1971; Stephenson *et al.*, 1981; Bartels and Elliott, 1985; Naylor *et al.*, 1985). They are also seen in colloidal suspensions (Troshin, 1948). And they are reported in gels (Collins and Edwards, 1971). Since none of these specimens contains a functional membrane, the measured potential difference between inside and outside must arise from the bulk properties of the gel/cytoplasm itself.

The existence of such gel/cytoplasm potentials is well recognized and their presence is generally ascribed to a "Donnan" mechanism. In other words, they are thought to arise out of an interaction between the proteins' fixed charges and the ions' mobile charges. Irrespective of mechanism, the point is that specimens devoid of a membrane and studied even under fairly arbitrary conditions can manifest potentials up to 50 mV (Collins and Edwards, 1971).

Our approach to a mechanism bypasses the formalism of the Donnan mechanism (*cf.* Overbeek, 1956) which applies specifically to *mobile* ions. We consider instead what transpires in situations in which ions are largely protein-adsorbed (Chapter 6).

Consider first the magnitude of charge contributed by the relevant players. Total protein charge can be estimated in a back-of-the-envelope calculation. By knowing each amino acid's fractional content (surprisingly uniform across proteins), and charge, along with average amino-acid mass (126.4 Daltons), one can estimate the charge per unit protein mass. Per kilogram of protein the figure is 1.6 mol negative charge and 1.01 mol positive charge (Wiggins, 1990). Thus, excess negative charge of about 0.6 mol/kg is borne by typical cell proteins. Additional negative charge comes from glycoproteins, nucleic acids and other non-protein constituents, but their contribution is inconsequential because their numbers are relatively smaller.

Compare this excess negative charge with the charge carried by the cell's ions. The concentration of positively charged ions, mainly potassium, is about 160 mM. With a cell-water content of 3 kg/kg dry weight of protein, this figure translates to approximately 0.5 mol potassium per kg dry protein. This positive charge is partially offset by negatively charged ions, mainly chloride, at 0.2 mol per kg. Thus, the cell's ions, largely potassium and chloride, present a net positive charge (0.3 mol/kg). This is not sufficient to balance the proteins' excess negative charge (0.6 mol/kg). The cytoplasm, by this calculation, remains negatively charged (Fig. 7.1).

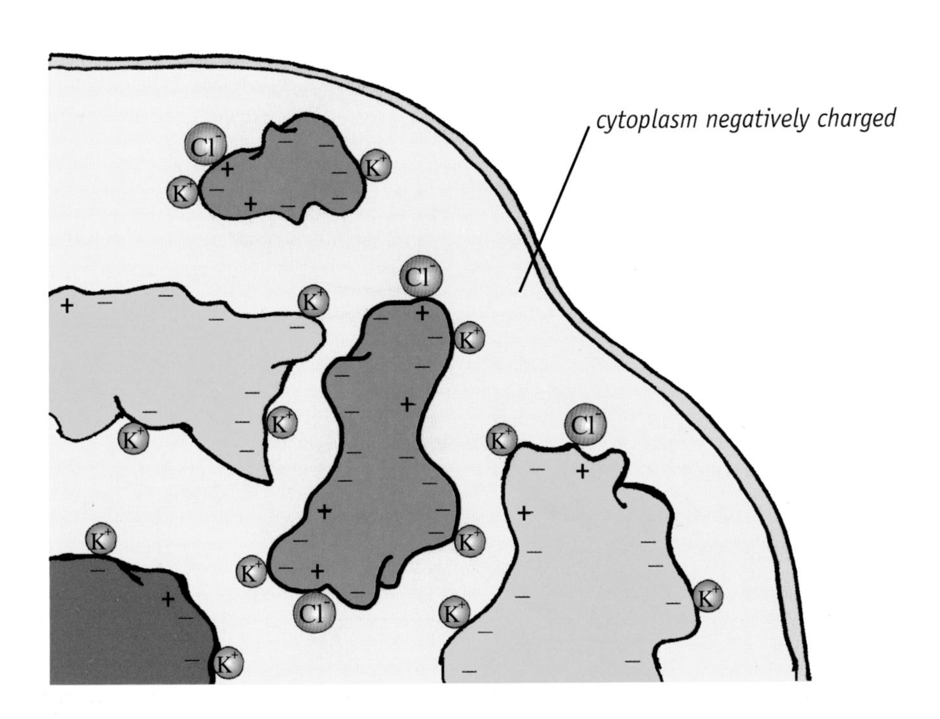

Figure 7.1. *Charge status of the cytoplasm. Because cation charges do not balance protein charges, the cytoplasm remains negatively charged.*

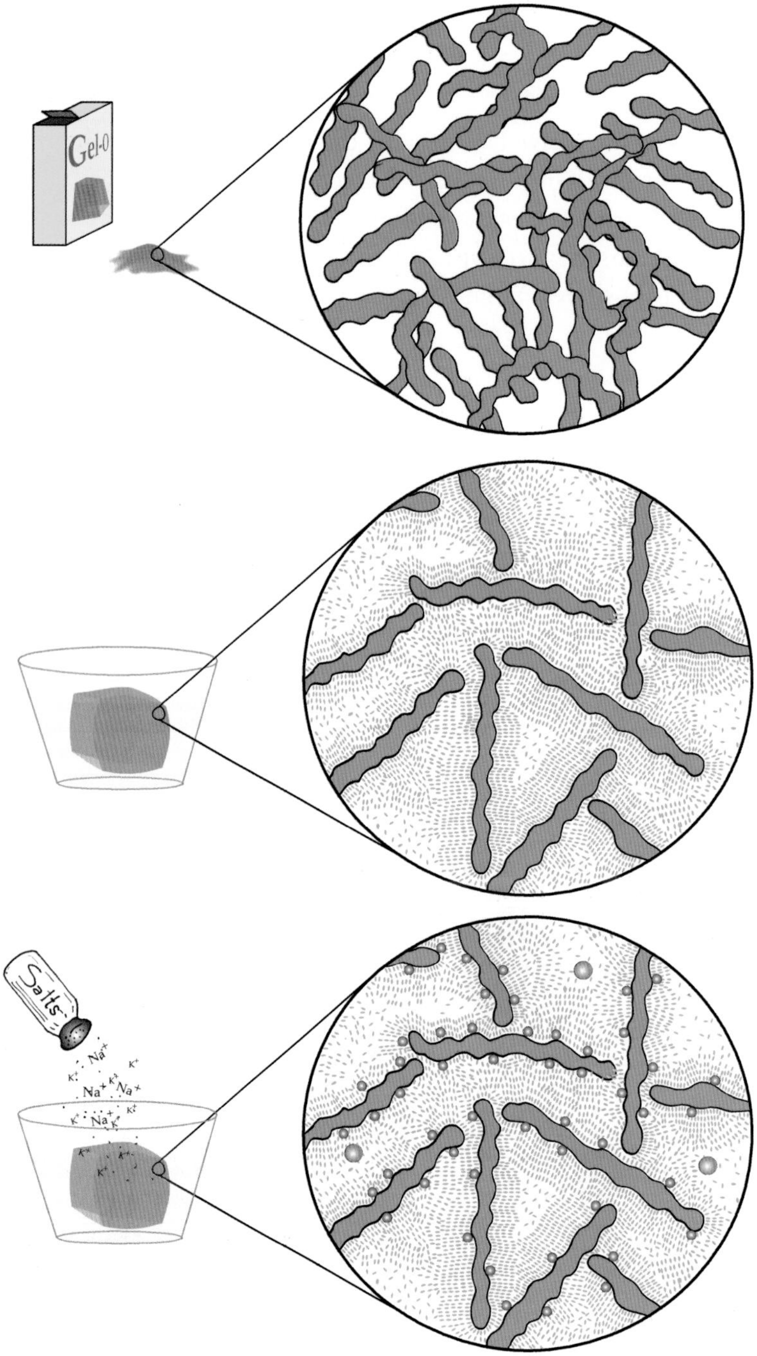

Figure 7.2. *Hypothetical protein network (top). Immersion into water hydrates the protein into a gel. Addition of salt results in ion adsorption.*

To appreciate the consequence of this net charge, imagine a hypothetical network of protein initially devoid of any water or ions (Fig. 7.2). Now add water molecules. Because of their attraction to the proteins' surface charges, vicinal water dipoles will adsorb, orienting themselves layer upon layer over the protein surface. The influx will continue until the drawing potential of the proteins is exhausted—or put another way, until the system's free energy has reached a minimum. At that stage the matrix will be fully hydrated—much like a hydrated gel.

Next, add ions. Immerse the protein gel in a bath containing physiological concentrations of NaCl and KCl, omitting ions of less quantitative significance to keep matters simple. The sodium ions will remain largely excluded because of their low solubility in structured water. Potassium ions, being more soluble, will penetrate; and because of their high protein affinity they may displace adsorbed water, clinging to the protein's surface in high concentration. Chloride, being negatively charged, will avoid the matrix's negatively charged environment and remain largely excluded.

Nothing is particularly startling in this scenario—all elements follow from previous chapters. But there is one new quantitative feature. If this hypothetical protein matrix resembles the one inside the cell, the quantity of potassium that is able to partition into the system will be insufficient to neutralize all of the protein's excess negative charge; the matrix will remain negatively charged.

Now, stick a microelectrode into this gel-like matrix. Given the excess negative charge, the microelectrode should register a negative potential, and that is precisely what is seen. Microelectrodes stuck into anionic gels or demembranated cells routinely report

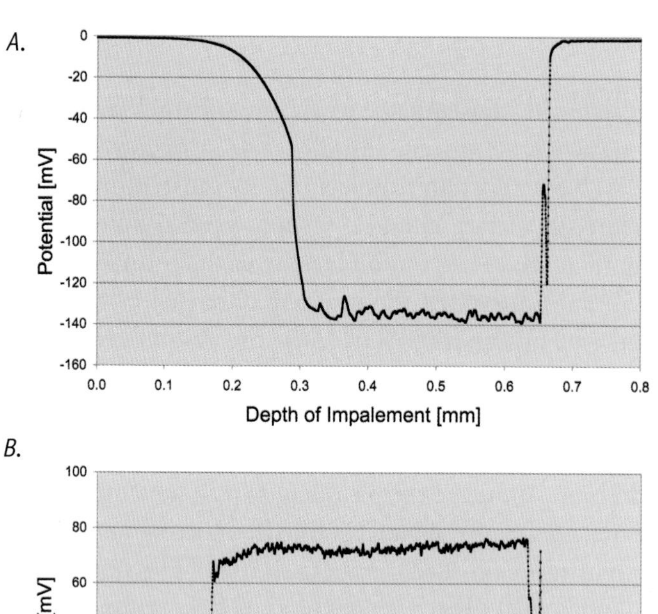

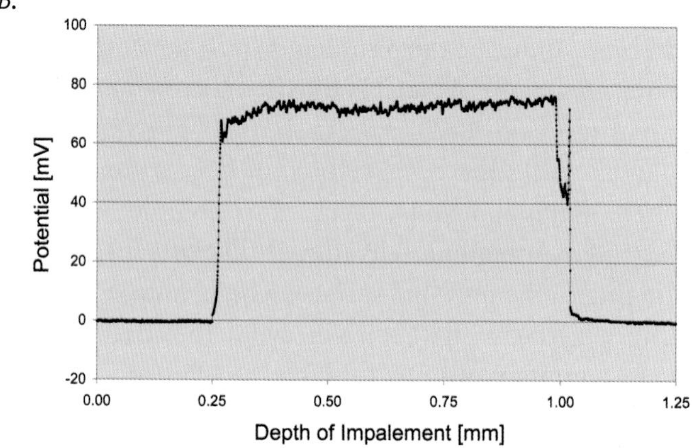

Figure 7.3. *KCl-filled microelectrodes stuck into gel strips at slow constant velocity and then withdrawn.* A: *Typical anionic gel, polyacrylamide/ polypotassiumacrylate shows negative potential.* B: *Typical cationic gel, polacrylamide/ polydiallyldimethylammoniumchloride shows positive potential. Courtesy of R. Gülch.*

negative potentials relative to the outside (see above). A recent example, supplied to me by Prof. Rainer Gülch, is shown in Figure 7.3. It confirms gel potentials on the order of 100 mV or more.

If negative potentials arise from excess negative charge, then charge and cytoplasmic/gel potential should go hand in hand; agents that affect matrix charge ought to affect potential in a similar way. Consider the effect of salt. When the KCl concentration in the bath is lowered, fewer potassium ions are available to diffuse into the protein matrix. The matrix will therefore be less neutralized. With more residual fixed charge, the potential magnitude should increase. Conversely, if the salt concentration is increased, the protein's charges will be more fully neutralized and the potential magnitude should diminish toward zero. These expectations are confirmed—not only in the bare cytoplasm but similarly in gels (Collins and Edwards, 1971; Stephenson *et al.*, 1981).

Cell Potential

We now move from cytoplasm to cell. Having established that the cytoplasm has a net charge and corresponding potential, we consider whether this charge and potential are the same in the intact cell (Ling, 1955). We consider, in other words, whether the membrane is electrically inconsequential, as early chapters imply.

To check this expectation, one may compare the cytoplasmic potential with the whole cell potential. This is not quite a simple matter of ripping off the membrane and measuring the voltage before and after: the solution chosen to bathe the exposed cytoplasm needs to be carefully adjusted to mimic the "solution" that naturally surrounds the cytoplasmic proteins; otherwise, the bath may impact the intracellular milieu and compromise the comparison. Nevertheless, the "ballpark" figure of 50 mV cited earlier for cytoplasmic potential is not far from the cell potential. And the cytoplasmic potential measured in the "cut-end" experiment where the membrane is functionally eliminated is approximately the same as the potential measured when the cell is intact (Ling, 1988).

An additional test is to compare cellular and cytoplasmic potentials in situations in which cytoplasmic charge is altered—the two potentials ought to vary similarly if the membrane is electrophysiologically transparent. We consider alterations of charge mediated by changes of ion content, and then by protein content.

The effect of potassium has already been considered. By adsorbing to the protein's negative surface charge, potassium ions diminish net cytoplasmic charge and decrease cytoplasmic potential. Cell potential is correspondingly diminished: As intracellular potassium is elevated by increasing extracellular potassium, the cell potential decreases (Fig. 7.4).

As for sodium, little or no effect is anticipated. Because of its exclusion from the cytoplasm, sodium's neutralizing power is feeble, so any change of external sodium concentration should go largely unnoticed by the cytoplasm. In the intact cell, experiments show a similar insensitivity. Decreasing external sodium concentration from its characteristically high level toward zero has practically no effect on the measured cell potential (Hodgkin and Katz, 1949; Draper and Weidmann, 1951).

Finally, consider chloride ions. Because of their negative charge, chloride and other anions are ordinarily in very low concentration inside the cytoplasm. Chloride is therefore a minor contributor to the cytoplasmic potential. Effects on cell potential are similarly negligible. Changes of external chloride mediate only transient effects, the steady-state potential remaining unchanged (Hodgkin and Horowicz, 1959).

Another way of varying cytoplasmic charge is through variation of protein charge. Proteins containing less net negative charge ought to give rise to a smaller cytoplasmic potential and a predictably smaller cell po-

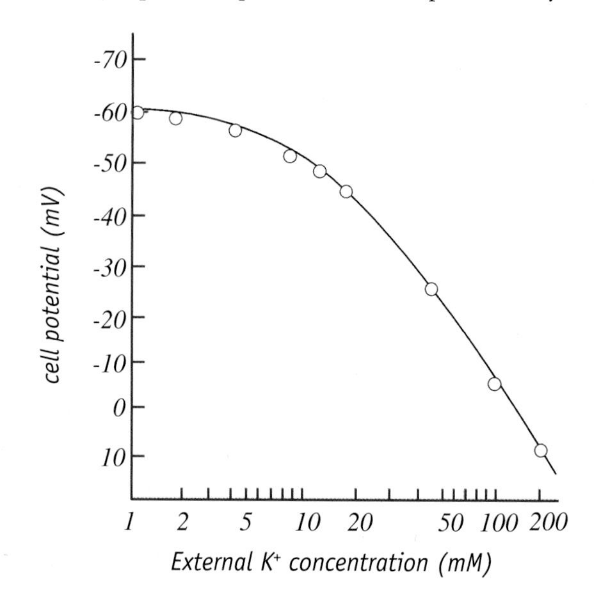

Figure 7.4. *Effect of variation of extracellular potassium concentration on cell potential. After Ling (1992).*

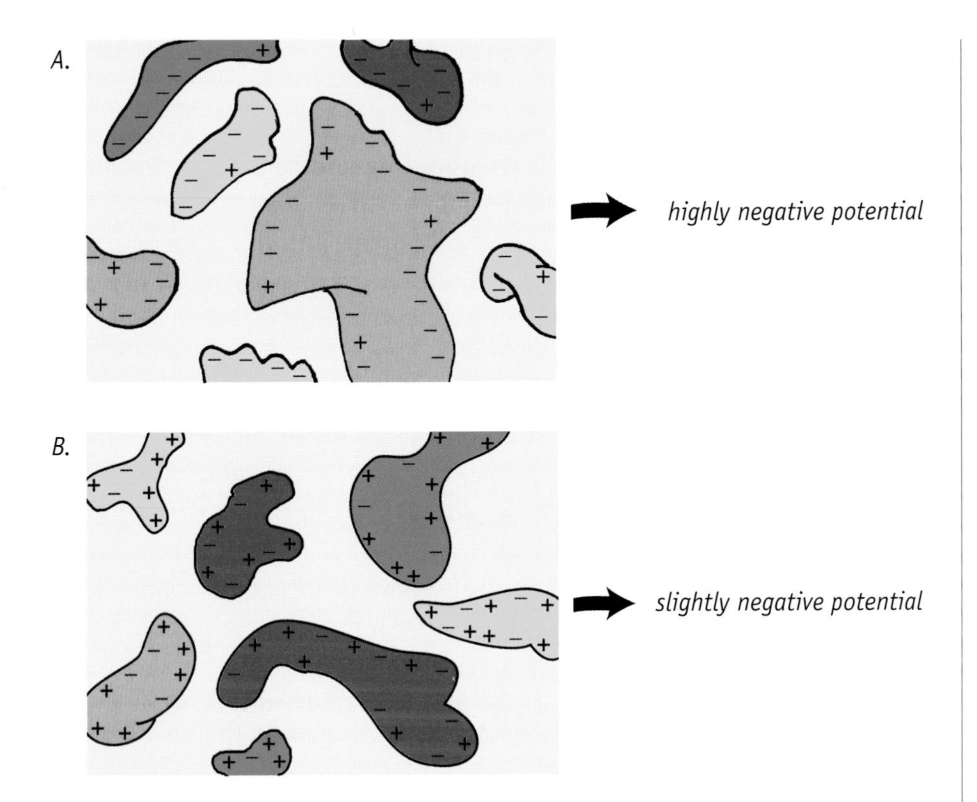

Figure 7.5. *(A) High negative charge on proteins. (B) Abundance of positive residues leads to low net negative charge.*

tential (Fig. 7.5). Ordinarily, this prediction would be difficult to test because cellular protein makeup is complex, and it is not easy to tally the net charge of one cell type and compare it to another. Even though cell potential varies widely among different cell types (between *ca.* -10 mV and -100 mV), computation of the respective protein charges is not straightforward. But there is an exception.

The exception is the red blood cell whose cytoplasm is dominated by a single protein, hemoglobin. Hemoglobin in turn is fairly atypical in its high density of positively charged lysine and histidine residues. Hemoglobin's net charge is therefore only slightly negative, relative to, say, myosin or actin whose net charge is highly negative. Under normal physiological conditions then, the red-cell cytoplasmic potential is predicted to be only marginally negative, far less than other cells we have been considering. Indeed, experimentally measured red-cell potentials are typically only -10 mV (Ruch and Patton, 1965) compared to the -70 to -90 mV found in most muscle cells.

In sum, whether the charge is introduced by adding cations or adding positive residues, the effect is the same: positive charge diminishes the (negative) cytoplasmic potential and cell potential in the same way. The cell potential may derive exclusively from the cytoplasm's excess negative charge.

Readers with physiological background may be wondering how a mechanism as simple as this could conceivably go on to produce an action potential—the positive-going spike of voltage characteristic of excitable cells. The action potential is a subject important enough that it will consume an entire chapter (Chapter 10). Suffice it to say that albeit of small magnitude, action potentials can be elicited from cells whose membranes have been stripped free (Natori, 1975); the membrane is not required. Large swings of potential can on the other hand arise from structural changes within the cytoplasm: For example, solvated vs. gelled regions of amoebae show potential differences on the order of 40 mV (Bingley, 1966). As we shall see later, cytoplasmic structural changes will play a central role in action-potential genesis.

Revisiting Holey Membranes

Finally, we close the loop by revisiting an issue that had earlier seemed paradoxical—concern over the anticipated ravages of membrane disruption that turned out to be unfounded (Chapter 2). Poking the membrane full of holes by gun, dagger or sword did not lead inexorably to cell death. The cell could survive, and the electrical potential was unaffected.

We now better understand why. If the cell's essential properties, including its charge and electrical potential, arise from the bulk properties of the cytoplasm, membrane violation should be relatively inconsequential.

As the human survives a tear in its skin, the cell survives a tear in its membrane. It is not the sheath that counts—but what lies within.

CONCLUSION

This chapter concludes the book's second section. For those readers who wish to delve more deeply into these concepts, I commend to you the works of Troshin, Szent-Györgyi, and Ling. The book by Troshin (1966) gives an especially readable account of solute partitioning. The volume by Szent-Györgyi (1972) is more philosophical, offering a brief but wonderful account of structured water's significance for life. The books by Ling (1962, 1984, 1992) are comprehensive and focus far more deeply than I could on water structure, ion-adsorption, cell potentials, and related matters. The 1992 book also reflects on why evidence of the kind presented in the previous several chapters has been ignored. Although the approaches of these three scientists differ from mine in some aspects, the similarities are legion. Indeed, it is on the shoulders of such giants that the current work stands.

It goes without saying that the paradigm offered here is orthogonal to convention—worlds apart from current views which place most of the action in the membrane instead of in the cytoplasm. To appreciate the full flavor of the difference, I commend to you once again to standard texts such as the ones by Alberts *et al.* (1994) and also Hille (1992). Comparing explanations of the cell potential provides excellent ground for evaluating the respective merits of the two approaches.

In the next section the foundational paradigm that we have derived springs to life as we explore how the cell carries out its functional tasks.

SECTION III

AN HYPOTHESIS FOR CELL FUNCTION

Building on previous chapters, this section advances the hypothesis that cell function resembles gel function. It examines the central role of the phase-transition in gels, and considers the potential for a similar role in cells.

Sky and Water I, 1938, by M. C. Escher

CHAPTER 8

8 PHASE TRANSITION: A MECHANISM FOR ACTION

Cells do not remain idle—they have tasks to perform. Cells manufacture substances and transport these substances from place to place. They react to stimuli. From time to time they may also divide. And specialized cells move, signal, secrete, contract.... The list goes on.

If such diverse tasks were mediated by equally diverse mechanisms, the chances of gaining a foothold of understanding in the space of a few chapters would be akin to the chances of mastering the world's languages in a day. On the other hand, if a common denominator of function were to exist, then the chances might fall within the realm of possibility.

Common paradigms do apply elsewhere in nature. In the field of genetics, for example, the common paradigm is the double-helix, whose essential simplicity has nucleated astonishing spinoffs, both practical and scientific. In the field of physics, the search for a unified force paradigm initiated by Einstein continues to this day. Such quests are driven by the notion of parsimony—the presumption that once the proper underlying principle has been identified, a seemingly complex array of phenomena will fall into place simply. It is this quest for simplicity that drives the search for a common cell-action paradigm.

The framework of our search lies in the cytoplasm's similarity to the polymer gel. The physicochemical correspondence has been documented in previous chapters: both entities are built of cross-linked polymers; both contain structured water; both exclude solutes; both exhibit sizeable electrical potentials; and both have the ineffable "feel" of a gel. Recognizing this correspondence, we treat the cell as a gel and ask whether such treatment can bring us toward a common paradigm for cellular action.

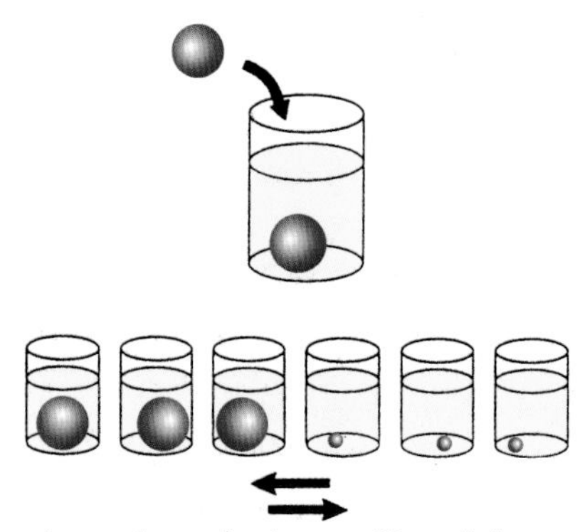

Figure 8.1. *Phase-transitions are triggered by subtle shifts of environment. After Tanaka et al. (1992).*

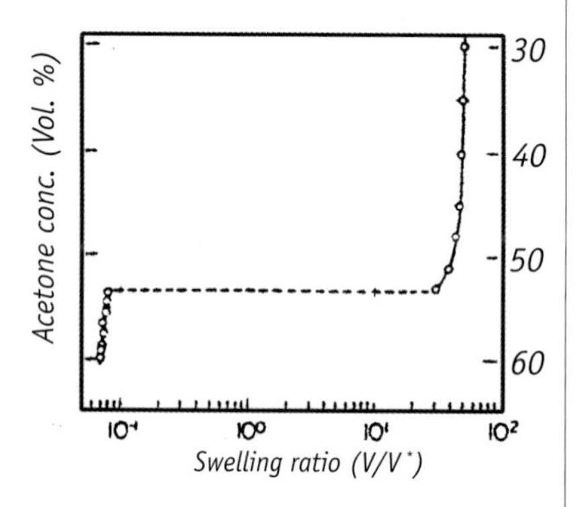

Figure 8.2. *Effect of change of solvent composition on gel volume. After Tanaka et al. (1992).*

Gel Action

How, then, do gels act? Until now, the gel has been treated as an autistic mass devoid of any characteristic responsiveness. However, gels are typically schizophrenic: in much the same way as water can change state from solid to liquid, gels can undergo equally remarkable transformations. A Jello-like mass may transform into a liquid. A supple, water-filled gel may contract into a dry, tightly condensed mass. The respective states often bear little more resemblance to one another than water does to ice. And, transformation from one state to the other can be mediated by nothing more than a subtle shift of environment (Fig. 8.1).

Such tranformations are labeled phase-transitions because of their massive nature and because they can be triggered by a small environmental shift, much like the water-ice transition. An example is shown in Figure 8.2. Here the transition is induced by a shift of the organic - aqueous solution ratio. When the ratio just crosses a threshold, gel volume changes by two orders of magnitude. Since the discovery of the phase-transition several decades ago by the late Toyoichi Tanaka of MIT, following the theoretical prediction of Dusek and Patterson (1968), phase-transitions have been exploited by chemists and engineers for practical applications ranging from disposable diapers to artificial sensors and muscles. Phase-transitions are the polymer gels' premier vehicle of action.

To appreciate the scope of the phase-transition's potential, consider Table 8.1, prepared originally by Professor Allan Hoffman, a luminary of the polymer-gel field. The table reveals two notable features. The first is the diversity of triggering agents (left panel). These include pH change, temperature change, chemical agents, mechanical force—agents that are among the most typical cell-action effectors. The second is the array of responses (right panel). Many of these, including changes of shape, volume, permeability and electrical potential are again just the kinds of actions the cell employs to mediate function.

If the cell does have a common denominator of function, then, the phase-

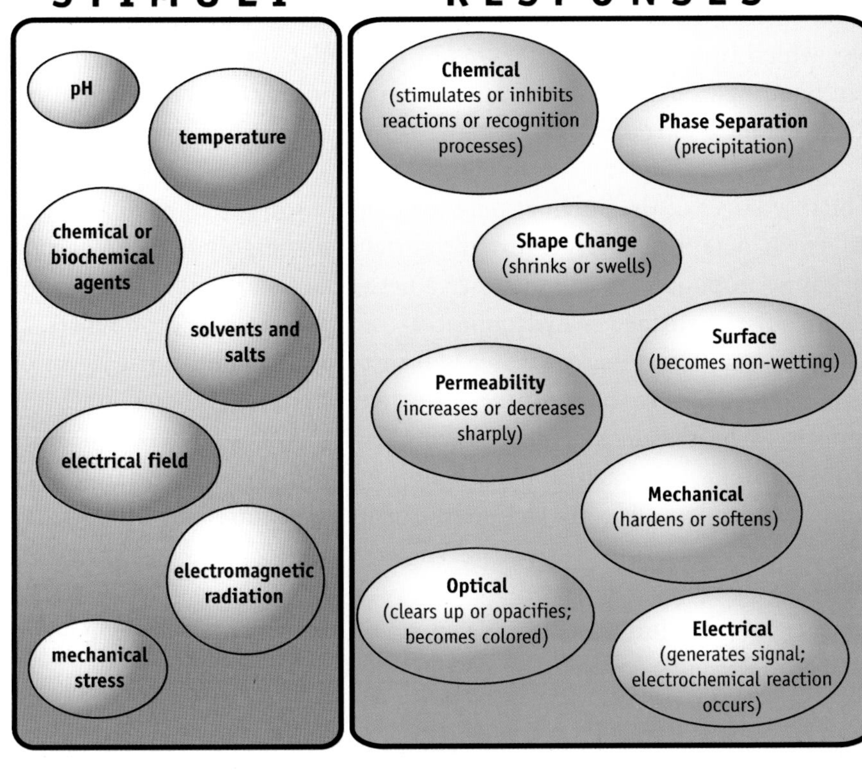

Table 8.1. *Typical stimuli and responses of polymer hydrogels. After Hoffman (1991).*

transition is a candidate worthy of consideration. The subject of phase-transitions will consume our attention for the remainder of this book as we explore essential cellular processes and see whether the phase-transition provides an adequate explanation. In this chapter we set the stage. We consider the basic features of the phase-transition along with a few representative examples.

Gel-Transition Basics

The framework of the gel is the polymer matrix. The matrix is created by long polymeric strands typically cross-linked to one another either chemically or physically. Chemical cross-links tend to be stable with heat, pH, and solvent composition, and do not ordinarily undergo transition. Physical cross-links are formed by intermolecular interactions of types such as coulombic, dipole-dipole, van der Waals, and hydrophobic, which bring water-shy regions of molecules together, or by hydrogen bonding. Such bonds are as a rule susceptible to environmental influence and tend to disrupt—leading to structural transitions.

The interstices of the polymer network contain solvent. The quantity of solvent is determined largely by the surface of the polymer. If the surface is hydrophobic, it will organize a network of clathrate water (Chapter 4). If the surface is hydrophilic as in most biological polymers, it will attract water: adherent water will in turn draw additional layers of water, and so on. In determining the extent of layering, the major factor appears to be the density of charged (Lewis) sites: surfaces with more charged sites will structure more water, while surfaces with fewer sites will structure less. Highly charged surfaces are reported to structure up to 600 water layers (for review, *see* Vogler, 1998).

The gel's water content will thus depend on the nature of the polymeric surface. Gels built around a framework of uncharged polymers may have a water-to-polymer volume ratio typically 5:1 or 10:1, while charged polymers can draw a ratio up to 3,000:1 (Osada and Gong, 1993). Layering of water may not be the only driving force for hydrophilic water retention. Osmotic forces may also draw water when counter-ions are present around surface charges—but the fact that a water-to-polymer ratio of 10:1 is easily achieved with uncharged polymers implies that this factor may be of secondary relevance.

In essence then, polymer gels and cellular gels share basic structural features. Both are built of polymeric matrices that may contain substantial surface charge; and both matrices are invested with appreciable amounts of organized water, whose extent can be evaluated through various physico-chemical means.

Transitions in gel structure entail massive shifts of both polymer and solvent. The shift of polymer, often from an extended to a contracted state, is well recognized, whereas the accompanying solvent shift is less broadly appreciated. Figure 8.3 shows the fate of relaxation time T_2, which reflects water's freedom of motion. As the temperature is raised in solution of the common polymer acronymed PNIPAM, the T_2 value increases progressively, implying a progressive increase of motional freedom. The control sample of pure water (bottom) shows similar behavior. Then, at the critical gelation temperature the T_2 value sharply decreases: water suddenly becomes restricted as the system gels. A progressive increase of mobility follows further increase of temperature. Although the extent is difficult to quantitate, the sharp inflection indicates a sharp change of water structure on either side of the transition.

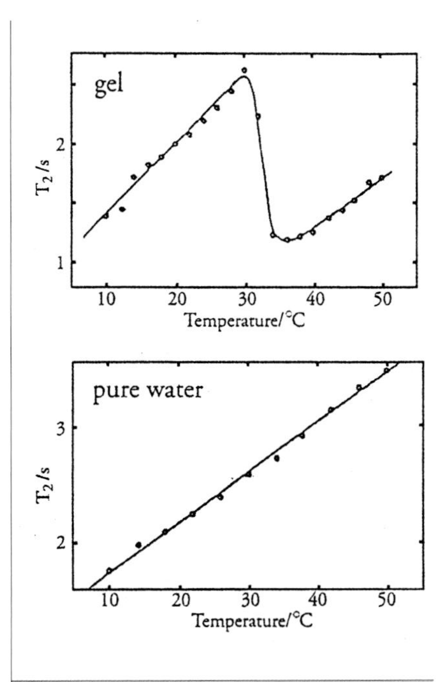

Figure 8.3. *Effect of temperature on the T_2 value in PNIPAM gel (after Yasunaga* et al.*, 1996). Note the abrupt reduction in T_2 at the point of gelation.*

Solvent-organization change during transition is also implied by measurements of the rate at which substances can penetrate through the gel. Tokita and Tanaka (1991) measured the rate at which pressurized water could be forced through a polyacrylamide gel. As the gel underwent transition (but prior to any volume change), the penetration rate increased by a factor of ~1,000. The proffered interpretation was an increase of spatial non-uniformity: with some matrix spaces opening while others closed, a disproportionately higher penetration rate could be achieved through the larger spaces. However, the thousand-fold change of penetration rate seems at least as easily explained by a water-structure change similar to the one implied in Figure 8.3. Destructuring sharply reduces viscosity and effectively opens the gel's diffusion spaces.

Alteration of the state of water is key to understanding how global physical change takes place. Consider a gel whose polymer undergoes a long to short transition (Fig. 8.4). The long configuration is stable because the polymer's hydrophilic surfaces thirst for water—the polymer extends itself to maximize the number of water contacts, thereby minimizing the system's energy. The short state is also stable. Here the polymer is folded, with surface charges now contacting one another instead of water. To the extent that charged surfaces are no longer exposed, vicinal water will be destructured. Thus, the two states differ radically, and one component of the difference lies in the state of water.

TRANSITION COOPERATIVITY

Now consider the transition between these two states. Suppose a gel that is in equilibrium in the long state is exposed to one of the triggering agents listed in Table 8.1. Exposure edges the system just past its critical point. Once this happens, the transition proceeds toward the shortened state with the inevitability of a sneeze (Fig. 8.1).

Why so? Why is the progression of change as inevitable as it is?

Inevitability presupposes some kind of cooperativity—a change that increases the propensity for additional change in the same direction. Cooperativity typically arises out of competition between two or more forces, *e.g.*, two magnets pulling on an iron ball suspended exactly half-

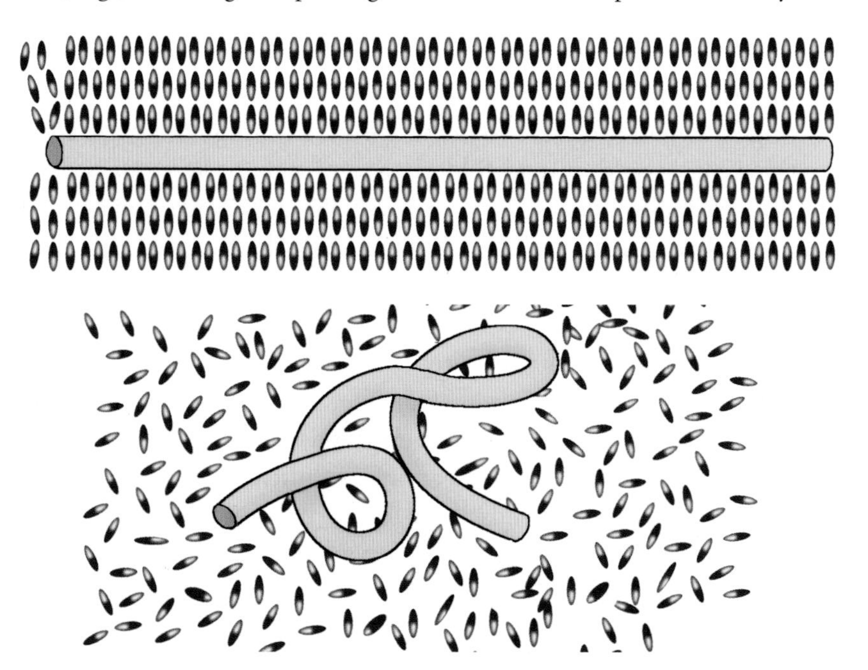

Figure 8.4. *Long vs. short configurations. In the long state the polymer is surrounded by structured water. In the short state water is randomly oriented.*

way between. Slight lateral perturbation confers an edge, which ensures full swing to one or the other magnet. In the case of the polymer strand, the competing forces arise from the polymer's attraction to water and to polymer. Either one can dominate. Consider a strand dominated by polymer-water interactions initially, and edged toward polymer-polymer interaction by an environmental shift. Once the number of polymer-polymer interactions passes a critical threshold, the proclivity for more polymer-polymer interactions must somehow increase if the transition is to proceed with inevitability.

Cooperative behavior of this sort may be realized in different ways in different gels, but the underlying mechanisms almost certainly involve water as well as polymer. The water molecule is unusual in that it can

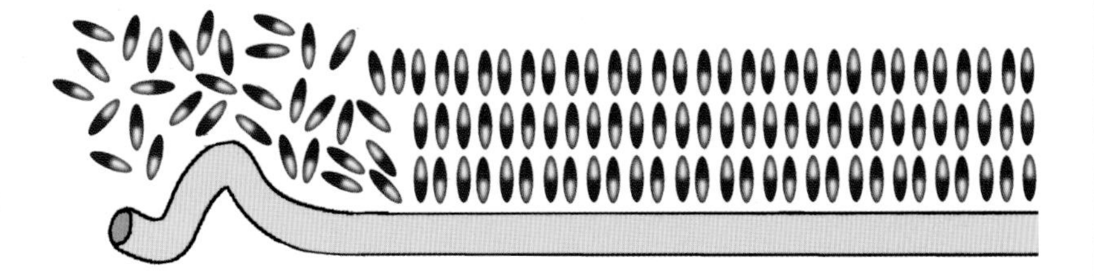

Figure 8.5. *Propagation of phase-transition by water depolarization and polymer condensation.*

form hydrogen bonds with up to four other waters, creating a strongly interlinked hydrogen-bonded network. Such bonding is known to be cooperative—once a single hydrogen bond forms, the proclivity to form additional hydrogen bonds increases (Stillinger, 1980). Perturbations of water structure may therefore propagate (Watterson, 1997; Vogler, 1998), with local water destructuring spreading to neighboring regions, and thereby facilitating polymer shortening, additional destructuring, more polymer shortening, *etc.* (Fig. 8.5).

Figure 8.6. *Analogy showing how elements can be induced to organize cooperatively.*

The polymeric component also contributes to cooperative propagation, and this may be realized in several ways. One mechanism is based on classical induction theory set forth by G. N. Lewis early in the last century and extended in some detail by Gilbert Ling (1962, 1992). Known as the "Association-Induction" hypothesis, this theory posits that in a carbon chain such as that of a protein or polymer, local structural change produces an electron-cloud shift, which induces a similar cloud shift and structural change in the next region, *etc.* Thus, the transition propagates along the polymer. This propagation mechanism is analogous to what happens when a magnet is brought near an array of nails loosely strung to one another with bits of string (Fig. 8.6). The first nail is magnetized, which magnetizes the second, *etc.*, until many or all nails are recruited

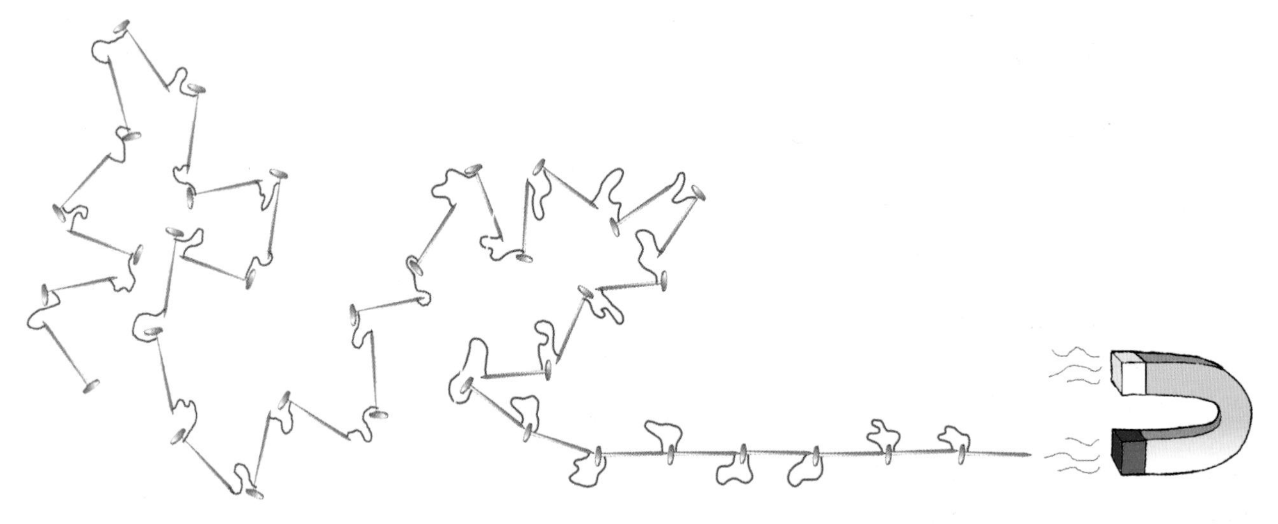

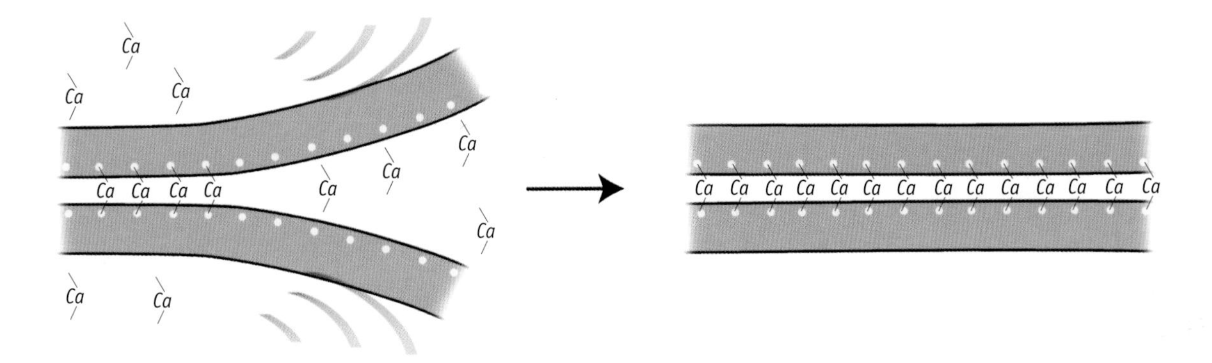

into the new configuration. In such a way the structural change propagates.

Another cooperativity mechanism involves divalent cations such as calcium, which are prevalent in the cell. Because of their two positive charges, divalents can grasp nearby anionic sites and hold them tightly cross-linked to one another. Imagine a network of polymeric strands undergoing thermally induced fluctuation, exposed to increasing levels of calcium. When any two surfaces come transiently close to one another, calcium may link their anionic sites. This linkage constrains the strands' flanking regions to remain in closer proximity, increasing the likelihood of additional cross-linking. The process proceeds in zipper-like fashion, the polymer transitioning from its extended, hydrated state, to a condensed state largely devoid of water (Fig. 8.7).

Figure 8.7. *Calcium and other divalent cations can bridge the gap between negatively charged sites, resulting in zipper-like condensation.*

Through cooperative mechanisms such as these, a modest shift of environment can induce massive action. The polymer gel may condense and release most of its water. Or, conversely, the contracted polymer may undergo reverse transition, imbibing huge volumes of water and expanding cooperatively back toward its expanded state. Of course it is not always so simple. Some polymers show intermediate states of quasi-sta-

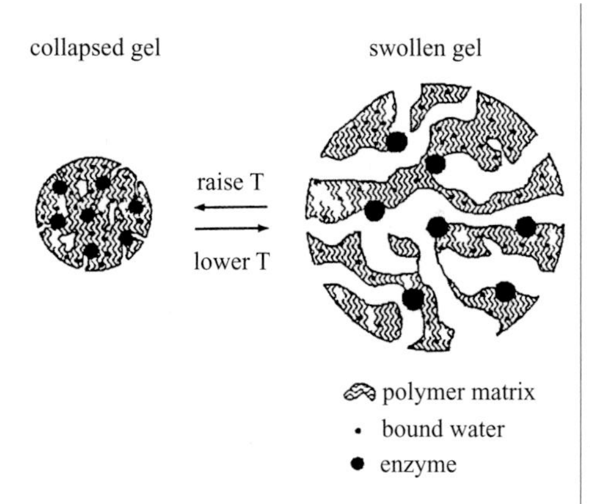

Figure 8.8. *Controlled release of enzyme or drug from a gel microsphere. After Hoffman (1991).*

bility (Annaka and Tanaka, 1992), and polymer affinity for water is not necessarily a yes or no affair. Nevertheless, the cooperativity mechanisms described above, involving both polymer and water, can be thought of as generic.

Armed with these essentials, we consider several examples of phase-transitions that have been developed for practical use. We consider gels producing largely chemical responses, gels producing largely mechanical responses, and protein-like gels that produce chemomechanical responses. As we proceed, we bear in mind the potential for these transitions to apply within the cell.

Chemical responders

A good example of a chemical responder is the so-called gel-valve, used by the pharmaceutical industry for drug release. In common application the valve surrounds a drug-containing capsule. Because of the gel's high density, substances will ordinarily not pass; but when the gel undergoes structural change that "opens its pores," substances can pass. Opening is designed to be intelligent. For example, swallowed medications can ordinarily be attacked by the acid environment of the stomach before they are able to reach their target site in the intestine. By employing a valve that opens only in the more alkaline environment of the small intestine, the drug is protected from the ravages of stomach acid—a clever scheme that ensures efficacious release. The scheme works because transitions are commonly pH-sensitive (Table 8.1).

Drug release can also be designed to respond to the presence of a particular chemical. A good example is the release of insulin in response to glucose (Albin *et al.*, 1985). The insulin is encased within the gel membrane, which also contains amine groups and glucose oxidase. Ambient glucose diffusing into the membrane is converted to gluconic acid by glucose-oxidase catalysis, thereby lowering pH, increasing the amine groups' ionization and repulsion, and triggering massive swelling and insulin release. In this way insulin is released only when glucose is present.

This artificial system mimics what nature itself accomplishes through the pancreatic beta-cell.

Yet another genre of intelligent drug-release is based on the gel microsphere. Here, an entrapped enzyme or drug is released as the microsphere undergoes transition and swells (Fig. 8.8). A specific example is the artificial mast cell—a negatively charged proteoglycan matrix loaded with histamine, which can respond to immunological challenge by undergoing phase-transition and swelling (Yuan and Hoffman, 1995). This release process is quite generic—it can deliver diverse bioactive substances including lysozymes (Yuan and Hoffman, 1995) and drugs (Verdugo *et al.*, 1995).

Apart from substance release, phase-transitions are useful also for substance recovery (Cussler *et al.*, 1984). A shrunk polyacrylamide gel is immersed in an aqueous bath containing the solutes to be recovered. Upon pH-sensitive transition, the gel imbibes bathwater up to 20 times its weight. Because of size-based solute exclusion (Chapter 5), dissolved solutes larger than a critical size are largely not taken up; they are left behind in the bath to accumulate in higher concentration for easier recovery. The process is useful, for example, for recovering antibiotics from a fermentation broth.

The examples above illustrate shifts of solutes toward or away from transitioning gels. Solute shift—or transport—is a basic feature of cell function. Whether transport is achievable through a phase-transition is a question that the next several chapters will consider.

Mechanical Responders

We now move to mechanical (actually, chemo-mechanical) responders—a vivid example of which is the "gel-looper" (Okuzaki and Osada, 1993). The looper is a gel strip that hangs in solution from a supporting ratchet bar by two metal loops (Fig. 8.9). The gel transports itself along in re-

Figure 8.9. *The "gel-looper" moves from left to right along a sawtooth ratchet as a result of cyclic phase-transitions in the gel. After Osada and Ross-Murphy (1993).*

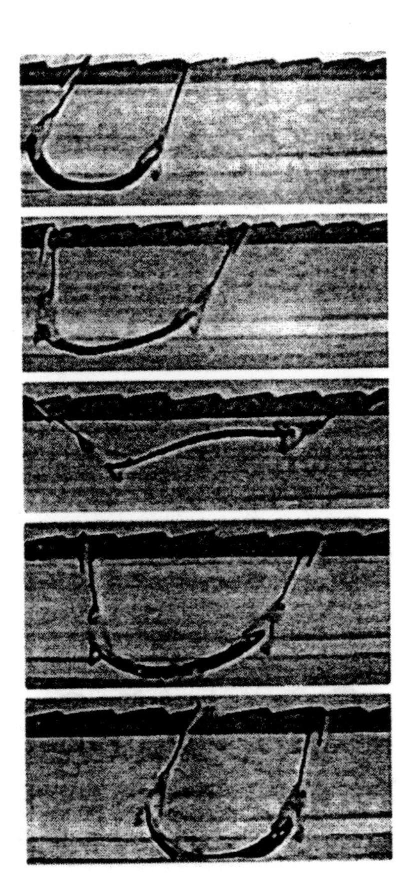

Figure 8.10. Gel golf. The principle is the same as that illustrated in Figure 8.9, but here gel bending is used to strike a small sphere. After Osada and Ross-Murphy (1993).

sponse to an alternating electric field. When the proximal electrode is positive, the gel edge near the electrode shrinks, curling the gel and pulling its trailing end along the ratchet's teeth. When the polarity is reversed the gel begins to straighten and its front end inches forward. In this way the gel propels itself inchworm-like along the supporting ratchet.

This same bending action can be used to play "gel golf." See Figure 8.10. Here the gel's bending action is used to strike a microsphere, which advances to an incline where it falls down a precipice—into the hole, as it were.

Although these applications may seem more fanciful than useful, they illustrate how even minor shape changes can be harnessed to achieve substantial translations. Particularly in the gel-looper, repeated inchworm-like movements can generate translations of appreciable magnitude. The mechanism anticipates snake-like motions, called reptation, which are

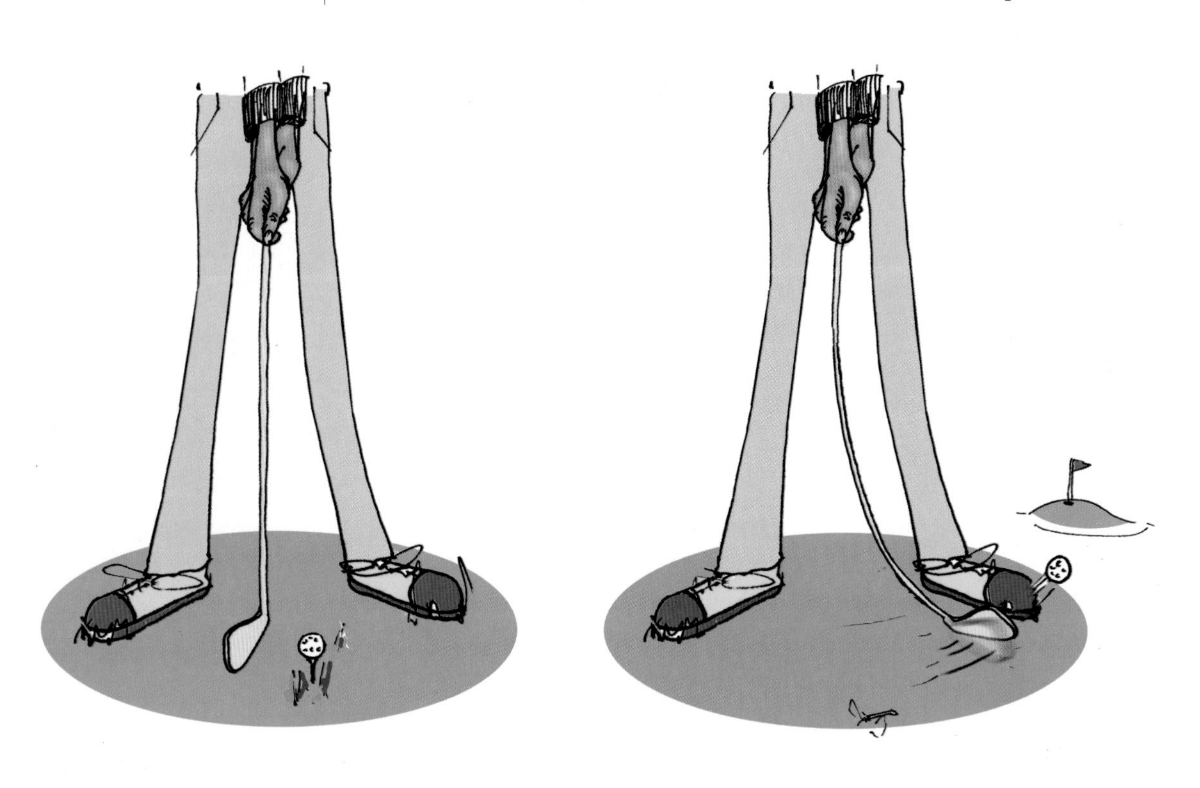

in fact characteristic both of artificial polymers and protein filaments (Yanagida *et al.*, 1984; Kas *et al.*, 1994). Reptation could drive a protein filament along a surface or past another filament (Chapter 14).

Another example of a mechanically responsive gel is the artificial pressure receptor (Sawahata *et al.*, 1995). Gels co-polymerized from polyacrylic acid and polyacrylamide are particularly sensitive to mechanical force. Pressure induces a change of pH and electrical potential (Figure 8.11). Whether this kind of mechanism could be responsible for pressure reception in the cell is a question the sensor's developers raise.

Finally, gels can self-oscillate. When certain conductive gels are placed in a steady electric field, sustained current oscillations are detectable from electrodes plunged into different regions of the gel (Miyano and Osada, 1991). Oscillations are common features in bio-electrical systems such as natural pacemakers. Although generally attributed to membrane cur-

Figure 8.11. *Mechanical pressure can generate an electrical potential and pH change. After experiments by Sawahata* et al. *(1995).*

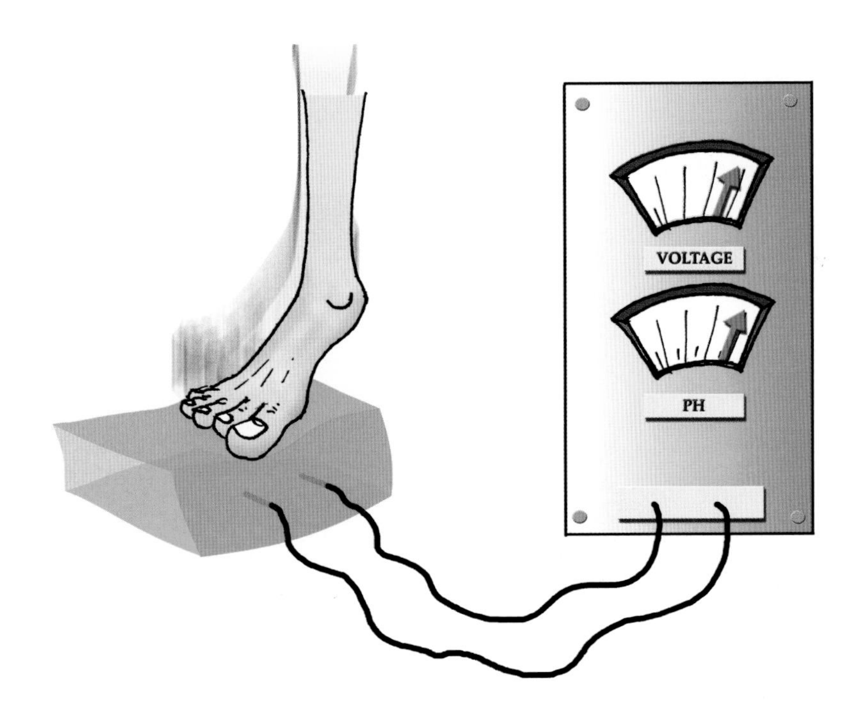

rents, such oscillations can evidently arise in gels devoid of any membrane, implying that they could as well arise from the gel-like features of the cytoplasm (Chapter 10).

PHASE-TRANSITIONS IN PROTEIN-LIKE GELS

Distinct from the polymer gels described above is a class of gel built of protein-like polymers. The best known of these may be the elastomeric gel based on a repeating motif in elastin (Urry, 1993). This gel undergoes reversible transition between a longer and a shorter state (Fig. 8.12). In the longer state the water is ordered; in the shorter state water is disordered, and induced disordering can facilitate the long-to-short transition. The transition generates enough mechanical power to function as an artificial muscle—work per unit volume is similar to a comparable-sized unit of muscle. And the gel can be cycled through transitions again and again without noticeable functional degradation.

Figure 8.12. *Elastin-like polymer that undergoes transition between long (top) and short (bottom) state. After Urry (1993).*

Triggering of this elastomeric transition can be effected by a multiplicity of agents. A common one is temperature. Temperature sets water mol-

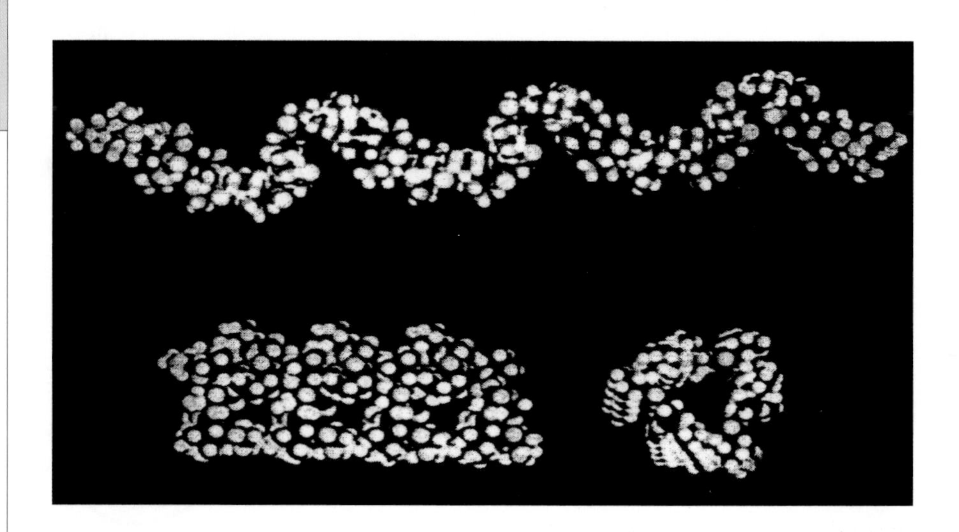

ecules in motion, destabilizing their structure and facilitating the shortening transition. Shift of pH is another trigger, along with addition of salt, exposure to organic solutes, increase of pressure, and exposure to light following chromophore incorporation (*see* Table 8.1). A trigger of impressive potency is phosphorylation: for every 300 molecules of kinase included in the gel, one phosphate is all that is needed to trigger massive transition. Such sensitivity mirrors what takes place in the cell, where phosphorylation of a small number of sparsely distributed residues can commonly trigger massive action.

A second example of a protein-like gel is the one based on cross-linked collagen. When immersed in high salt, this gel shortens to about half its length—undergoing transition from a helical structure to a random coil whose equilibrium length is shorter than that of the helix. Machines built around this principle were constructed by the renowned physical chemist Ahron Katchalsky and his colleagues. The machines were ingenious devices in which multiple shortening-lengthening cycles could be achieved by bringing a belt of cross-linked collagen into contact alternately with solutions of low and high salt (Steinberg *et al.*, 1966). Like the elastin gel, the collagen gel's production of work per gram was similar to that of a contracting skeletal muscle.

Phase-transitions are also seen in "artificial protein" gels built of polymers arranged in alpha-helices, beta sheets and other common protein motifs. Such gels are used increasingly for drug-delivery applications (Aggeli *et al.*, 1997; Petka *et al.*, 1998; Wang *et al.*, 1999).

In sum, semi-natural gels produce useful functions. Elastin-like and collagen-like gels undergo phase-transition, and in so doing they generate mechanical work that can be useful as an artificial muscle, or actuator. In the elastin gel, complementary functions such as superabsorbency and tissue scaffolding are also realized (Urry, 1993). And, gels built of polymers with protein motifs are finding application in drug-release systems. Such successes with semi-natural gels bode well for the possibility that natural gels within the cell may function comparably.

CONCLUSION

The phase-transition's potential has not escaped the attention of cell scientists. Standard texts consider the gel-like nature of the cytoplasm and frequently go on to deal with the cytoplasmic "gel-sol" transition, which is a common member of the phase-transition set. In the absence of a theoretical framework, however, biologists have been left with little more than a vague notion of the phenomenon, with which little could be done.

The polymer-gel field, on the other hand, embodies a growing framework of understanding. Driven in part by the potential for practical device design, chemists and engineers have spent much time exploiting the phase-transition, and gaining a foothold of understanding in the process. As a result, a rich array of products has recently come on the market. These products are based mainly on large-scale transitions of volume, shape, water-content, electrical potential, permeability and/or ion-content (Table 8.1).

The relevance of these transitions for the cell is not difficult to envision: contraction arising out of shrinkage; motility arising out of shape change; transport arising out of solute separation; action potentials arising out of permeability change; *etc.* The concept is in fact less radical than you may think. In the chapters that follow it will become clear that phase-transitions have already been suggested to apply in a diverse number of biological arenas. Such suggestions have eluded wide-scale recognition, however, largely because they have remained scattered among sub-disciplines, and thereby lack cohesion.

Of the phase-transition's attractive features, one of the most prominent is the prodigious response that can be elicited by a subtle environmental shift such as a slight change of temperature or pH. Such amplification should be enhanced further in the biological context by the high degree of order commonly present: Whereas synthetic gels are typically built of tangled polymers with little apparent order, cellular organelles such as the ciliary axoneme and muscle sarcomere exhibit extraordinary supramolecular order, with X-ray diffraction patterns showing regularity down to ~1 nm. With structural regularity, the triggering threshold should be the same everywhere, and the response should therefore be decisive and rapid.

Another attractive feature of the phase-transition hypothesis is the implicit notion that cellular processes are not necessarily anything special—cells would operate by the very same physical and chemical principles that govern ordinary non-biological systems, and not by special inventions of the celestial committee. The notion of continuity across the living - non-living boundary seems logical because the boundary is fuzzy. Is the seed living? What about the virus? Presupposing similar principles operating on either side of the boundary provides an appropriately seamless transition between living and non-living. It also forces us to accede to the notion that the cell does not function by special mechanisms: Although the cell may seem special to us, its mechanisms may be fully humdrum and completely orthodox.

A third attractive feature is the potential for cell function to be governed largely by a single common underlying mechanism. Life would be simple indeed. The objective of science after all is not to add layers of interpretational complexity with each discovery, but to collapse layers of apparent complexity into a few simple governing principles; parsimony should prevail. If the adequacy of the phase-transition is confirmed, there is high potential for simplicity.

On the other hand, one needs to be mindful of simplicity's seductive allure. It will be a matter of testing whether the phase-transition is merely a fancifully attractive notion or whether it can really account for known phenomena in a more naturally direct way than can current mechanisms. In the chapters that follow we do just that—we consider whether the phase-transition provides direct explanations for basic cellular processes such as secretion, communication, transport, motility, division, and contraction.

SECTION IV

APPROACHING CELL DYNAMICS

This section pursues details of active cell function. Building on previous chapters, it explores the possibility of a common underlying basis for cell action—the phase-transition.

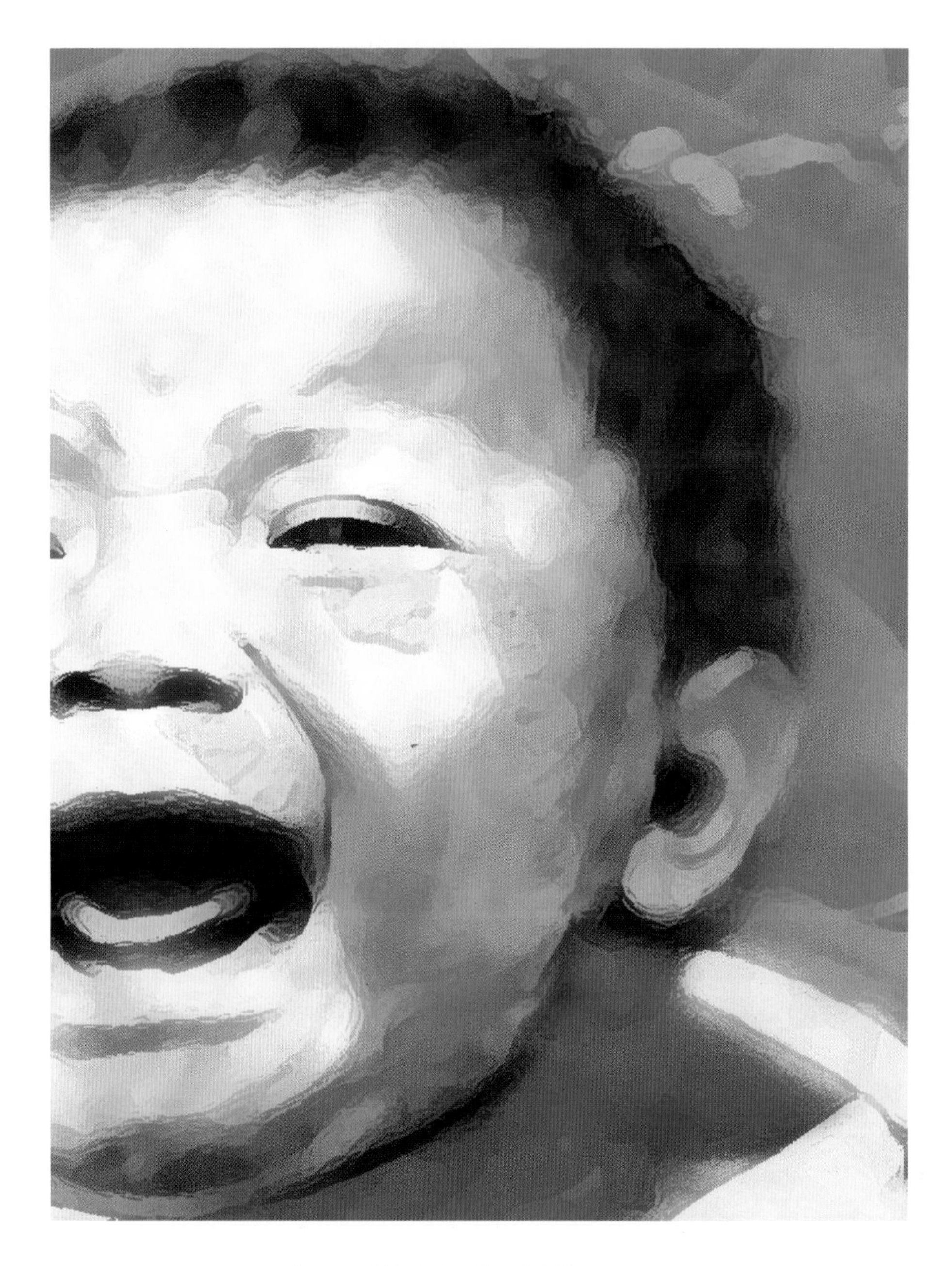

Teary, 2000, by David Olsen

CHAPTER 9

SECRETION

We deal first with the cell's main vehicle for external communication—secretion. Secretion entails the cellular discharge of chemical substances, which may include enzymes, hormones, neurotransmitters, or special agents depending on cell type. Substances to be secreted are packed into small vesicles and shipped to sites near the cell periphery. There they stand poised for release, awaiting a chemical or electrical stimulus.

A good example is in the nerve cell. When activated by an action potential, the cell's terminal region secretes a neurotransmitter that diffuses across a short gap to a muscle cell or another nerve cell, where it triggers an action potential. In this way the chain of information is passed from cell to cell. Another example is the mast cell. The mast cell contains packets of histamine, which are discharged when antigens bind to a cell-surface receptor. Discharged histamine diffuses locally and into the systemic circulation, where it can defend against invading pathogens.

Secretion is considered first because of its essential simplicity. Although staging events may be intricate, the focus here is on discharge—release of chemical packets from the cell. The question is whether this simple action could involve some kind of phase-transition.

STRATEGIES

According to current views, the secretory vesicle is a kind of soup surrounded by a membrane—a miniature of the cell itself according to the

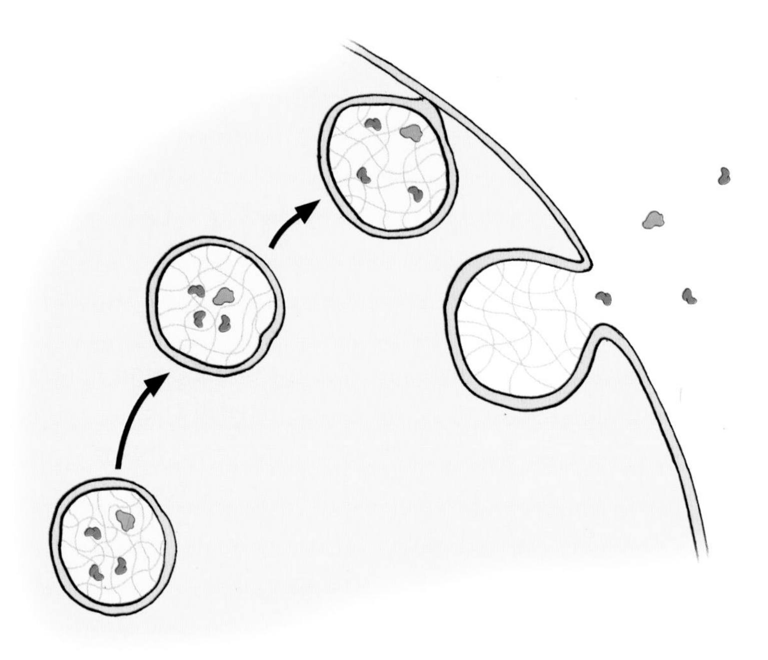

Figure 9.1. Classical view of the secretory discharge process.

classical view. The soup contains the substance to be secreted embedded in a matrix of polymers. For discharge, the vesicle docks with the cell membrane. Cell and vesicle membranes fuse, opening the interior of the vesicle to the extracellular space and allowing the vesicle's contents to escape by diffusion (Fig. 9.1).

Although attractive in its apparent simplicity, this mechanism does not easily reconcile with several lines of evidence. One concern is that discharge is often accompanied by dramatic vesicle expansion. Some expansion is anticipated because the concentrated solutes in the soup can draw water osmotically: if water diffuses in before solutes diffuse out, the vesicle might expand transiently. The observed expansion, however, can be practically explosive. Isolated mucin-producing vesicles undergo a 600-fold volume expansion within 40 ms (Verdugo *et al.*, 1992). Vesicles of Nematocysts, aquatic stinging cells, are capable of linear expansion rates of 2,000 μm/ms (Holstein and Tardent, 1984). Such phenomenal expansions imply something beyond mere passive diffusion of water.

A second concern is that the expansion responds anomalously to solvents (Fig 9.2). Vesicle contents can be expanded and re-condensed many times by exposure to various solutions. But the required solutions are not those expected. When condensed vesicle matrices are exposed to low osmolarity solutions—even distilled water—they remain condensed although their thirst for water should be enormous (Fernandez *et al.*, 1991; Verdugo *et al.*, 1992). But if small amounts of sodium (<10 mM) are dumped into these low osmolarity solutions, the matrices expand instantly.

A similar anomaly applies to re-condensation of the expanded matrix. Adding enough solutes to the bath should withdraw water and precipitate condensation. But expanded matrices show almost no response even when the bathing solution is augmented by 1M sucrose; only ~2% shrinkage is observed (Curran and Brodwick, 1991). Osmotic mechanisms again do not suffice—vesicle contents do not behave like chicken soup.

Given such anomalies, it is no surprise that investigators have begun looking elsewhere for mechanisms. Several groups including those of Pedro Verdugo at the University of Washington and Julio Fernandez at the Mayo Foundation have looked increasingly for explanations within the realm of the phase-transition.

Figure 9.2. *Condensed vesicle matrices immersed in low osmolarity solutions do not adsorb water (left). However, when solutions are augmented by small amounts of sodium (right), the matrices expand fully.*

Why the phase-transition? One reason is that the vesicle is built of a dense polymer network that abruptly expands. A second reason is that expansion and discharge require a critical shift of environment: as temperature or solution composition edges past a threshold, matrices abruptly expand or condense. In the case of mast-cell matrices and mucus-producing goblet-cell matrices, thresholds for expansion lie within a window as narrow as 1% of the critical value (Verdugo *et al.*, 1992). Hence, the phase-transition's signature criteria are satisfied.

So too are other features anticipated of phase-transitions (Table 8.1). These include: substantial volume change, shifts of solvent, changes of consistency, and even triggering by an electrical current (Nanavati and Fernandez, 1993).

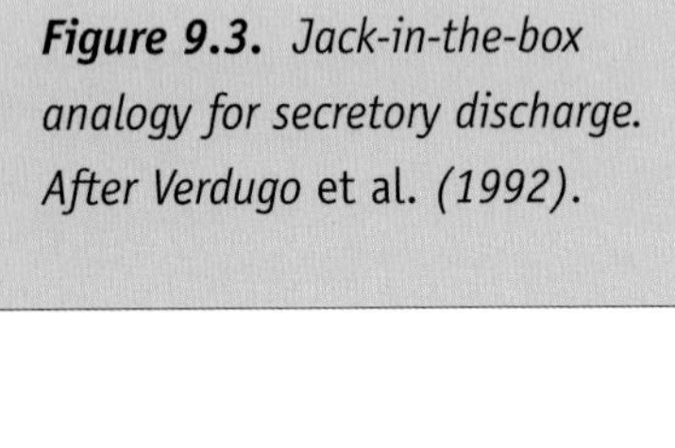

Figure 9.3. *Jack-in-the-box analogy for secretory discharge. After Verdugo* et al. *(1992).*

The phase-transition mechanism also seems strategically sensible. Secretory cells specialize in high payloads. In the nerve terminal for example, the acetylcholine concentration approaches 2M—enough to fuel intense barrages of neurotransmitter discharge. Concentrations of such magnitude would ordinarily confer an osmotic drive intense enough to flood the cell with water. But this soggy scenario is averted by using the phase-transition to pack molecules into condensed aggregates and thereby diminish the osmotic draw. The condensed matrix can later re-expand by a reverse transition that allows hydration forces to explode the matrix and expel the neurotransmitter molecules.

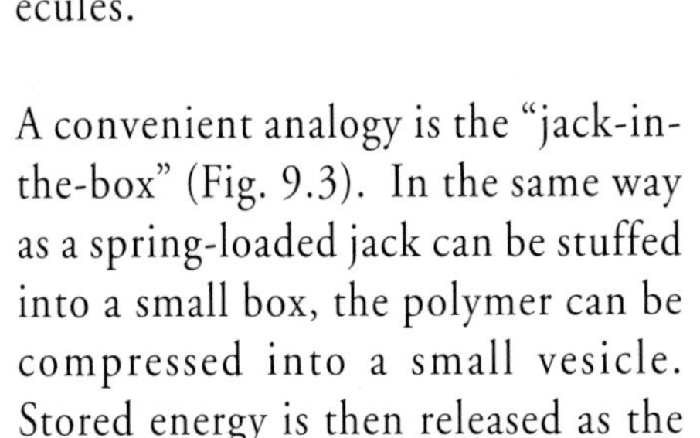

A convenient analogy is the "jack-in-the-box" (Fig. 9.3). In the same way as a spring-loaded jack can be stuffed into a small box, the polymer can be compressed into a small vesicle. Stored energy is then released as the jack pops out, or as the vesicle explodes. The stored energy is dissipated largely as increased entropy, or disorder (Chapter 15)—similar to the popping of popcorn.

As an explanatory principle, then, the phase-transition mechanism has some potential. It can account for vesicle expansion, and it can explain why osmotic considerations do not suffice for explaining the response to solvents. Whether this mechanism is adequate can only be determined by addressing basics such as: What keeps the polymer matrix condensed? How is the expansion triggered? And how, exactly, could expansion be tied to chemical discharge?

Condensation

In exploring condensation, it is tempting to nominate the vesicle membrane as protagonist. The membrane could keep vesicle contents packed as the box keeps the jack packed. But there is a problem. Secretory matrices can condense even in the absence of a membrane: by exposure to appropriate solutions (see above), vesicle matrices can be expanded and re-condensed many times. No membrane is required. Nor is it clear how a membrane could force the enclosed contents to condense.

A more plausible condensation mechanism follows from the matrix's physicochemical nature. The matrix is built of charged polymers. Most commonly, the polymers are randomly entangled, but they can also be semi-ordered, as in goblet-cell vesicles, or even crystalline, as in insulin-containing pancreatic beta-cell vesicles. Although matrix architecture varies, what remains invariant is the polymers' strongly anionic character. High negative charge is the rule.

The matrix condenses because its negatively charged sites can be cross-linked to one another (*see* Fig. 8.7). Divalent cations are effective cross-linkers because of their high charge density (Katchalsky and Zwick, 1955): like two grasping hands, the two positive charges can link negatively

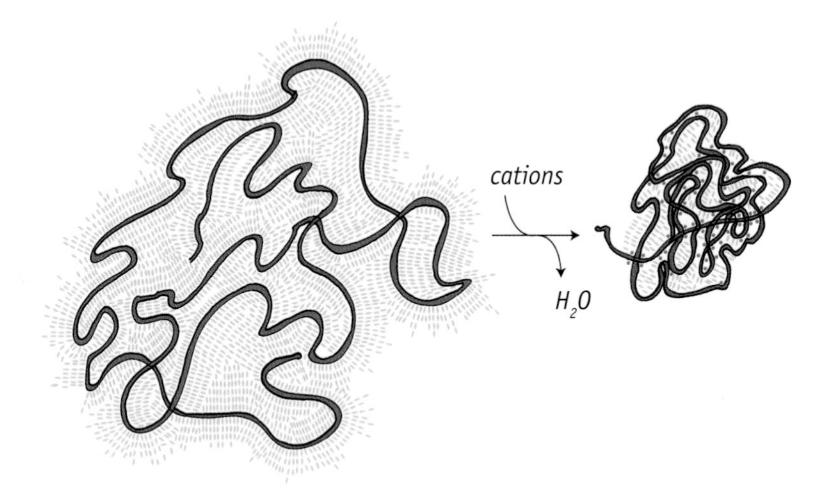

Figure 9.4. Mechanism of condensation. As divalent cations cross-link the negatively charged matrix, adsorbed water is forced out.

charged polymeric surface sites into close proximity. As one might expect from such a mechanism, cross-linking effectiveness strictly follows valency: monovalents are ineffective, divalents are effective, and trivalents are especially effective (Curran and Brodwick, 1991). An example of this hierarchy is the behavior of histamine. At neutral pH, histamine behaves as a monovalent cation, whereas at low pH (3 – 4), it becomes divalent. Divalent histamine easily condenses the polymeric network, whereas monovalent histamine does not (Fernandez *et al.*, 1991).

As cross-links condense the polymer network, adsorbed water must be forced out (Fig. 9.4). The matrix then becomes dry, granular, condensed—and relatively easy to transport (Chapter 11). It remains stable because the divalents are firmly held; they exchange slowly within the matrix and very little with the space outside of the matrix (Rabenstein *et al.*, 1987).

An example of such a condensed network occurs in the mast-cell vesicle. The anionic polymer is heparin, a giant glycosaminoglycan bound to a protein core. Depending on pH, heparin contains two to three negative charges per disaccharide repeat unit, conferring an extremely high negative charge on the polymer. The cross-linker is histamine—a divalent cation. Histamine brings heparin's anionic sites into coalescence and condenses the network. It holds the network in its condensed state until some agent triggers expansion.

EXPANSION

To trigger the expansive transition, divalent cations need to be cut loose. Once a few cation bridges are broken, repulsive forces arising from the surfaces' negative charge will push the strands apart, allowing them to satisfy their natural thirst for water. Water dipoles will build one upon another, wedging polymers farther apart (Fig. 9.5). As flanking bridges are broken by this wedging force, still more water dipoles will adsorb, and so on. By this highly cooperative unzipping process the vesicle rapidly expands and solvates, allowing its chemical contents to be released to the environment.

Expansive mechanisms of this ilk are not without precedent. Dry, highly

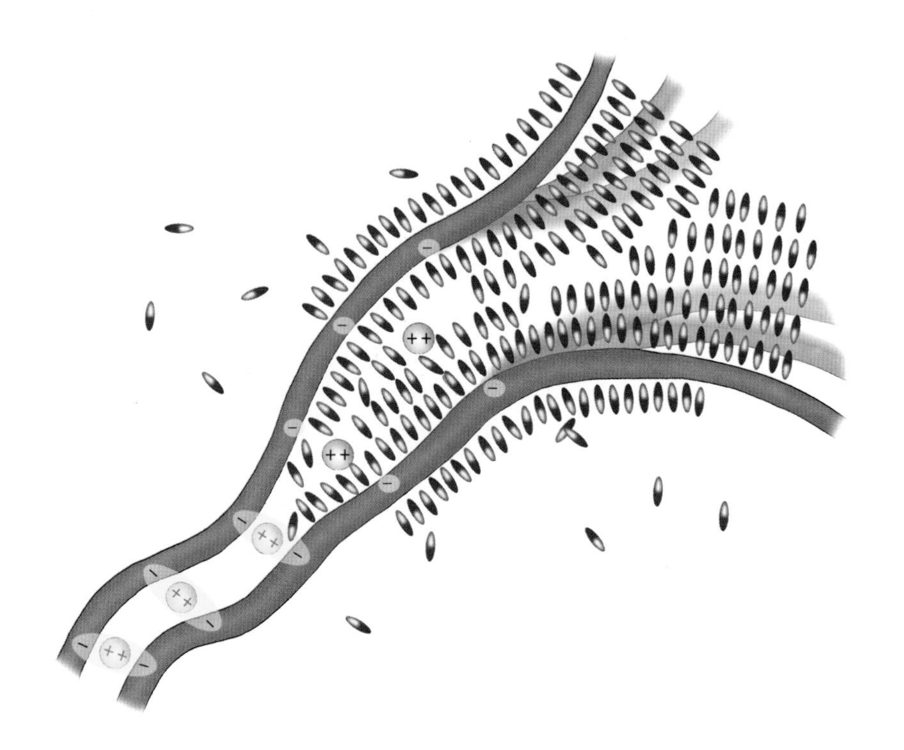

Figure 9.5. *Mechanism of expansion. As breakage of divalent bridges allows strands to separate, water clings increasingly to anionic surfaces. This pushes surfaces farther apart, thereby breaking more bridges and further expanding the matrix.*

charged synthetic polymers immersed in water routinely expand by 3,000 times or more through adsorption (Osada and Gong, 1993). This figure is quantitatively similar to the expansion of a vesicle matrix that is, say, initially half water and undergoes a 500-fold volume increase—water content will have increased by 1,000 times.

Expansion is breathtaking because the adsorptive potential of the polymers' hydrophilic surface is so large. Water is drawn in layer upon layer. The drawing process ultimately terminates when the surfaces' water-drawing potential is exhausted or when the polymer matrix is strained to its physical limit. These factors are ordinarily not severely restrictive, however, so hydration is typically prodigious. Water molecules within the matrix remain organized: Those closest to the surface are most highly structured, while more distant layers are progressively less structured. If polymer strands are not permanently cross-linked and the outer layers' structure is weak, the discharged matrix may dissipate.

What happens to the displaced cations? If the matrix dissipates, the divalent cations may simply diffuse away. If the discharged matrix does not dissolve, the cations will be forcibly expelled by virtue of their low solubility in structured water—solubility is restricted because of the cations' large hydrated radii (Chapter 6). As water organizes around polymers, the cations will therefore be ejected.

Therein lies the punch line. With rare exception, the divalent cations responsible for keeping the matrix condensed—calcium, histamine, adrenaline, mucin, serotonin, *etc.*—are the agents to be secreted. Cut loose from cross-linkage, these cations are freed to carry themselves to their respective targets. Cross-linkers become agents of communication.

TRIGGERING

Triggering of discharge takes place following a series of staging events that include vesicle transport to the cell periphery (Valentijn *et al.*, 1999).

Staging is beyond the scope of this chapter, although it may involve mechanisms of cytoskeletal transport such as those considered in the next several chapters. Here the question is the ultimate trigger—why does the vesicle explode when exposed to the extracellular space?

A logical candidate would be some feature of the new environment. Present in abundance in the extracellular space are monovalent cations such as sodium. Concentrated monovalents can displace divalents (Chapter 6). When they do, the gels abruptly expand in the converse way that expanded gels condense when exposed to divalents (Katchalsky and Zwick, 1955; Tanaka, 1981). This flip-flop paradigm is exemplified in Figure 9.6. The gel either expands or contracts over a narrow window depending on the monovalent/divalent ratio. This paradigm appears to be a general feature of polyanionic gels including those involved in secretion (Uvnas and Aborg, 1977; Verdugo *et al.*, 1992). Thus, monovalent ion substitution is a plausible candidate for triggering divalent expulsion.

Among monovalents, sodium is far more abundant in the extracellular space than potassium, and is therefore the likely candidate. High potassium inside the cell, however, raises a question of premature discharge. Such catastrophe is averted because the intracellular vesicle is enveloped by a lipid membrane—and even if this lipid barrier were absent or in-

Figure 9.6. *Monovalent-divalent switch paradigm. Addition of divalent calcium or magnesium to solutions containing sodium causes poly (sodium acrylate) gels to shrink, abruptly. Withdrawal of divalents triggers expansion. From Tasaki and Byrne (1992).*

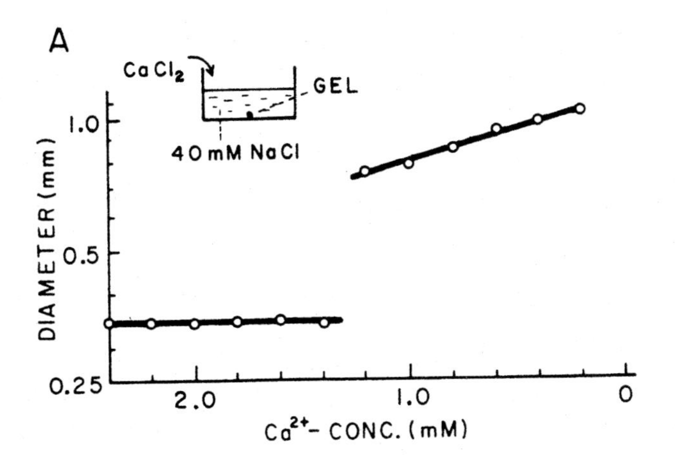

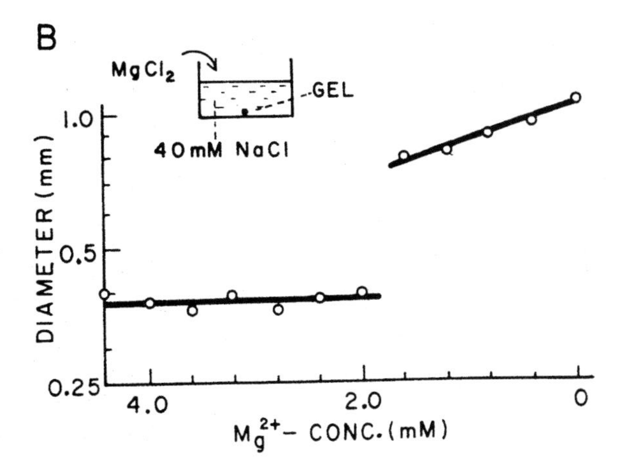

complete, potassium is largely bound to proteins and therefore inaccessible (Chapter 3). Thus, discharge is reserved for the extracellular space, where abundant sodium can displace calcium and trigger the expansion.

The mechanism described above ultimately rests on affinity differences between monovalents and divalents. Sodium can displace divalents because its affinity for negative surface charges is higher. Divalents have lower affinity (except when anionic bridging is involved). This is because their concentrated charge builds larger hydration shells, which require more energy to remove (Chapter 6). According to the Hofmeister series, affinities are anticipated to be: $K^+ > Na^+ > Ca^{++}$ and other divalents > trivalent cations. Indeed, Na^+ and K^+ displace Ca^{++} for matrix expansion (Uvnas *et al.*, 1985); and for condensation, the trivalent lanthanum is even more effective than Ca^{++} or Mg^{++} (Curran and Brodwick, 1991). The mechanism is therefore quite orthodox.

CONCLUSION

The proposed secretory mechanism rests on the physicochemical nature of the vesicle matrix (Fig. 9.7). All vesicles studied to date contain negatively charged polymers cross-linked by a multivalent cation. Because vesicles are formed in the presence of the cation (calcium, histamine, adrenaline, mucin, *etc.*), the respective matrix is condensed into a tight packet. The packet is then shipped to the cell periphery in anticipation of an action potential or some other release signal. As vesicle contents open to the extracellular space, sodium displaces the divalent. Sodium cannot cross-link, so the network gains freedom to expand. Rapid expansion is facilitated by association of water, whose layered structure excludes the divalents, which are thus expelled and freed to act.

While all of these steps seem consistent with available evidence, two features of the mechanism seem especially impressive. First, the agent used to condense the network is the very agent responsible for action. The only known exception to this rule is the insulin-release system, where

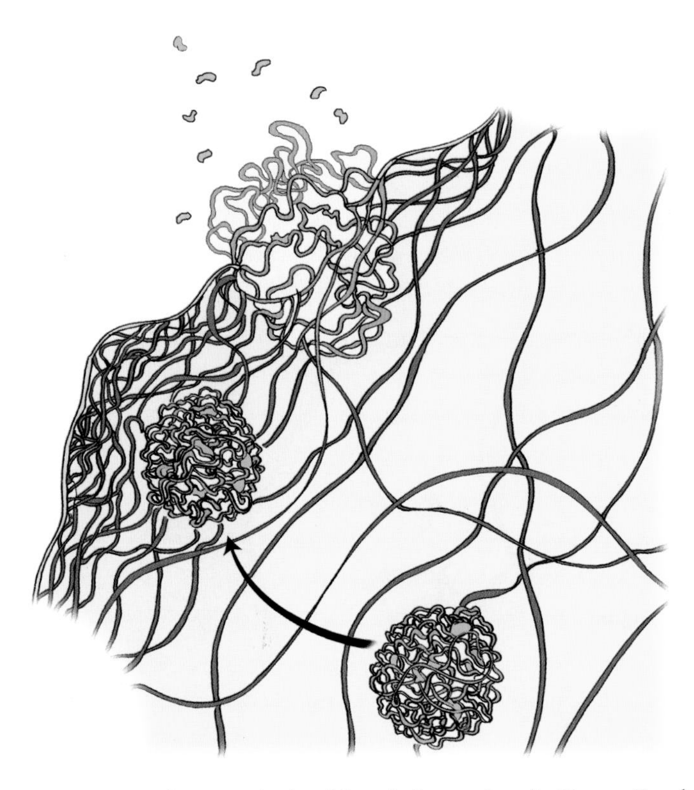

Figure 9.7. Release of divalent cations to the extracellular space by the phase-transition mechanism.

the active agent is the matrix itself and the cation is Zn^{++}. Exploiting the same element for condensation and action is a neat trick of nature. And so too is condensing these cationic agents into packets in order to avoid the ravages of osmotic flooding.

The phase-transition mechanism of secretion remains largely unfamiliar to biologists, but engineers and chemists have begun exploiting it for artificial drug-release (Chapter 8; *see also* Kiser *et al.*, 1998). Hence, the principle is beyond theoretical—it works. Whether it is the working principle in the cell is the question at hand, and the observations above lend substance to such a possibility.

The next chapter explores another matrix-expansion process—the action potential. Here again a textbook mechanism has been in place for many years, and here again several groups have pursued an alternative model based on the phase-transition.

"Hey baby, what's your phylum?"

CHAPTER 10

THE ACTION POTENTIAL

Oddly but aptly named, the action potential is the transient electrical spike that galvanizes the cell into action. It triggers processes ranging from secretion to contraction.

Figure 10.1. *Classical model explaining events leading to the action potential.*

Textbook explanations of the action potential stem from the well-known experiments of Hodgkin and Huxley in the early 1950s. Hodgkin and Huxley studied the electrical properties of the squid giant axon. By shoving a long electrode into the core of the axon and positioning a complementary electrode outside, they could impose voltage steps across the membrane and measure the resulting currents. Their observations were seminal. They led to a model in which the action potential arose out of discrete membrane-channel currents (Fig. 10.1).

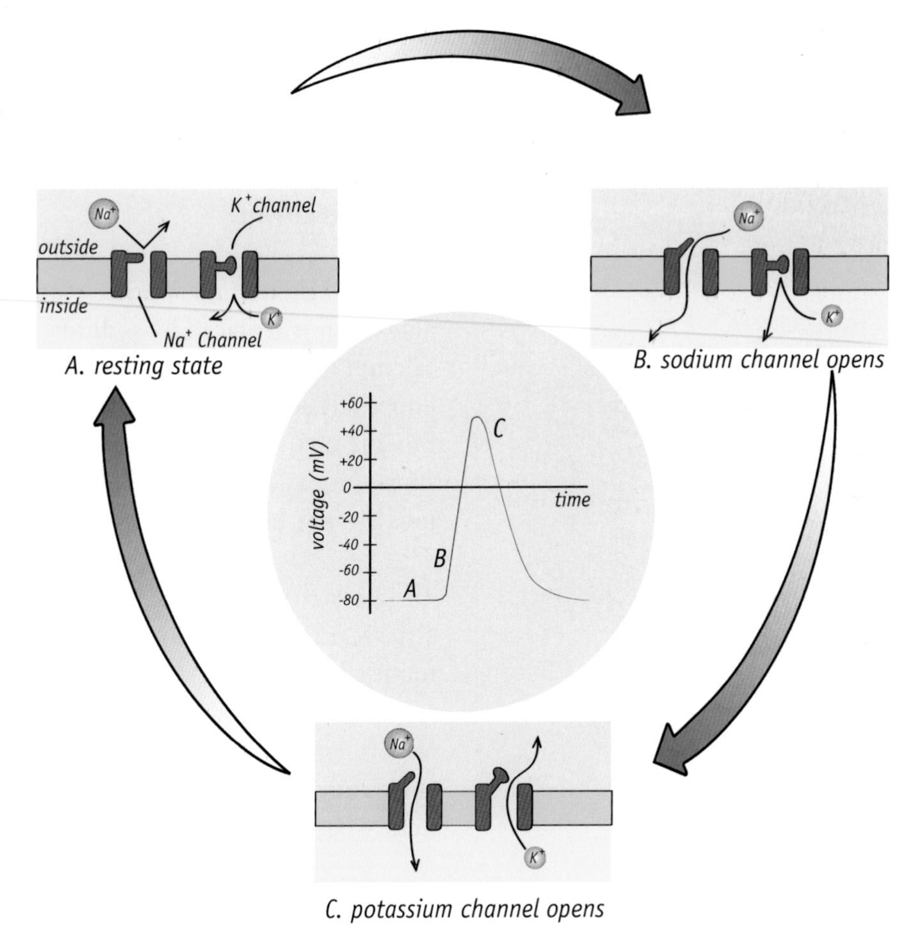

In this model, sodium channels open first, triggered by a chemical or electrical signal. Sodium flows down its concentration gradient, bringing positive charge into the cell and enabling potassium channels to open. Potassium then flows out of the cell. This returns the potential toward its negative resting level, at which time sodium and potassium are pumped across the membrane to restore the initial condition. The action potential propagates because local electrical events induce flanking channels to open, creating a wave that travels rapidly along the membrane. These processes are detailed further in standard cell biology books such as the one by Alberts *et al.* (1994).

Presupposed by this mechanism are central roles for sodium and potassium channels. Yet, when sodium is removed from the medium surrounding the axon, action potentials are not obliterated; sodium-free action potentials are documented in a broad array of experimental preparations (Meaves and Vogel, 1973; Inoue *et al.*, 1973; Tasaki, 1982, 1988). Frog nerve cells and crustacean muscle fibers, for example, produce robust action potentials in media containing no sodium at all (Osterhout and Hill, 1933; Hagiwara *et al.*, 1964). Depending on ionic conditions, such action potentials can be of normal amplitude, although their duration is generally extended. Thus, the positive-going spike can be elicited when sodium is absent.

Similarly, there is no specific requirement for potassium. When internal potassium is replaced by sodium, excitability persists for hours as long as calcium is present in high enough concentration (Tasaki, 1982). Action potentials are again of longer duration and about half normal amplitude. But the characteristic voltage reversal evidently does not require potassium, just as the upstroke does not require sodium. The roles of these ions are not unique.

To imply therefore that the action potential arises out of uniquely specific features of sodium channels and potassium channels seems ill-founded. The underlying mechanism must involve something less narrow. On the other hand, sodium and potassium (or their surrogates) do flow in and out of the cell during the action potential and the challenge is to understand how. If such flows do not depend on particular channels, what can be the responsible agent?

The Peripheral Cytoskeleton

Lying just inside the cell membrane is a dense polymer-gel matrix known as the peripheral cytoskeleton (Fig. 10.2). The presence of such a matrix had been unknown during the Hodgkin-Huxley era when experimental axons were routinely "rolled" to extrude the cytoplasm and presumably leave only the membrane. What in fact remains is the combination of membrane plus contiguous cytoskeleton (Metuzals and Tasaki, 1978)—the latter 100 times thicker than the former. "Transmembrane" currents flow through both. Classically measured currents could as well arise out of the dynamics of the cytoskeleton as the dynamics of the membrane.

Figure 10.2. *Freeze-fracture image of the peripheral cytoskeleton, obtained by John Heuser. (From Alberts* et al.*, 1994).*

The peripheral cytoskeleton is composed mainly of cross-linked actin filaments and microtubules. In the nerve cell, the actin filaments cluster at distinct foci, while enmeshed microtubules run axially just beneath the membrane. Both are present in high density, and both are endowed

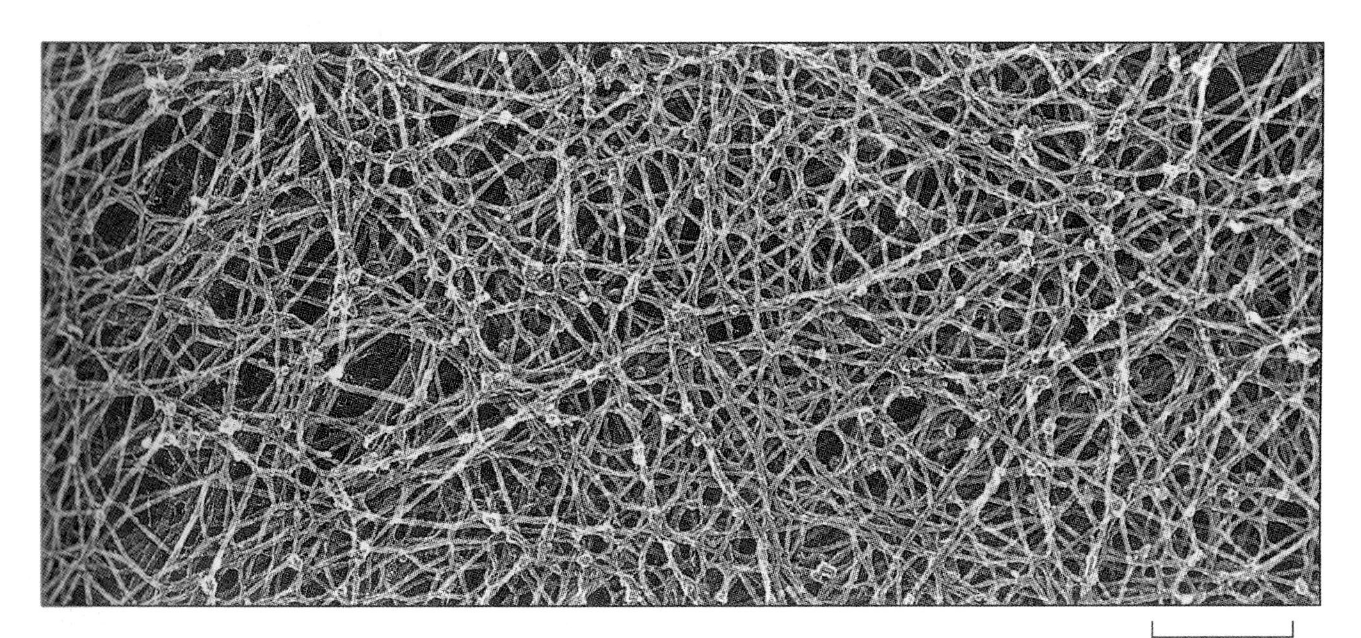

with high negative surface charge (Tsukita *et al.*, 1986). Thus, the cytoskeleton may be thought of as a shell of negative charge that envelops the cell. Additional islands of charge pervade the cytoplasm (Chapter 7), but the peripheral shell is continuous. A microelectrode stuck inside the cell must therefore register a decidedly negative potential (Fig. 10.3).

If the cell potential arises largely from the cytoskeleton, then any change of cell potential implies a respective change in the cytoskeleton. The latter is plausible—actin filaments and microtubules both undergo physical transitions described in detail in the next two chapters. If such transitions expand the cytoskeletal matrix in the same way as transitions expand the secretory matrix (Chapter 9), the increased water content could

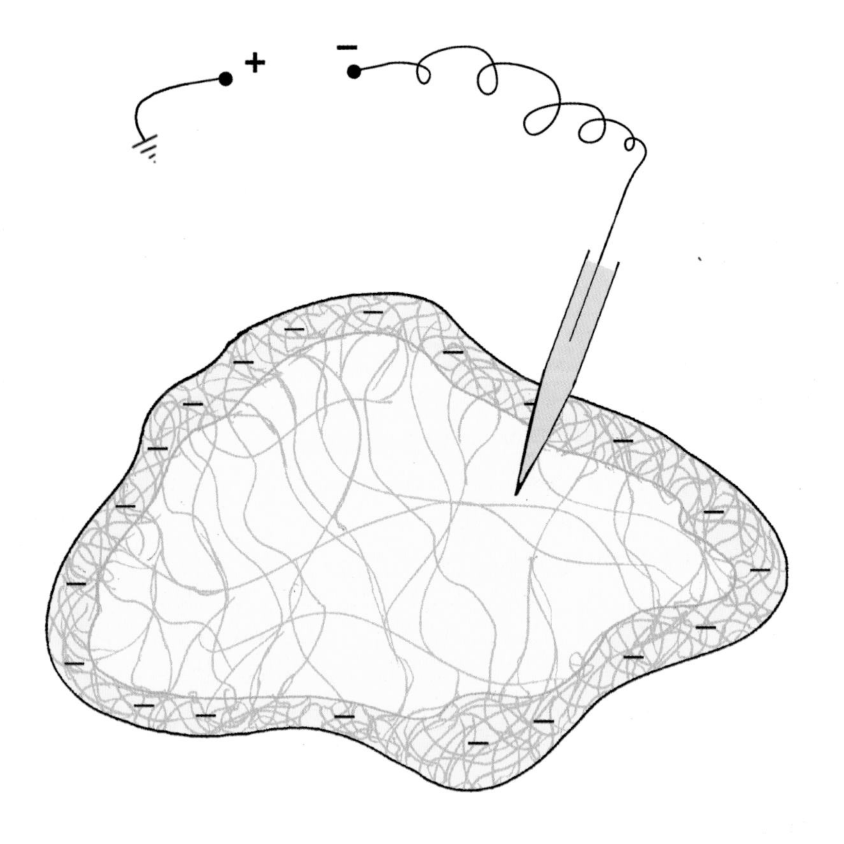

Figure 10.3. *The peripheral cytoskeleton encircles the cytoplasm with a ring of negative charge, leading to a negative cell potential.*

allow ions to pass. Sodium could flow inward, down its concentration gradient, and potassium could flow outward. The prospect of a cytoskeletal-gel phase-transition therefore seems worthy of exploration.

The phase-transition hypothesis for action-potential generation has been pursued extensively in the laboratories of two scientists: Ichiji Tasaki at the NIH, and Gen Matsumoto in Japan. Schooled in the laboratory of Sir Alan Hodgkin in the early 1950s, Tasaki gained world recognition in classical electrophysiology and then broke from the ranks. His views, founded mainly on the results of simple, direct experiments, have implicated the cytoskeletal gel phase-transition as the unique mediator of the action potential. Matsumoto's work details the specific cytoskeletal changes that accompany the action potential, and shows that elimination of the peripheral cytoskeleton eliminates the action potential. For detail beyond what is practical to present here, I recommend a review by Matsumoto (1984), as well as a book and a more recent review by Tasaki (1982, 1999). At age 90, Tasaki continues to pursue this phenomenon on a daily basis.

INVOLVEMENT OF THE CYTOSKELETON?

Linkage between the action potential and the peripheral cytoskeleton is implied in maneuvers that impair the peripheral cytoskeleton. In the squid-axon cytoskeleton, microtubules are packed at a cross-sectional density of ~$100/\mu m^2$ just inside the membrane, and thin out substantially with distance inward (Endo *et al.*, 1979). This density gradient is obliterated by the drug colchicine, which dissolves mainly the peripheral microtubules (Matsumoto *et al.*, 1982; 1983). At concentrations that just eliminate the spatial gradient, there is a concomitant loss of electrical excitability. Similar linkage is seen with a half-dozen other microtubule-depolymerizing agents (Matsumoto *et al.*, 1984). Microtubule depolymerization also impacts heart excitability (Gomez *et al.*, 2000).

The possibility remains, however, that such parallel effects are merely coincidental. To check this possibility, disrupted microtubules were repolymerized. The core of the squid axon whose excitability had been impaired by exposure to colchicine was perfused with a solution containing microtubule- and microtubule-cross-linking proteins under conditions favorable for assembly. This solution restored the matrix and also restored the lost excitability (Matsumoto *et al.*, 1979; 1984).

The linkage between cytoskeletal integrity and excitability implies that action potentials could be impacted by most any agent that solubilizes gels. Consider the following. When axons are internally perfused with KCl, action potentials will wane progressively and vanish within 25-40 minutes. This decline is impacted in a telling way by ion substitutions (Tasaki *et al.*, 1965). If fluoride replaces chloride during the period of decline, action potential amplitude is promptly restored. Bromide substitution accelerates the decline. Examination of a large number of anions showed that restoration efficacy followed the classical Hofmeister series (von Hippel and Wong, 1964). Thus, F > HPO_4 > glutamate > SO_4 > acetate > Cl > NO_3 > Br > I > SCN. The anions toward the left had the highest tendency to restore action potential amplitude, and are the most stabilizing. Those toward the right most strongly hastened the action-potential decline and have the highest tendency to solubilize gels.

Without an intact cytoskeleton, then, action potentials cannot occur. Action potentials seem somehow to be linked to the peripheral cytoskeleton's integrity. Even in normal, untreated axons, when the existing cytoskeleton is further stabilized by polymerization-promoting agents such as Taxol or dimethyl sulfoxide (DMSO), excitability is enhanced (Matsumoto *et al.*, 1984). Thus, the action potential seems inextricably linked to some feature of the cytoskeleton.

A CYTOSKELETAL PHASE-TRANSITION?

An array of cytoskeletal transients accompanies the action potential. One of these transients is in form birefringence, an optical measure of structures lying predominantly in one direction—longitudinally in the axon (Metuzals and Izzard, 1969). Because the form-birefringence transient persists after the axon has been surgically flattened to squeeze out the cytoplasmic core, the source has been localized to the longitudinal structures of the peripheral cytoskeleton (Sato *et al.*, 1973). Thus, the peripheral cytoskeleton undergoes physical change during the action potential.

Coincident with this change is a transient liberation of heat (Fig. 10.4). Heat liberation during the action potential was first reported long ago (Abbott *et al.*, 1958) and has been repeatedly confirmed (Tasaki and Iwasa, 1981; Tasaki *et al.*, 1987, 1989). The signal is biphasic—heat is released during the action potential's rising phase, and absorbed during its recovery. The absorption phase can be seen most prominently when internal potassium is replaced by tetraethyl ammonium, a maneuver that extends the action potential duration by a factor of 100 to 1000. A brief suprathreshold heat pulse applied during this extended electrical plateau can trigger a swing all the way back to the resting potential (Spyropolous,

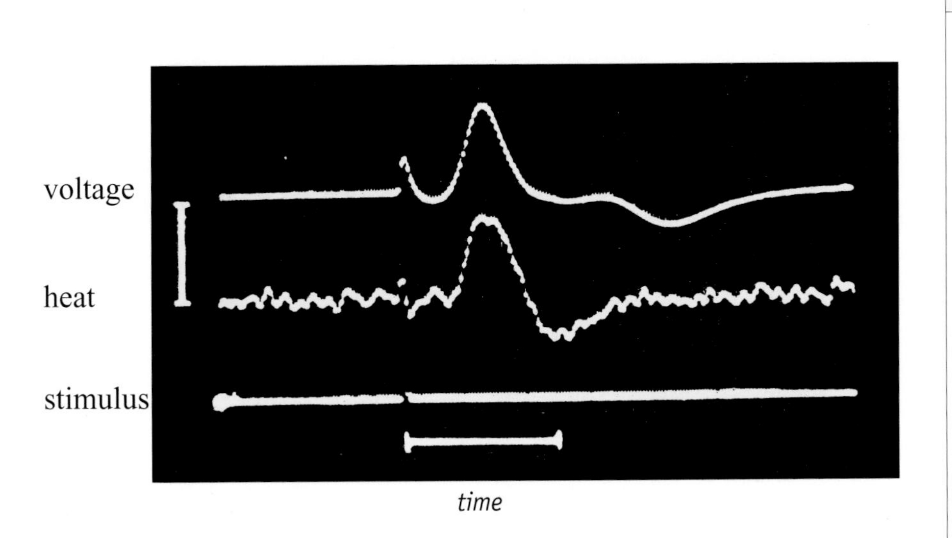

Figure 10.4. *Heat liberation accompanying the action potential in garfish olfactory nerve. Action potential (top), thermal response (middle) and stimulus artifact (bottom). Vertical bar: 4 x 10^{-3} deg/sec; horizontal bar: 20 msec. From Tasaki (1988).*

1961); *see* Figure 10.5. This action shows that relaxation is a heat-absorbing process. It also illustrates the process's critical character.

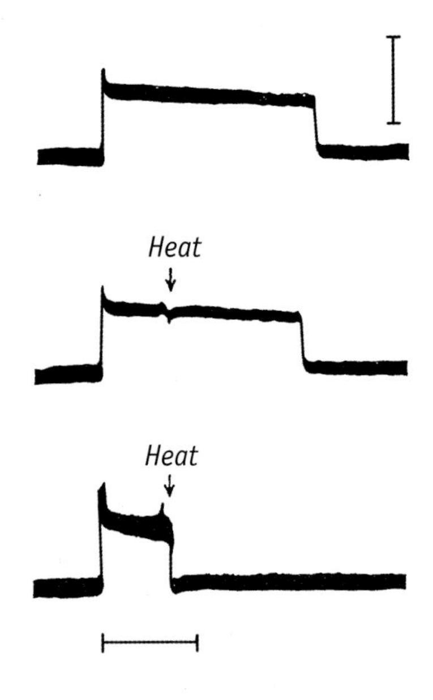

Figure 10.5. *Pulse of heat applied during the extended action potential in a Ni-treated node of Ranvier. Subthreshold pulse exerts only transient effect (middle panel). Slightly larger pulse triggers potential swing all the way to baseline value (bottom panel). After Spyropoulos (1961).*

Accompanying these physical changes is a transient cellular expansion (Fig. 10.6). Transient swelling coincident with the action potential is observed not only in the squid axon but in a series of specimens using a variety of measurement techniques (for review, *see* Tasaki, 1999). It implies transient accumulation of water inside the cell. Hence, the action potential is accompanied by water adsorption and heat liberation, and these dynamics are coincident with some structural change that takes place in the peripheral cytoskeleton. The critical nature of the heat trigger implies that the structural change could well be a phase-transition.

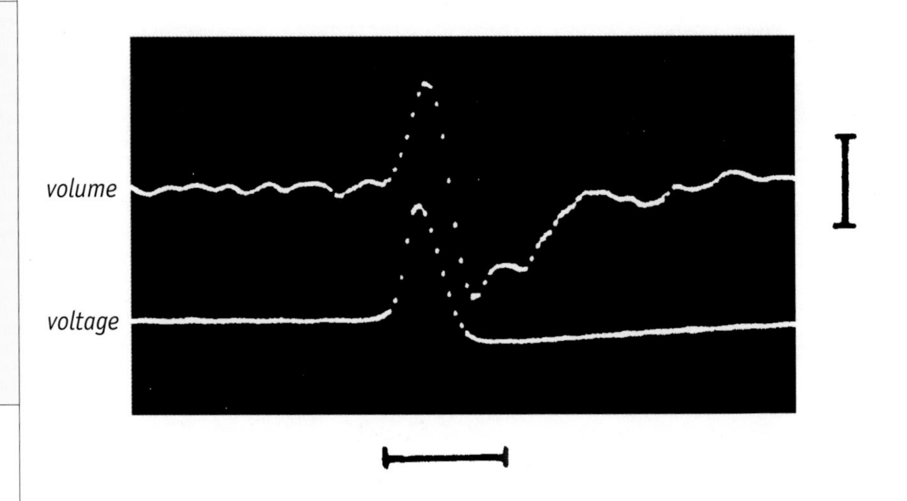

Figure 10.6. *Transient cellular expansion measured using a tension transducer (upper trace) in squid axon. Corresponding action potential shown in lower trace. Bar: 10 μg expansion force. From Tasaki (1988).*

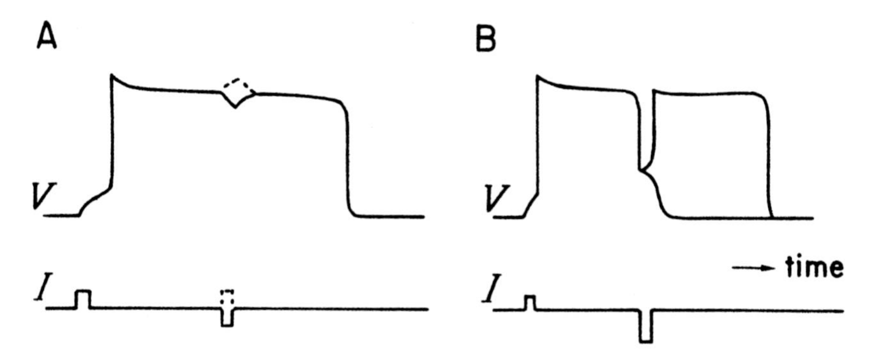

Figure 10.7. *Effect of repolarizing pulse applied during an extended action potential in squid axon. Subthreshold pulses elicit only transient changes (A, and top trace of B). Suprathreshold pulse (B, lower trace) elicits a swing to baseline. The result confirms electrical bistability. From Tasaki (1988).*

Phase-transition implies a flip between two states. Identifying any such flip within the brief interval of an action potential is a formidable challenge, but experiments carried out on cells with extended action potentials (as above) confirm bistability. Such experiments employed a brief electrical repolarizing pulse (Fig. 10.7). Repolarizing pulses of sub-threshold value had no special effect—the potential quickly returned to its plateau value. But when the pulse amplitude exceeded a threshold, the potential could swing precipitously to its baseline value, just as it did with the heat pulse (Tasaki and Hagiwara, 1957). Thus, the potential is stable at its plateau level and stable at its baseline level, but unstable in between.

Electrical bistability has been confirmed repeatedly. In one experiment (Tasaki and Byrne, 1994) the contents of the perfusion bath were varied progressively: as the sodium/calcium ratio crossed a threshold, the axon potential took a sudden step (Fig. 10.8). Similarly critical was the effect

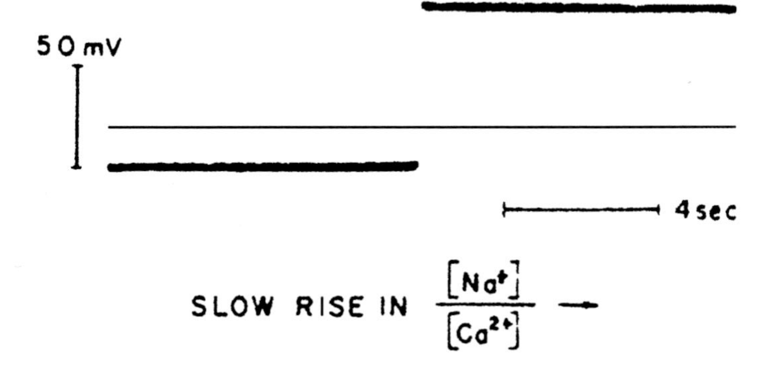

Figure 10.8. *Effect of monovalent/divalent ratio on electrical potential in squid giant axon. From Tasaki and Byrne (1994).*

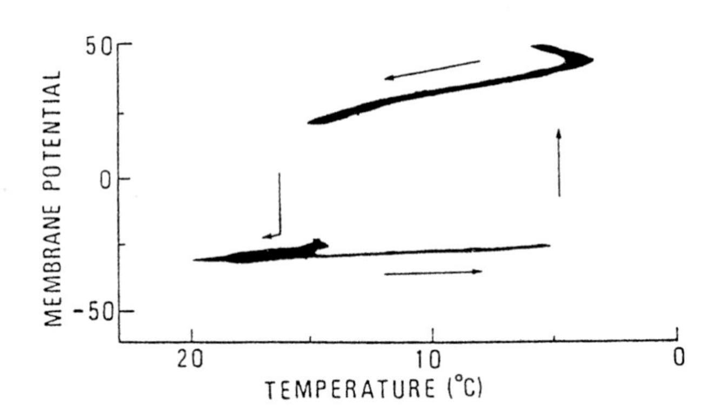

Figure 10.9. *Effect of cyclic changes of temperature on cell potential in squid giant axon. From Tasaki (1999b).*

of temperature (Fig. 10.9): As temperature of the bath surrounding the squid axon was lowered just past 15°C, the intracellular potential took a reversible step of 50 mV upward (Inoue *et al.*, 1973).

Taken together, all of these observations imply an electrically consequential transition between two structural states of the cytoskeleton. One state corresponds to the baseline potential, the other to the plateau potential. Somehow, a structural transition in the peripheral cytoskeleton appears to create the voltage swing commonly known as the action potential.

IONIC MECHANISM

In pursuing the mechanism of any such transition, an important clue is the ion requirement. Sodium and potassium cannot be critical because action potentials can be generated, as we have noted, even when these ions are absent. Calcium, on the other hand, is critical. When calcium is removed from the extracellular solution, action potentials are immediately abolished. For generation of action potentials, many experiments confirm the requirement for calcium or a divalent cation surrogate (for review, *see* Tasaki, 1988).

A logical way of approaching the transition mechanism, therefore, is to probe the role of calcium. In the secretory granule, calcium condensed the polymeric matrix (Chapter 9). A similar theme could play in the peripheral cytoskeleton. If calcium maintained condensation of the cytoskeletal lattice, then calcium displacement by a monovalent, which cannot cross-link, could trigger the reported expansion—the same as in the secretory granule. The lattice would then expand, and because of its more open nature the expanded lattice would have higher conductance. Ions could pass, and the cell potential could change.

A difference between the peripheral cytoskeleton and the secretory vesicle, however, is the organization of polymers. In vesicles, polymers are generally tangled. In the cytoskeleton, polymers are covalently cross-linked to one another so the matrix cannot expand indefinitely. As the matrix does expand it stores elastic energy, some of which can ultimately be returned to the system. Any such energy transfer does not necessarily imply a perpetual motion machine—only an indication that stored energy could be put to use in restoring the initial condition and reversing any conductance change (see below).

A plausible way in which the action potential could be initiated, then, is by replacing calcium with a monovalent. Classically, sodium is thought to enter the cytoplasm through a localized, receptor-mediated permeability increase. In the proposed model, sodium ions flow into the peripheral cytoskeleton and begin displacing calcium. Replacement loosens the network, enabling it to adsorb water and expand. As it expands, permeability is increased, allowing for more sodium entry, further Ca displacement, additional expansion, *etc.*—like ripping open a zipper (Fig. 9.5). Meanwhile, the entering sodium drives the cell potential to a more positive value, establishing the action potential's rising phase. This mechanism is not dramatically different from the classical view except that sodium entry has a function beyond merely changing the potential.

The cell potential soon reverses, and an obvious mechanism to consider is a reversal of the phase-transition. Water must exit and calcium must return to condense the cytoskeleton. Something must drive the change:

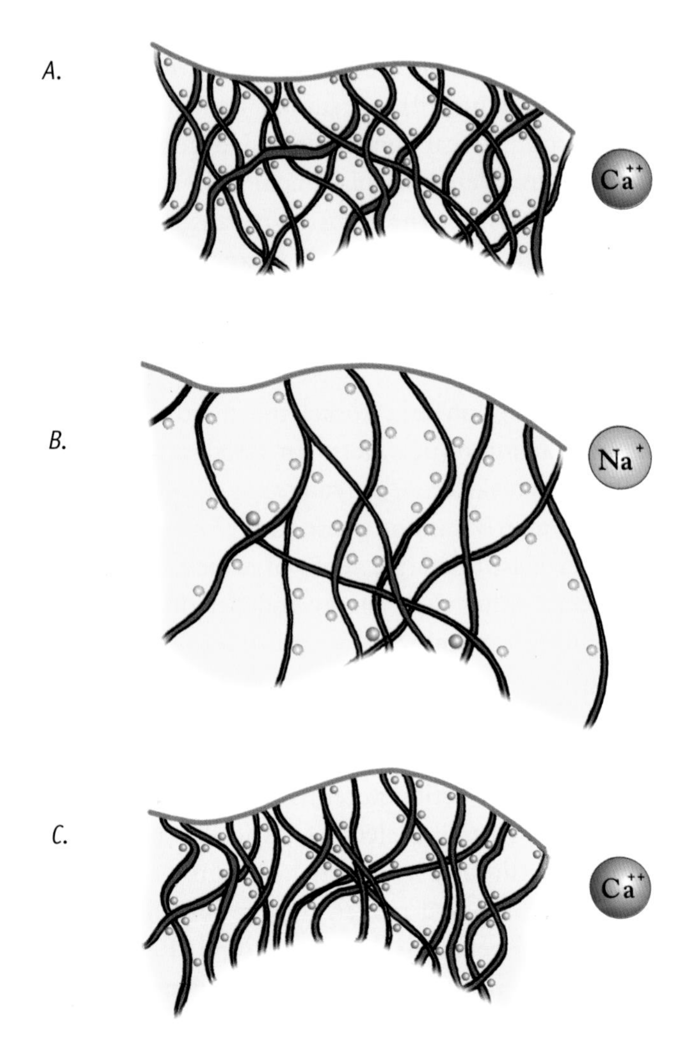

Figure 10.10. *Structural dynamics and the action potential. Initially (A) the network is collapsed as strands are bridged by calcium. The network expands as sodium replaces calcium (B). But increasing sodium eventually neutralizes surface charge, weakens water structure and allows the polymer-retractive force to collapse the network (C), at which stage calcium may easily bridge the strands once again.*

one speculation is that the retractive force of the expanded and strained cytoskeleton could initiate condensation. With retraction, closely apposed polymers may then more readily accept calcium bridges, perpetuating condensation by closing the zipper.

For such a mechanism to work, the expansive force of adsorbed water must give way—for only then could the matrix retract. In fact, water "melting" is a seemingly inevitable consequence of the progression of events (Fig. 10.10). During the rising phase of the action potential, protein surfaces adsorb water. They also adsorb ions in competition with the water (Chapter 6). Water adsorption is limited by matrix expansion, but sodium continues to flow down its concentration gradient and into the permeabilized network. Adsorbed ions nullify surface charge and thereby erode water structuring (Luck, 1976). At some point, the increase of adsorbed sodium should impair water structuring to the point that the network's retractive force wins out against the expansive water force, and the network begins to collapse. With anionic sites brought into closer apposition, calcium can displace the bound sodium, restoring both volume and cell potential toward their initial values.

An advantage of this mechanism is its parsimony—recovery arises from the same events that generate the rising phase. Sodium entry initially triggers expansion and permeability increase; the permeability increase permits more sodium to enter, which ultimately destabilizes water structure and permits the network to collapse. Recovery is therefore inevitable. It explains why, when failure does occur, the potential rarely hangs up at a positive value.

The speculative aspect of this mechanism is the anticipated collapse of water structure at the peak of the action potential. Water's involvement in recovery is supported by an energetic argument. Since recovery occurs spontaneously, the free energy change must be negative, by definition. Free energy consists of two terms—enthalpy and entropy. Thus, $\Delta G = \Delta H - T\Delta S$, where ΔG is the free energy change, ΔH is the enthalpy change, T is the absolute temperature, and ΔS is the entropy change. ΔH is positive because heat is absorbed during recovery (above). To offset this, ΔS must be very positive (increasing entropy)—implying a substantial loss of structure during recovery. This loss need not arise from water destructuring *per se*, but the required size of the change implies the involvement of many molecules, and the number of water molecules greatly outweighs all others. Hence, the likelihood of water involvement is high.

Recognition that water movement plays a central role in generating the action-potential may bring closure to the unsettled question of general anesthetic action. Anesthetics eliminate action potentials. In seeking a mechanism, Pauling (1959) noted that many general anesthetics have no covalent chemistry at all, and form no hydrogen bonds. A common feature of these anesthetics, he noted, was that they surround themselves by water clathrates, and are therefore likely to act through this route. For action-potential generation according to the proposed mechanism, water must be free to move. Any water constrained by clathrate formation will be preoccupied, and thereby unable to enter easily into the cytoskeletal lattice. In sufficient concentration near the axon, such water could eliminate the action potential altogether.

Finally, it is worth returning briefly to the classical patch-clamp experiments (Chapter 1) to see how those results might fit with the proposed phase-transition mechanism. Envision a patch—membrane plus cytoskeleton—held by a pipette. Suppose a suprathreshold potential difference is imposed across the patch. The pulse of current passing through the patch has been taken as evidence of a single channel opening and closing. The "channel" could well correspond to the cytoskeletal actin cluster. In the axon, such clusters are distributed on a micron scale—sufficiently far apart to permit independent phase-transitions, yet close enough to allow a typical 1-μm patch to contain one or a few. Thus, quantal currents are anticipated—one quantum per cluster. Such quantal currents should be self-arresting: monovalents entering the cytoskeleton in sufficient number will shut down ion flow in the same way as they shut down the action potential. Thus, patch currents occur as a pulse rather than as a step.

In fact, self-arresting patch-clamp responses should be produced by any kind of phase-transition. The phase-transition need not arise physiologically—it could for example reside in the seal between specimen and pipette, where pockets of water may undergo phase-transition. As accumulating monovalent ions impair water structure and collapse the pocket, the current-producing transition will self-arrest. It is therefore understandable why quantal currents can be observed in the broad array of circumstances illustrated in Figure 1.3, even circumstances in which channels are demonstrably absent.

In a Heartbeat

A tidy way of tying together the central points of this and the previous chapter is to consider a situation in which the action potential arises directly out of a secretory event. This occurs in the synapse, where neurotransmitter secreted by one cell triggers an action potential in a contiguous cell. It may also occur in the cardiac pacemaker—the cell cluster that triggers the heartbeat.

Pacemaker cells are small spherical entities lodged in a compact cluster at the base of the heart's right atrium. These cells undergo spontaneous cyclic changes of potential (Fig. 10.11). When depolarization crosses a threshold, the heart is triggered to contract.

These spontaneous potential changes have been thought to arise out of a complex summation of channel currents (Noble and Bett, 1993). On the other hand, a characteristic feature of pacemaking cells is a conspicu-

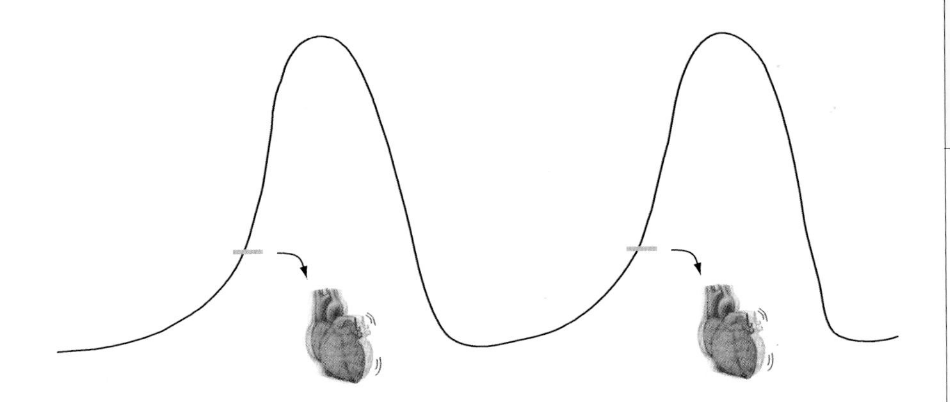

Figure 10.11. *Time course of electrical potential in cardiac pacemaking cells. When depolarization crosses a threshold, the heart is signalled to contract.*

ous abundance of secretory vesicles. These vesicles contain high concentrations of catecholamines (Trautwein and Uchizono, 1963; Challice, 1966, Cheng, 1971, Miyagishima, 1975). In fact, the catecholamine (adrenaline, epinephrine) concentration in pacemaking cells of primitive hearts is as high as in some nerve terminals. When these catecholamines are experimentally depleted, pacemaking stops; when they are restored, pacemaking recommences (Tuganowski *et al.*, 1973; Noda and Yugari, 1973). Thus, catecholamines are required for pacemaking.

Noting the catecholamine requirement, I suggested some years ago that the agent responsible for spontaneous depolarization in pacemaking cells could be the binding of catecholamines (Pollack, 1977). After all, nearby nerve terminals discharge catecholamines that bind to pacemaker cells and accelerate depolarization, making the heart beat faster. Why couldn't catecholamines secreted from the pacemaker cells themselves do the same (Fig. 10.12)? Discharged catecholamines would bind to the same or another pacemaker cell, triggering a phase-change that increases local per-

Figure 10.12. *Proposed mechanism of pacemaking. Catecholamines released from a pacemaking cell bind to the same or another pacemaking cell, promoting depolarization. The more the pacemaking cell depolarizes, the faster catecholamines are discharged. This dynamic results in the pacemaking potential's exponential rise (Fig. 10.11).*

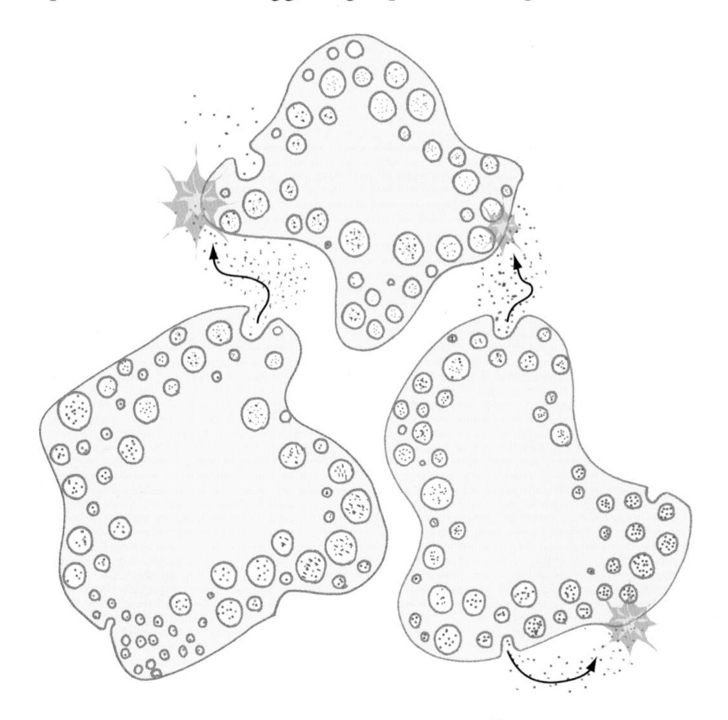

meability and allows cations to enter and incrementally depolarize the cell. The more depolarized the cell, the higher the catecholamine-discharge rate (same as the nerve terminal), and hence the faster the depolarization. Such positive feedback could explain the pacemaker potential's exponentially rising shape (Fig. 10.11).

Although this mechanism is speculative, it accounts for several elementary facts that current models do not address. It explains why pacemaking cells are densely packed with catecholamines. It accounts for the absolute requirement of catecholamines for pacemaking. It explains why epinephrine-induced pacemaking can be associated with cyclic volume changes (Tasaki, 1998). And it ties together the known effects of catecholamines discharged from neurons with catecholamines discharged from internal sources: both promote pacemaker depolarization by the same route. The mechanism thus removes the pacemaking process from the realm of the unique and places it squarely within the realm of ordinary synaptic action.

CONCLUSION

The message of this chapter and the previous one is that the cell's communication mechanisms may be governed by phase-transitions. Secretion and action-potential generation rest on the dynamics of negatively charged polymeric matrices—condensed by divalent cations and decondensed by monovalent cations. In the secretory matrix the decondensing transition goes to completion, the network expanding fully and in some cases irreversibly. In the cytoskeletal matrix the transition does not go to completion because matrix polymers are covalently cross-linked and the matrix cannot expand indefinitely; the transition is therefore easily reversed. Thematically, however, the two mechanisms are similar—and it is notable that two such similar mechanisms arose from researchers largely unfamiliar with one another's approach.

Cellular Transit Authority

Route and usage information

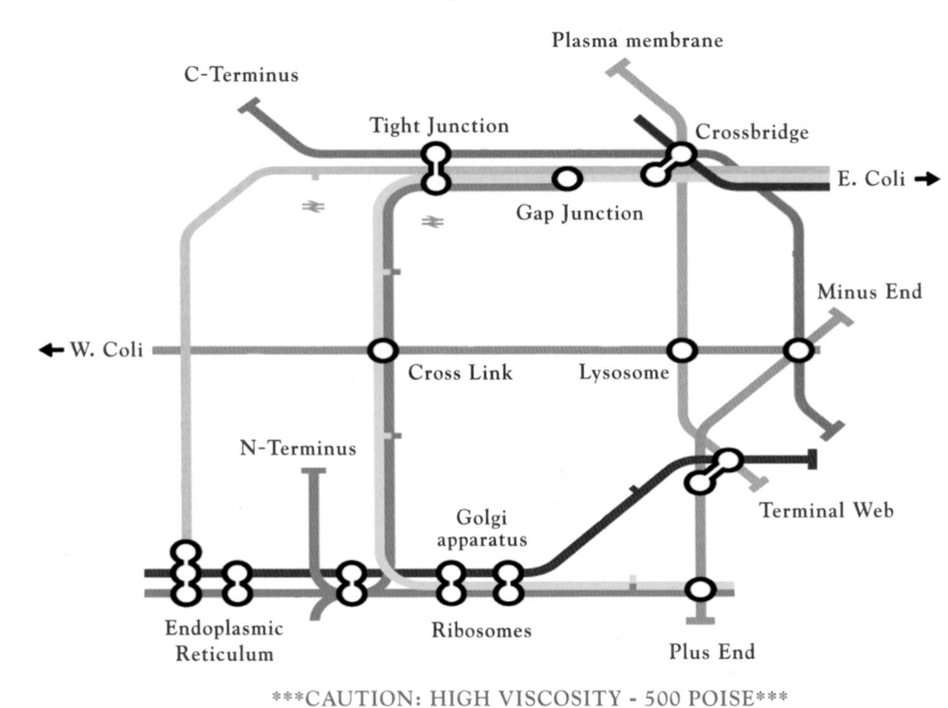

CAUTION: HIGH VISCOSITY - 500 POISE

Platform A
Lines departing for the actin cortex; see beta-sheet for departure times.

- Germ Line
- Phase Transit
- Second Messenger

Platform B
Plus-end directed lines; diffusion times to be announced.

- Metabolite Shuttle
- Axonal Transport
- Random Walk

Platform C
Lines departing for all extracellular destinations. Translation sequences may vary.

- Brownian locomotion
- NaK Transporter (DELAYED)
- Brush Border Express

In consideration of other patrons:

no apoptosing
no endocytosing
no secretion
fine for public meiosis: $500
absolutely no focal adhesion

Please place unused metabolites in the designated lysosomes.
Thank you for keeping your Golgi bodies to yourself.

We realize you have a choice of fundamental biological hypotheses; we'd like to thank you very much for riding with us.

- the management

11

CHAPTER 11

TRANSPORT

Having considered communication, we move next to a process at least as fundamental—transport. Without material transport, the cell cannot metabolize, move, secrete, or divide. Material transport is central to cell function.

In principle, the transport requirement could be met through diffusion. But diffusion lacks directionality; and it is slow. The cell relies instead on vectorial processes that transport substances actively from one place to another, the most primitive version of which is "streaming." Streaming is an archaic analog of blood circulation; it carries substances along as though suspended in a flowing stream. Streaming is a vehicle of choice in both plant and animal cells ranging from the most primitive to the most differentiated—from algae, amoebae and bacteria up through human neurons.

This chapter explores the underlying mechanism of streaming. We consider whether a phase-transition such as one of those presented in the last few chapters holds the key. The significance of this exploration extends beyond streaming *per se,* for the streaming phenomenon is primitive enough that whatever underlying principle is uncovered here may well be generic for related processes.

THE ORGANELLE

The streaming organelle is quite simple. It consists of a bundle of filaments containing three essential elements—actin, myosin and water.

Actin is the primary building block. Globular-shaped actin monomers about 5-nm in diameter link to one another to create a "strand of pearls" and two such strands entwine to form the filament (Fig. 11.1). Actin filaments rarely exist in isolation, and the streaming organelle is no exception. Through cross-linking proteins such as alpha-actinin, filamin and villin, filaments bundle into a parallel array.

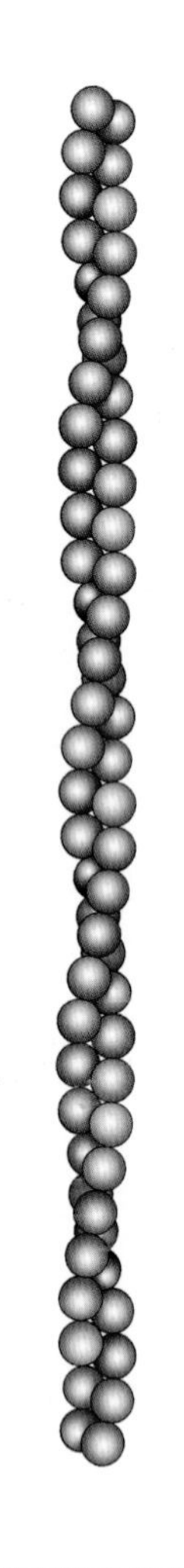

Figure 11.1. *Molecular structure of the actin filament.*

Actin is an exceptional protein both in ubiquity and function. Its presence has been documented in perhaps every eukaryotic cell examined to date, implying a central role in cell function. And whereas most proteins typically evolve into numerous function-specific isoforms, actin has remained resistant to change; relatively few viable isoforms exist. Such consistency implies a delicate function that even conservative structural change could render inoperative. Actin is not likely to be an inert rod.

The second component of the system is water. Water, as we know, will self-organize around protein surfaces. The actin surface appears to be especially adept, for actin filaments gel practically at the snap of a finger, implying high water-adsorptive capacity (Chapter 8). Even huge osmotic gradients cannot remove actin-gel water (Ito *et al.*, 1992). Moreover, the actin-filament bundle is confirmed by microscopic observations to be jacketed by a clear zone extending out as far as 1,000 nm from the bundle's surface, devoid of any particulate matter (Kamitsubo, 1972). Exclusion of solutes again implies high water-structuring capacity.

The tightness of water structuring is inferred also from the electron micrograph of Figure 11.2. Here globular actin was infused into a muscle. The globular actin polymerized into filamentous actin, forming a gel-like construct. The strength of the actin-water bond was apparently strong enough that the muscle-filament lattice was split apart, as shown.

Actin's proclivity to structure water is consistent also with its resistance to freezing. An aqueous solution of propanediol and glycerol will ordinarily freeze readily. When augmented by only 0.5% filamentous actin (plus 0.1% cross-linker alpha-actinin), the solution will not freeze even

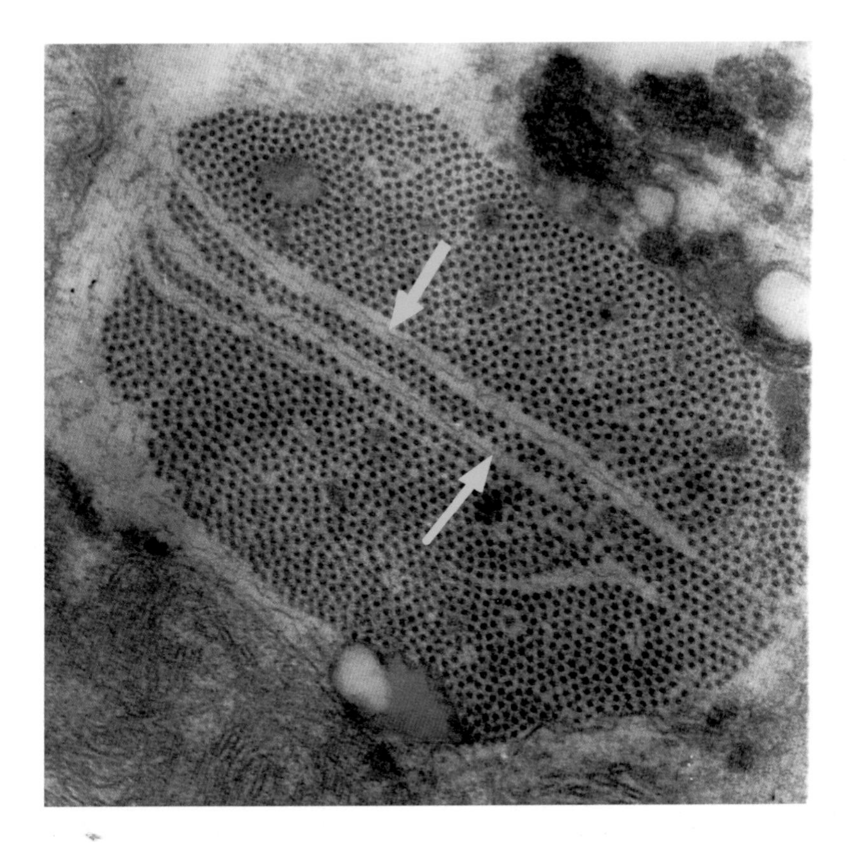

Figure 11.2. *Polymerization of actin creates a force strong enough to break apart the muscle-filament lattice (arrows). Micrograph courtesy of Károly Trombitás.*

down to liquid nitrogen temperatures (Prulière and Douzou, 1989). The filaments apparently organize the water into a configuration that is difficult to convert to ice. The small amount of actin required to achieve this resistance is telling.

Tight water structuring may account for some of actin's well-recognized biochemical features. Among proteins, actin is famous for its "stickiness"—it is said by biochemists to bind practically anything in sight. The number of actin-binding proteins identified by the mid-1980s totalled more than 60—over and above those solutes known to bind to actin non-specifically (Pollard, 1984). Since solutes are excluded from structured water, they will be forced to move in either of two directions: into bulk water, or to the protein surface—anything to avoid the structured water environment. Proteins with the most surface-adsorbed sol-

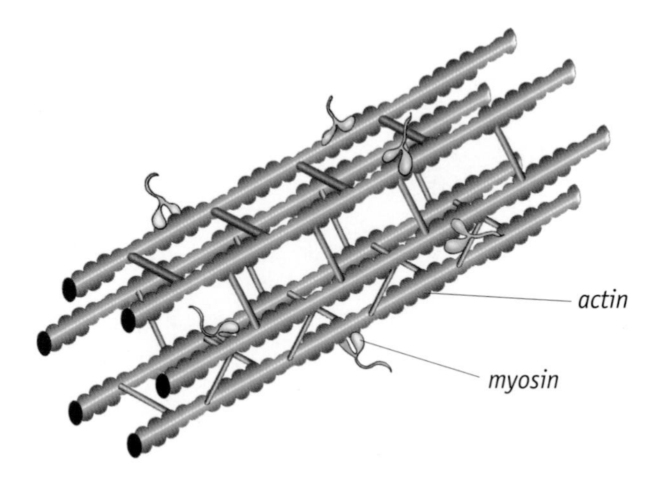

Figure 11.3. *Along with structured water, the streaming organelle consists of cross-linked actin filaments interacting with myosin.*

utes, therefore, will be those that structure water most extensively. Hence, actin's legendary stickiness.

Tight water structuring may also be tied to actin's resistance to mutation. Actin's molecular structure has remained uncommonly consistent over the eons. If actin function turns out to be linked to its water-structuring capacity (see below), then the protein's evolutionary consistency may be understandable, for water-structuring capacity is critically dependent on surface-charge distribution and hence, even modest alterations could defunctionalize the protein. This interpretation places water structuring at center stage—a possibility that will be considered.

The third component of the streaming system is myosin. Unlike actin, myosin appears in numerous isoforms and families, the most common of which are myosin-1 and -2. Myosin-2 is the larger filament-forming variety that appears in muscle and elsewhere, whereas myosin-1 and some of the higher-numbered myosins are diminutive forms. It is the diminutive form that predominates in the streaming system. Myosin appears to be required for initiation of streaming.

The system's essence then is quite simple: Actin filaments cross-linked into a parallel bundle; spaces within the bundle suffused with structured water; and myosin associated with actin (Fig. 11.3). The organelle is commonly situated in the cell periphery, where streaming is prominent. Even in isolation, this organelle system can propel materials along its length.

A PHASE-TRANSITION?

Materials could be induced to flow along the filament bundle through some kind of phase-transition. Phase-transitions mediate shifts of both solvent and solutes (Chapter 8). Hence, an actin phase-transition could be anticipated to produce some kind of solute shift, and if the transition were to propagate along the filament bundle, the solutes might shift vectorially.

That structural change does propagate along the actin-filament bundle is well known. Streaming bundles extracted from internodal cells of *Nitella* and *Chara* show vigorous undulations that propagate from one

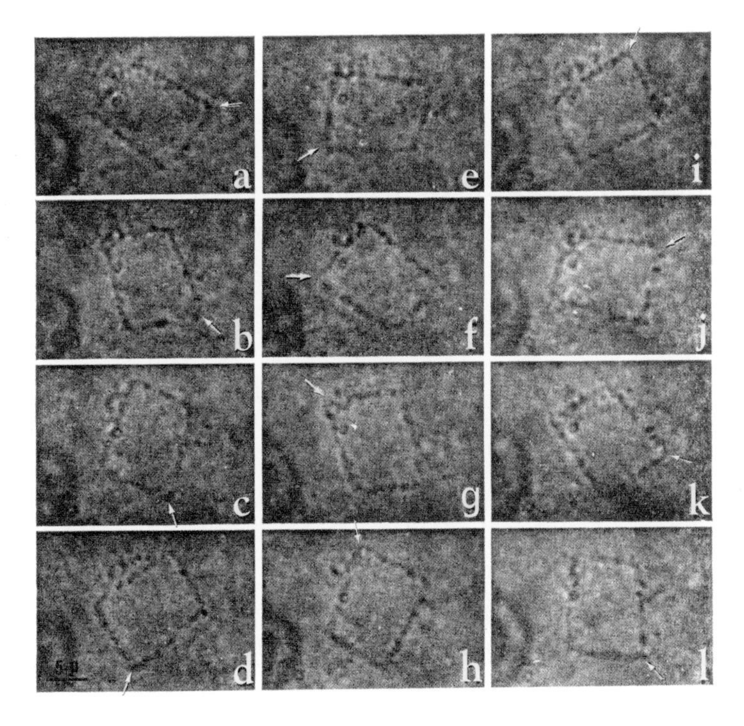

Figure 11.4. *Vertices of the actin-filament bundle propagate around the otherwise stationary polygon (arrows). From Kuroda (1964).*

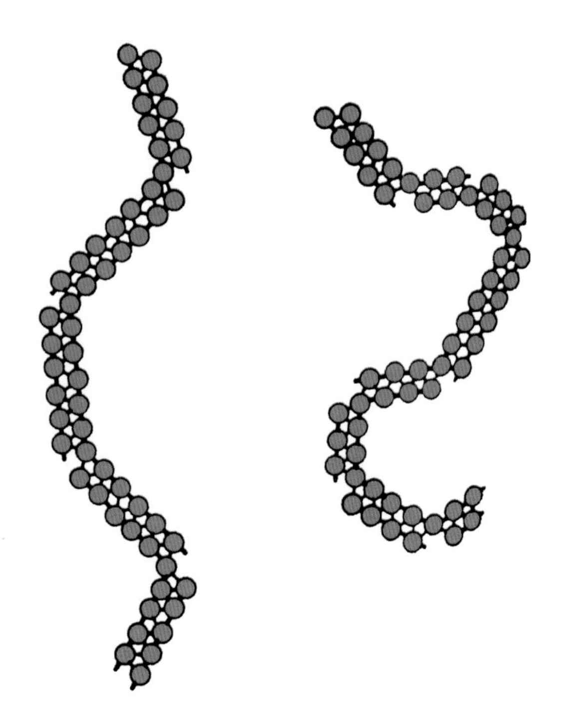

Figure 11.5. Long and short states of actin. After Hatano (1972).

end to the other (Kuroda, 1964; Kamitsubo, 1972). These undulations are particularly conspicuous in bent specimens that form closed polygons. Although the material is fixed in space, vertices of the polygon propagate ceaselessly along the perimeter at ~10 μm/sec (Fig. 11.4). Thus, a propagating undulation is directly visualizable.

The structural basis of such undulation was studied in the slime mold *Physarum* (Hatano, 1972). Actin filaments resynthesized from streaming-system proteins took on either of two structural configurations depending on the level of ATP and potassium—a linear configuration, or a shorter, more flexible configuration with multiple kinks (Fig. 11.5). This observation implied that the undulation might depend on local filament folding, which would locally shorten the filament. Indeed, isolated streaming bundles show spontaneous length oscillations on the order of ~10% (Kamiya, 1970), confirming their implied ability to shorten.

Undulations are also observable in single actin filaments such as those that make up the bundle. Known as "reptation" because of their snake-like character, the undulations are observed broadly: in filaments suspended in solution (Yanagida *et al.*, 1984); in filaments embedded in a gel (Käs *et al.*, 1994); and in filaments gliding on a myosin-coated surface (Kellermayer and Pollack, 1996). Such prominent undulations had been presumed to be of thermal origin. But that notion was brought under challenge by the observation that the undulations could be substantially intensified by exposure to ligands such as myosin (Yanagida *et al.*, 1984), as well as by ATP (Hatori *et al.*, 1996). These actions implied a specific structural change rather than a thermally induced change.

In fact, there is a long history of evidence for actin structural change. Molecular transitions were first noted by Oosawa and colleagues in the 1960s (Asakura *et al.*, 1963; Hatano *et al.*, 1967; Oosawa *et al.*, 1972). Actin monomers underwent a 10° rotation on exposure to myosin (Yanagida and Oosawa, 1978). Actin-conformational changes have since been confirmed repeatedly: in probe studies, X-ray diffraction studies, phosphorescence-anisotropy studies, and fluorescence-energy transfer studies—the latter showing an actin-repeat spacing change of 17% on exposure to myosin (Miki and Koyama, 1994). I draw your attention to Figure 11.6, a time series that illustrates the change from a straight filament to a wavy fila-

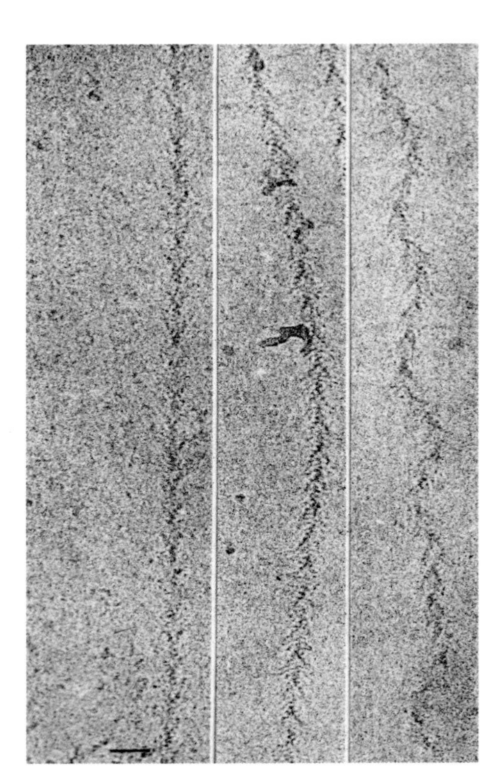

Figure 11.6. *Actin filaments decorated with S1 myosin fragments, flash frozen at various times (left to right) after exposure to ATP analog. Note progressive increase of filament waviness. From* Ménétret et al. *(1991).*

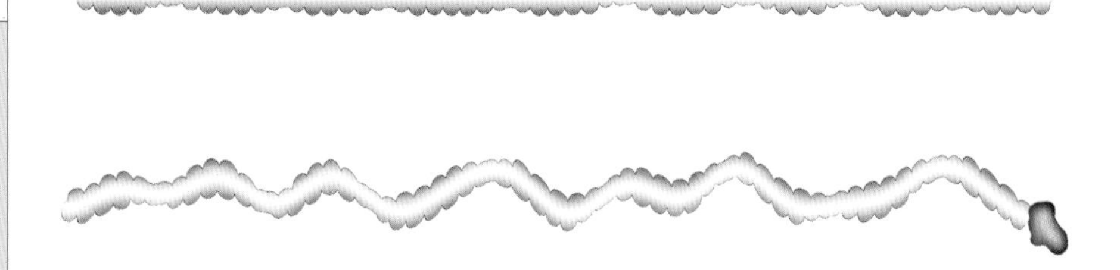

Figure 11.7. Binding of gelsolin to one actin monomer induces conformational change in the entire filament.

ment. Additional evidence for actin-structural change is considered in two recent reviews (dos Remedios and Moens, 1995; Schutt and Lindberg, 1998).

That such structural change propagates along the filament is implied in several studies. Gelsolin is a protein that binds to the actin filament at one end (the so-called "barbed" end). Yet the impact of point binding is felt along the entire filament (Fig. 11.7): Molecular orientations shift by 10° and there is three-fold decrease of the filament's overall torsional rigidity (Prochniewicz *et al.*, 1996). Thus, structural change induced at the end demonstrably propagates over the entire filament. Such propagating action may be responsible for the bend that progresses along the streaming bundle (Fig. 11.4). It may create the undulations commonly observed along single actin filaments. These undulations appear to propagate—as confirmed by cross-correlation of point displacements (deBeer *et al.*, 1998), and by tracking fluorescence markers distributed along the length of the filament (Hatori *et al.*, 1996b, 1998).

Could such a propagating structural transition drive solutes to stream along the bundle—as suggested early on by Szent-Györgyi (1972)?

WATER-STRUCTURE CHANGE?

A possible solute-driving mechanism could reside in the organizational features of water. Water in the structured state has high, gel-like viscosity, which impedes flow. If substances are to stream actively, a likely bet is a change of water structure.

Such change is a predictable accompaniment to the phase-transtion. Tight water structuring arises from an opportune distribution of surface charge (Chapter 4). If the phase-transition perturbs this distribution, said effectiveness must diminish and water will tend to destructure. In chemist's terms, the actin surface becomes less hydrophilic (more hydrophobic) and adsorbs commensurately less water. As the protein folds—either by protein-protein interaction or in the case of actin possibly by the formation of calcium bridges between nearby anionic surfaces (Tang and Janmey, 1998)—charge distribution is altered and fewer surface charges are exposed to the water (*see* Fig. 8.7). Thus, water desorbs.

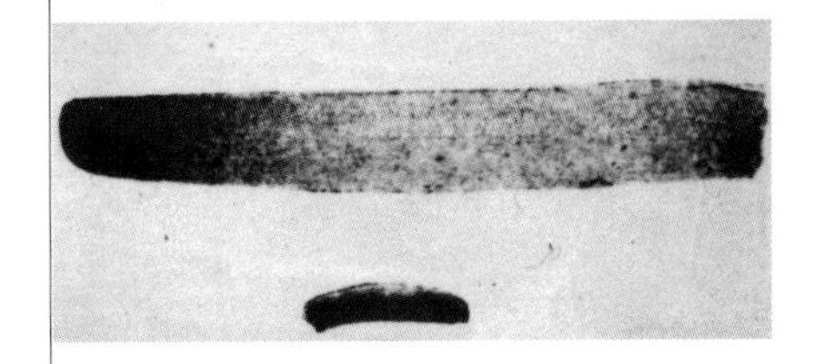

Figure 11.8. *Exposure of actin gel to critical level of ATP induces massive contraction and water release.*

Evidence for water desorption comes from the behavior of actin gels. Actin gels undergo massive all-or-none phase-transitions, and as this happens water exits (Fig. 11.8). The extent of water loss is impressive in such random gels, and also in more linearly oriented gels where the transition drops the water to protein ratio from 50:1 to 1:1 (Szent-Györgyi, 1951). The transitioned protein is practically dehydrated—much the same as in the peripheral cytoskeleton and the secretory vesicle.

If this evidence for water release seems too technical, consider the common experience of eating sashimi. Sashimi is raw fish muscle, which contains an abundance of actin. Just as actin gels retain water, so does sashimi—as do other fresh fish and meats. In Japan, it is possible to obtain sashimi that is particularly fresh, and my friend Professor Ogata will search doggedly for such delicacy. He relates that one measure of freshness is the chopstick test: sashimi that is genuinely fresh will contract to the touch. As it contracts, the water leaks out.

In sum, actin filaments have an unusually high propensity to structure water, and as the filaments undergo a phase-transition this capacity is lost. Structured water then desorbs, and reverts to bulk water. As the actin transition propagates along the filament, so also must the water transition. A propagating undulation must therefore encompass the composite change—including both the polymer and the associated water (Fig. 11.9).

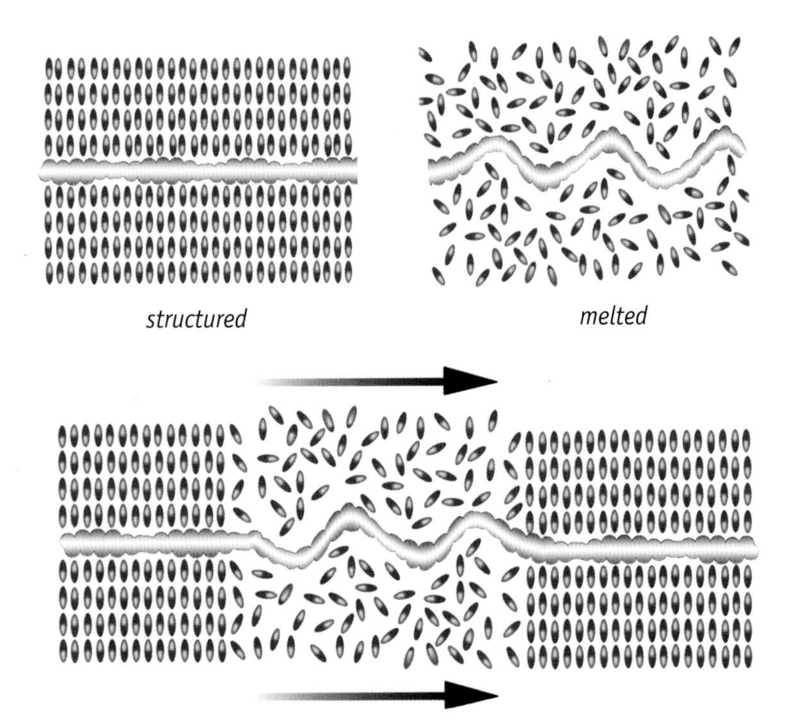

Figure 11.9. *Two states of actin. In the extended state, vicinal water is structured (left); in the folded state, it is unstructured (right). A propagating change (below) will involve both actin and water.*

TOWARD A MECHANISM

Could the scenario of Figure 11.9 lead to streaming? Consider an analogy. For the sake of illustration, imagine that structured water is replaced by ice. Suppose a bundle of robust 10-foot-long interconnected rods is immersed to fill a trough containing dirty water, and said trough is shipped off to the Arctic. The water will freeze. As the water freezes and ice crystals grow, the debris will be pushed out of the ice lattice and toward the nearest interface—either the rods' surface or the trough surface; and there it will remain indefinitely (Fig. 11.10A).

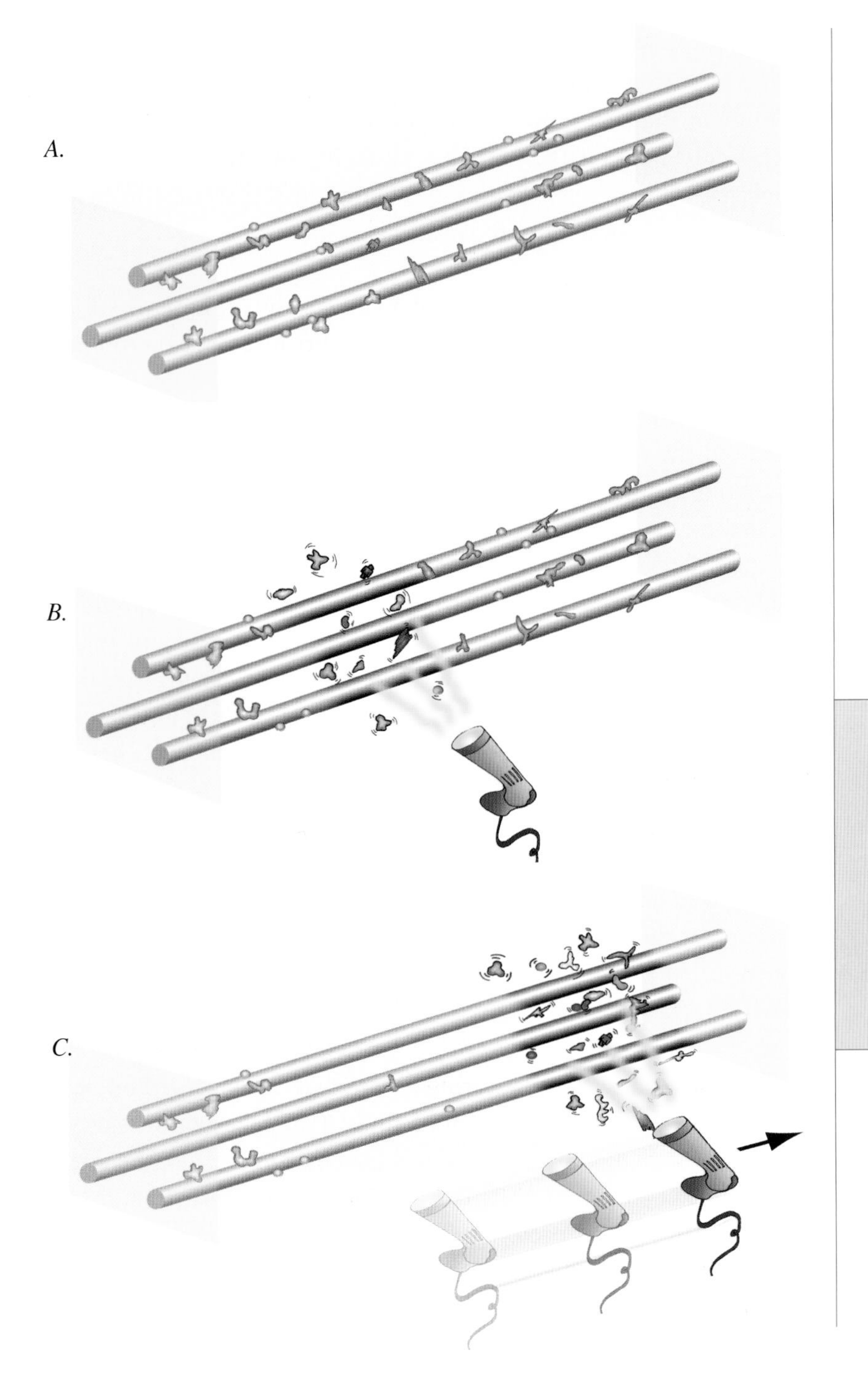

Figure 11.10. *Analogy of metal rods (actin filaments) immersed in ice (structured water). Solutes accumulate in the melted zone. As the melt zone translates, so do the solutes.*

Now suppose you ventured boldly to this remote experimental station armed with a heat gun and perhaps also a rifle to fend off polar bears. You aim the heat gun at some point along the frozen rod array and melt the ice (Fig. 11.10B). As the ice melts, the debris is released into the melted water. Turn off the heat gun and the arctic cold will drive the debris back toward the rods' surfaces.

Instead of merely holding the heat gun steady, suppose you draw it slowly along the array. The ice will melt, but because the experiment is carried out in the far reaches of the Arctic, the water will quickly refreeze in the wake of the moving heat gun. A refreeze front will follow immediately behind the thaw front, leaving a "window" of melted water with debris, moving along the array (Fig. 11.10C). As the window moves, so will the debris.

The analogy between this system and the actin system should, I hope, be clear. In the actin system it is not the moving heat gun that drives the melt but the phase-transition. The transition "melts" local water into bulk water, creating a window within which solutes or organelles can be suspended. As the window moves, so do the solutes. These solutes effectively stream along the actin filament array. One difference between the two systems is that structured water is less solute-exclusive than ice. Solutes may sometimes escape the clutches of the moving window and thereby remain stationary—until captured by a subsequent window. Thus, transport should occasionally cease, and then pick up again at full speed—a so-called "saltatory" mode that is in fact a common feature of the process.

The transport principle outlined here may seem unorthodox, but it has become standard in the realm of applied engineering: it is used in a process called "zone refining." The scenario is much like the Arctic experiment. A crystal rod is melted locally by heat exposure (Fig. 11.11). Impurities in the rod prefer the melted region to the crystallized region, so as the heater sweeps along the rod to create moving melt windows, the impurities will likewise translate, eventually accumulating at the rod's end for clearance. The method is extremely effective. With repeated sweeps, rod purification on the order of 1 part in 10^{10} is obtainable—which made practical the manufacture of the first germanium transistor (Pfann, 1962, 1967). The moving-melt-window principle is much the same as the one proposed for actin-based transport.

Given this similarity of principle, it may be worth looking toward this engineered counterpart for clues about energetics. In zone melting, the input energy is used to transform the crystalline rod from one state to another, solid to melted. If a similar principle were to apply for streaming, the energy would be used to "melt" the actin-water complex. Because water molecules outnumber actin molecules by more than 100,000 times (55 M vs. 200 – 400 μM), the water transition would likely dominate the energy landscape, unless it turned out to be energetically inconsequential. Much of the energy, in other words, would be used to effect the entropic change in water. Such a possibility will be considered in more detail in Chapter 15, by which time, the material of the intervening chapters can be brought to bear on the subject of energetics in general. By then, we will be in a position also to consider the role of ATP.

So far, the focus has been on solutes. The question arises whether it is solutes alone that are biologically transported, or solvent as well. Al-

Figure 11.11. *Principle of zone-refining, or zone melting.*

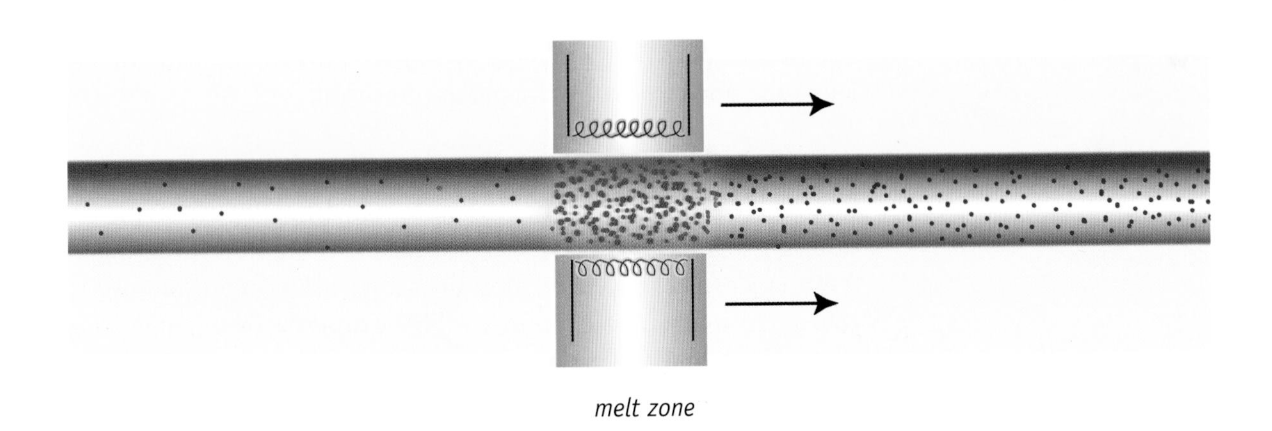

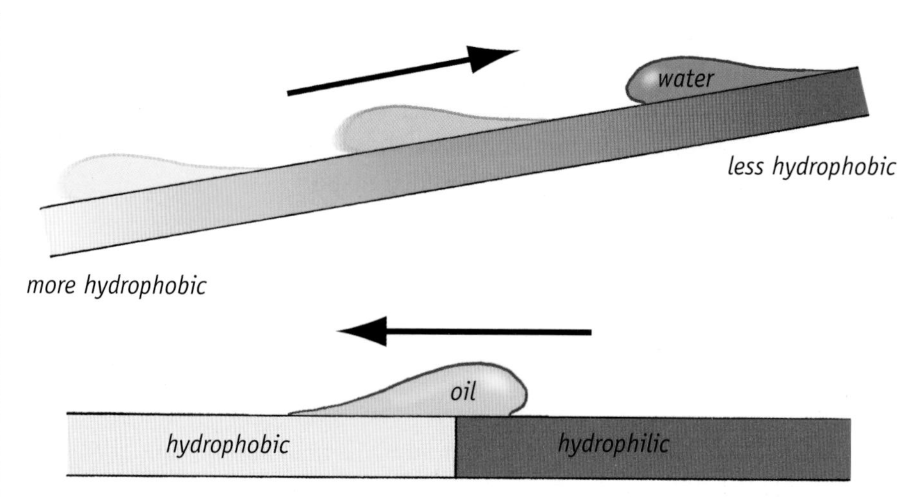

Figure 11.12. *Principle of fluid movement along a surface containing a hydrophilic-hydrophobic gradient.*

though "streaming" would seem to imply movement of fluid, water molecules cannot be seen by the optical microscope, so standard observations of streaming particles do not necessarily settle the issue of whether water flows. In fact, water does flow. This is confirmed in experiments on isolated streaming bundles. When the filament bundle is suspended in moist air, a growing droplet forms at the distal end (Wohlfarth-Botterman, 1964).

What drives the water to flow? Inevitably, water must move with the solutes because the solutes are themselves surrounded by substantial clouds of structured water (Chapter 6). As the solute moves so must the cloud. This passive mechanism may, however, not be the only one. Water can also be transported by a moving hydrophilic-hydrophobic boundary similar to the one created by the phase-transition.

The principle is best illustrated by two demonstrations (Fig. 11.12). The first involves a spatial gradient of surface hydrophobicity (top), set up by exposing a polished silicon wafer to the diffusing front of a silane vapor. A water droplet placed on this surface translates continuously toward the more hydrophilic region. The droplet can even move uphill (Chaudhury and Whitesides, 1992). A second demonstration (bottom) employs a discrete hydrophilic – hydrophobic boundary set up by butting hydrophilic and hydrophobic surfaces edge to edge. An oil droplet placed on the boundary translates abruptly toward the hydrophobic side (Suzuki, 1994). These demonstrations confirm that surface hydrophobicity gradients similar to those anticipated to propagate along the actin filaments (Fig. 11.9) are sufficient to drive fluid flow parallel to the filaments.

Finally, it is difficult to leave this topic without mentioning that successful functional devices have been built from re-assembled biological components of the streaming system (Fig. 11.13). Filaments suitably oriented can create a continuous flow stream (*A*). And in the same way that the force of a spinning propeller can move a boat by pushing on water, the reactive force of actin-filament water propulsion can turn a shaft to generate work (*B*). Whether any such mechanisms will replace the automobile engine is questionable. However, these demonstrations do illustrate that the streaming mechanism is in no way magical; it is a primitive process driven by a few simple biological components.

We next move beyond the mechanism of streaming to biological applications.

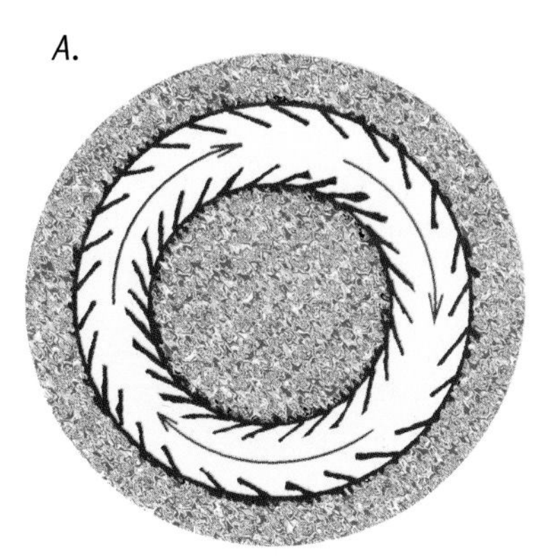

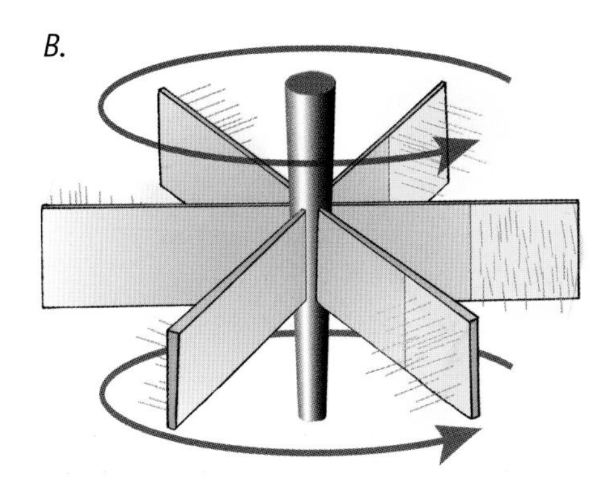

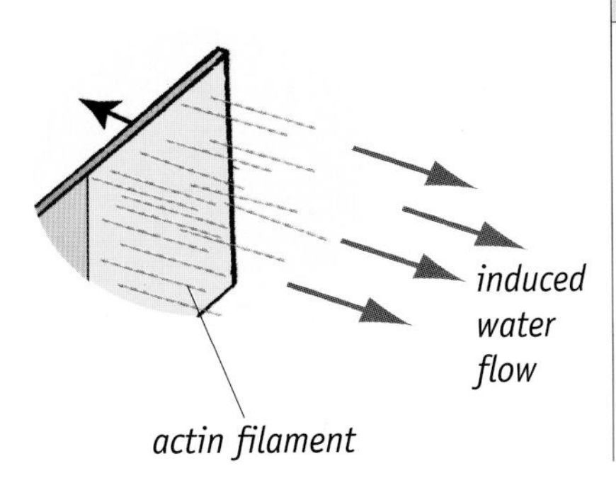

Figure 11.13. *Streaming machines built on the actin-streaming principle. A: (after Yano* et al., *1978). The gap between concentric cylinders contains actin filaments oriented as shown. Addition of ATP and heavy meromyosin induces flow in the direction of the arrow. B: (after Yano* et al., *1982). Actin filaments are oriented as shown. When ATP and heavy-meromyosin are added to the solution containing the paddles, the shaft turns.*

We look first at cell migration, and then at solute transport across the cell boundary. For each of these processes we consider whether the propagating phase-transition mechanism outlined above provides an adequate explanation.

CELL MIGRATION

Migration is a process common throughout phylogeny, from the amoeba in search of nourishment to the fibroblast migrating toward the site of a wound to deposit collagen. Migratory processes involve actin-filament bundles, although the manner of involvement is not well understood.

Consider migration in the amoeba. The amoeba will crawl in the direction most apt to provide sustenance. The sustenance source provides a chemotactic signal to which the cell responds by sprouting an appropri-

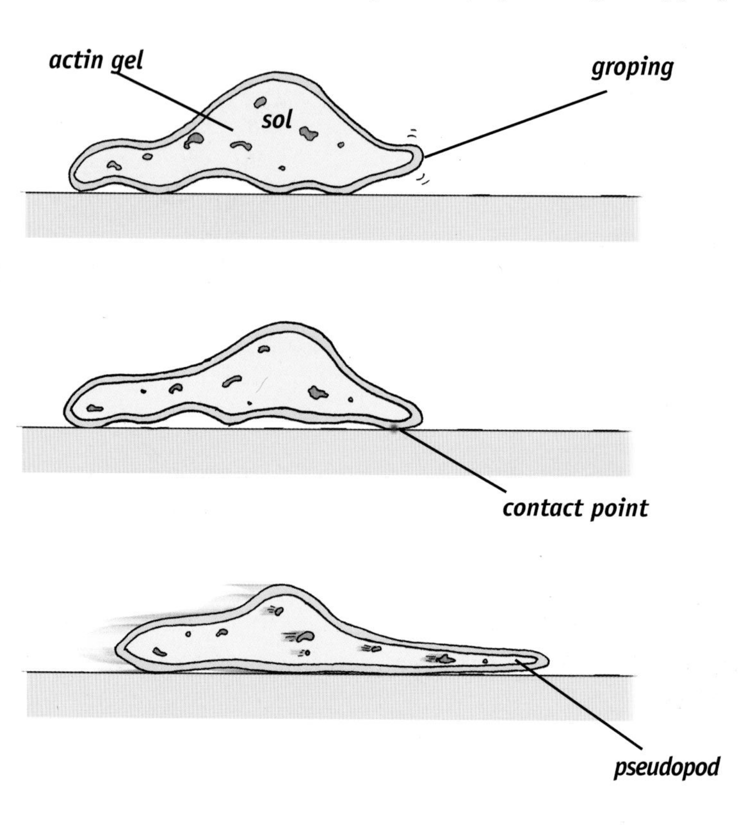

Figure 11.14. *The amoeba moves as mass is transferred from cell body to pseudopod.*

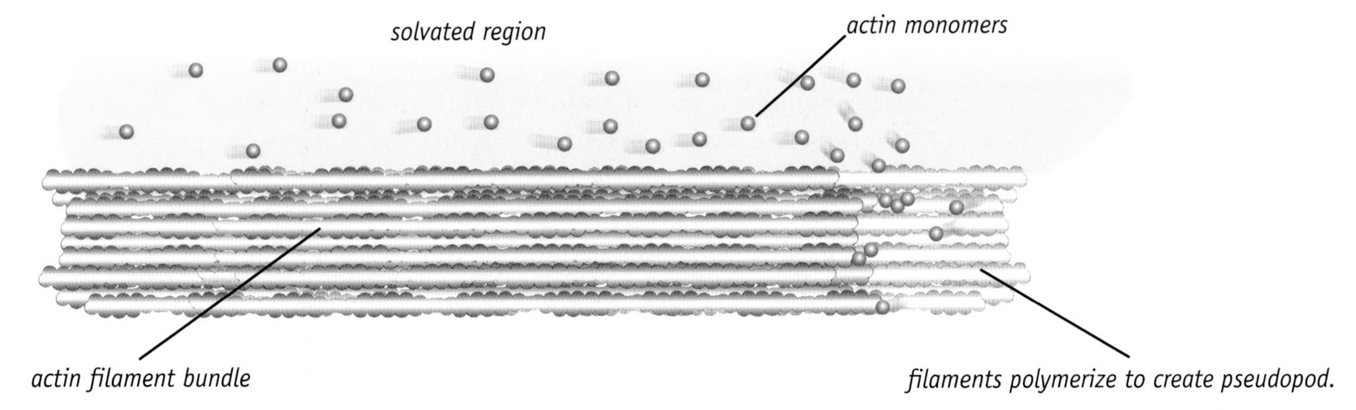

ately directed pseudopod (Fig. 11.14). The incipient pseudopod first gropes clumsily until it can cling. Then it proceeds to grow in length and girth by drawing much of the body of the cell into itself. Through mass transfer, the pseudopod grows until it becomes indistinguishable from the cell proper. The cell's center of mass effectively shifts to a locus closer to the site of nourishment.

Pseudopod growth requires a continuous supply of building materials from the cell body. Without "bricks and mortar" no growth is possible—hence no movement. To satisfy the demand for materials, the cell needs a transport mechanism, and suggested textbook mechanisms range from spatial gradients of osmotic pressure to the squeezing of cellular contents by regional contraction. Such *ad hoc* mechanisms seem less than necessary, however, as streaming in amoeboid (and many other) cells is a prominent feature that remains in evidence even when the cytoplasm is removed from the cell and inserted into glass capillaries (Allen *et al.*, 1960). In the amoeba, streaming is intimately linked to peripheral bundles of actin filaments. Streaming's presence is beyond question; the only question is whether the transport job done by the actin bundles fully obviates the need for any supplemental *ad hoc* processes.

Figure 11.15. *Actin filaments polymerize as monomers stream toward the plus-end.*

With the actin-transport mechanism, the pseudopod would receive its building materials in conveyor-belt-like fashion. Among materials, the most essential is the framework protein, actin. Actin is dissolved in monomeric form in the cell core—it accumulates there as the cell's posterior filaments are dissolved by gelsolin. Thus, abundant actin monomers should pervade the stream that flows along the bundle (Fig. 11.15). Flowing monomers arriving at the filaments' receptive (plus) ends can polymerize. The lengthening actin filaments will make up the framework of the incipient pseudopod, perpetuating the transport of constituents needed for additional growth and extension. This is a self-sustaining vectorial process that assures growth in a consistent direction.

Pseudopod growth is therefore an inevitable feature of the peripheral cytoskeleton. So long as a chemotactic or other signal is present to trigger propagation, the pseudopod will grow and the cell will migrate. For all this to happen, the cell requires little more than bundles of actin.

TRANSPORT ACROSS THE CELL BOUNDARY

An implication of this simple linkage between structure and function is that the mere presence of an actin-filament bundle should provide sufficient reason to suspect streaming. Many organelles contain actin bundles. Only rarely has it been suggested that such organelles might operate through a streaming mechanism (Oplatka, 1998).

An obvious case in which actin-filament bundles are prominent is the muscle sarcomere. When sarcomeric actin bundles are exposed to soluble myosin and ATP, streaming is immediately initiated (Tirosh and Oplatka, 1982). Thus, muscle actin supports streaming. In fact, it is muscle actin that drives the streaming machines of Figure 11.13. These observations imply that the water-propulsion principle could well be incorporated into the functioning sarcomere (Chapter 14)—either directly in the contractile process as Tirosh and Oplatka suggest, or indirectly to move required molecules into place.

Another venue in which actin-filament bundles are prominent is the microvillus. The microvillus is a tightly packed, cross-linked actin-fila-

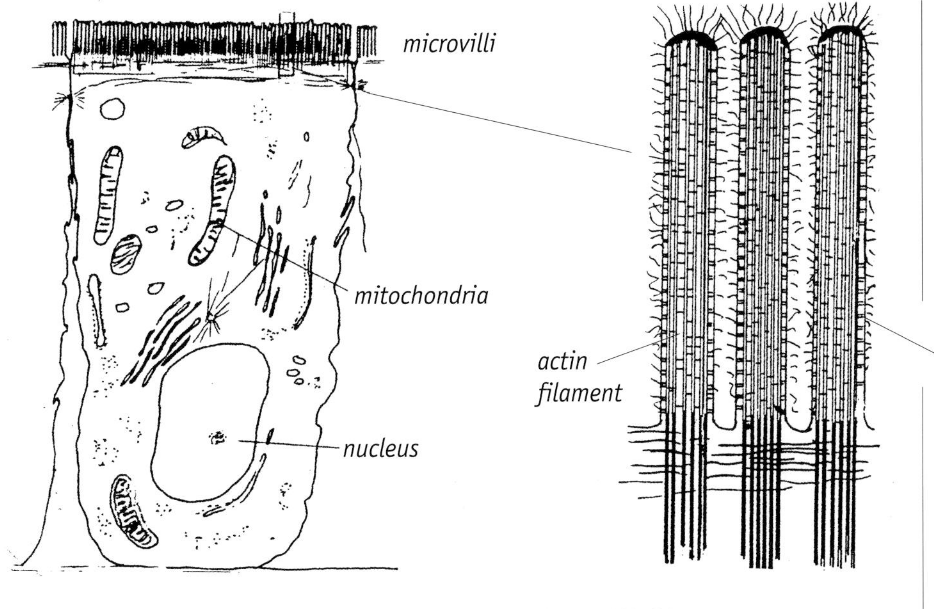

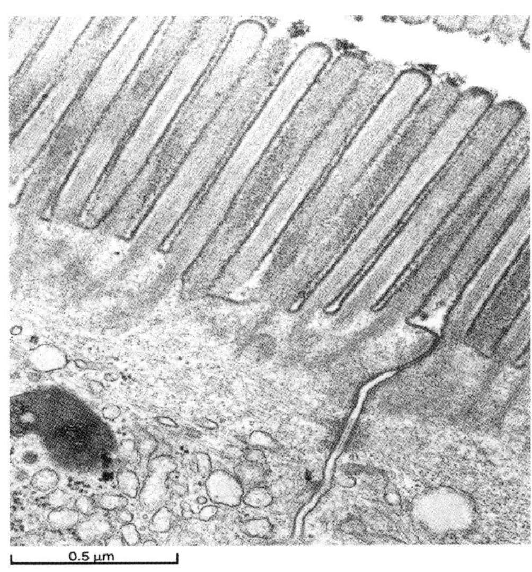

Figure 11.16. Schematic structure of the brush-border. After Bray (1992). Corresponding electron micrograph shown below, courtesy P. Matsudaira.

ment bundle that juts out of the cell to create a protrusion. The best-studied example is the intestinal brush-border, where clustered microvilli project like bristles on a brush (Fig. 11.16). Each bristle contains about 30 actin filaments cross-linked to one another by the proteins fimbrin and villin. Around each bundle's periphery lies an array of myosin-like proteins. Thus, all elements required for streaming are in place, although to my knowledge streaming has never been suggested.

The function of the brush-border is to facilitate nutrient exchange in the intestine. The consensus view is that the brush-border and other microvilli-based organelles operate by creating extra surface area to facilitate diffusion across the cell boundary. But this rationale is not entirely obvious, for any such advantage would be offset by the filaments' high packing density as well as the high viscosity of structured water within the dense array—both of which would impede diffusion.

An altogether different possibility is that the exchange is facilitated by streaming. Through active streaming along the bundle, substances could be transported across the cell boundary at rates far higher than achievable by diffusion. The array of actin filaments would then exist not as an incidental feature, but for the execution of a particular task. And the presence of myosin-like proteins around the periphery would be explained—as myosin is required for actin-based streaming. For situations like the intestine, where the main objective is high material throughput, streaming makes sense.

CONCLUSION

Among the most basic requirements of the living cell is the need to transport substances from one place to another. This need arises both within the cell and also across the cell boundary. Because the transport requirement is primitive, it is unlikely to rest on mechanisms of great complexity. The proposed transport mechanism satisfies the criterion of simplicity, in the sense that an actin-filament framework is largely all that is needed. Yet it is effective. Substances are transported vectorially, and they can be transported substantially faster than by diffusion. The fact that the system's underlying principles have been successfuly adopted in engineering applications lends support to their feasibility.

Merits notwithstanding, the actin-based mechanism is not the only transport mechanism employed by the cell. For more sophisticated applications, the cell has developed a commensurately more sophisticated system based on microtubules. But there is little need for concern. Nature does not invent a new working principle at every turn, so there is reason to expect that the working principle of this more highly evolved system is similar to the one we have just considered. It is on that premise that the next chapter builds.

The Red Stiletto, David Crow (photo by Harrod Blank)

CHAPTER 12

TRANSPORT WITH FLAIR

Prime among the limitations of the actin-based transport system is the absence of selectivity. Garden-variety substances such as proteins, organelles, ions, amino acids and sugars are generally transported indiscriminately (for an exception, *see* Mehta *et al.*, 1999). Bulk transport makes sense when the goal is to move everything. In other situations bulk transport makes less sense. In the neuron for example, the objective is to shift a neurotransmitter to the terminal release site without necessarily shifting the balance of cytoplasmic constituents—otherwise, the terminus would swell like a sprained ankle.

A second general limitation of the actin system is low efficiency. Even if a single molecule is to be transported, all parallel filaments generate the propagating undulation in concert. This is inefficient. Adding to this inefficiency is the fact that bulky loads cannot fit within the bundle and must therefore be transported along its periphery.

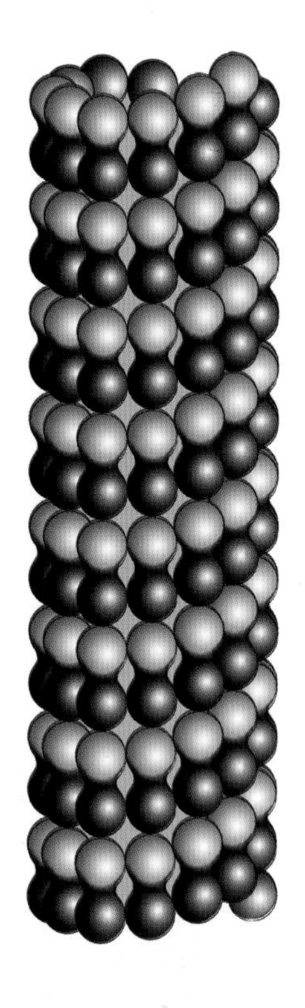

Figure 12.1. *Molecular structure of the microtubule. Basic element is the tubulin dimer, shown as coupled spheres.*

Given these limitations, it is no surprise that nature has developed a more selective system in which transport is linked to a single longitudinal element. This system is based on the microtubule, a long tubular polymer built of the protein tubulin (Fig. 12.1). Tubulin dimers, built of alpha and beta subunits, bind to one another to form long protofilaments, which in turn bind to parallel protofilaments to create the tubule. Microtubules ordinarily contain 13 parallel protofilaments around their circumference although 12 and 14 are not uncommon.

Unlike the actin system, the more highly evolved microtubule system transports a specified cargo. Each piece of baggage is grasped by an accessory protein, which then facilitates transport along the microtubule as myosin facilitates transport along the actin-filament array. Given this parallel, the possibility of a similar transport principle seems worthy of consideration.

WALKING MODELS?

Before seriously entertaining phase-transition models, we examine the currently accepted microtubule-transport hypothesis. This is based mainly on experiments carried out with the accessory protein kinesin, and to some extent dynein. Kinesin is a spindle-shaped protein containing three functionally distinct regions: a pair of myosin-like heads; an intervening stalk; and a fan-shaped tail (Fig. 12.2). The tail binds the specified cargo and the heads interact with the microtubule. Largely by analogy with myosin, the heads are presumed to contain internal motors. These motors are thought to drive the heads (operating more like feet) to saunter past one another along the microtubule and thereby transport the cargo.

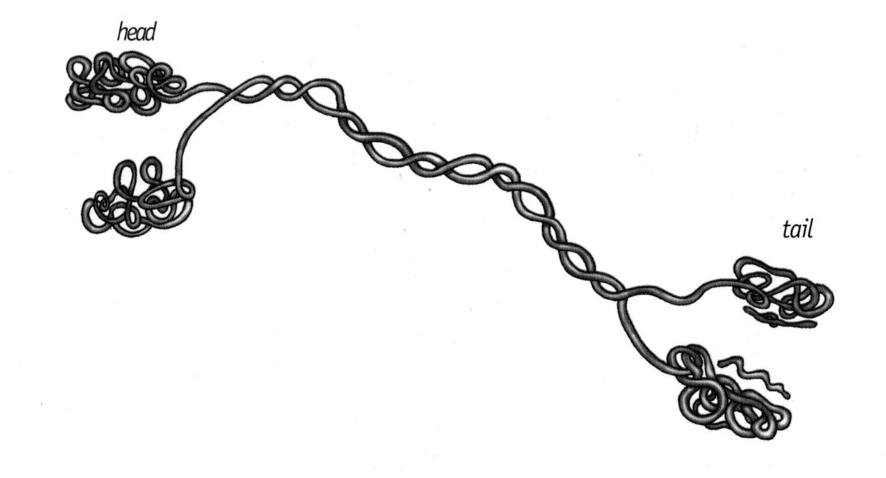

Figure 12.2. *Model of the kinesin molecule.*

Kinesin-translation studies have revealed two mechanistically relevant features. First, movement occurs along a straight line. Kinesin translates virtually without deviation along one of the 13 linear protofilaments that make up the microtubule (Fig. 12.1). Second, translation along this linear protofilament track occurs discretely, a step at a time (Fig. 12.3). The size of the step is equal to the spacing between successive tubulins, 8 nm. Occasionally, steps of two times or even three times 8-nm are observed.

To explain these 8-nm steps, the prevailing view is that the heads step from one tubulin to the next. Imagine one head settled on the *n*th tubulin and the other on the *n-1*th. The lag head then swings past the lead, settling on the *n+1*th. Then the *n*th swings to the *n+2*nd and so on. Each 16-nm swing produces an 8-nm shift of the center of mass. The problem is that the head is only 7-nm long (Kull *et al.*, 1996). The 16-nm stride

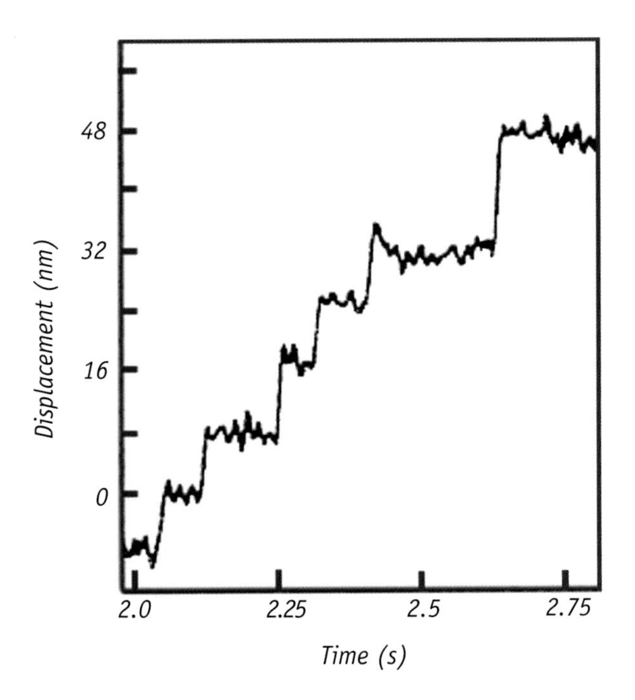

Figure 12.3. *Stepwise translation of kinesin along a microtubule. From Kojima* et al. *(1997).*

must therefore contain some element of leap (Fig. 12.4). The leap begs the zero-gravity question: Once the molecule leaps, why does it return so consistently to the microtubule surface instead of just diffusing away? And further, why does it return with 99% consistency to the very same protofilament (Vale *et al.*, 1996)?

Figure 12.4. *Walking model of microtubule transport. Top: Kinesin molecule undergoes motor action, which facilitates head swing. Bottom: Swing to another tubulin molecule, 16 nm away, requires a hop. Cargo shown as sphere.*

This "leap of faith" issue is even more serious for those kinesins such as the isoform KIF1A that contain only one head (Hirokawa, 1998). This single-headed isoform moves along the microtubule in distinct processive steps (Okada and Hirokawa, 2000). The same is true of single-headed isoforms of dynein (Sakahibara *et al.*, 1999). Unless some kind of supplemental mechanistic feature is invoked, it is not obvious how a striding mechanism could operate in molecules containing only one head.

A second and independent concern is whether hefty loads could be moved by so diminutive a motor. In the above-mentioned kinesin-translation

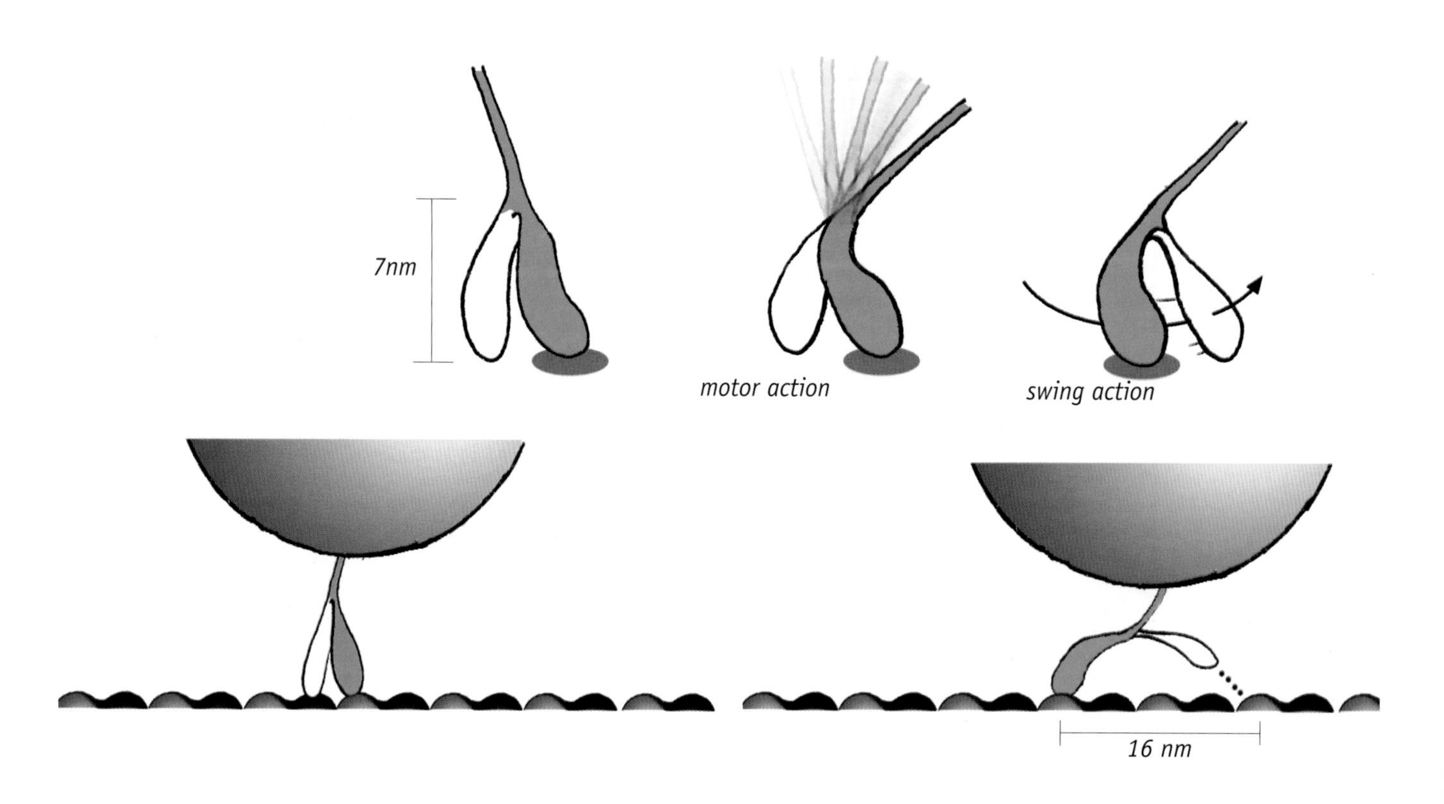

Figure 12.5. Size perspective. The biological motor needs to haul a substantial load in a relatively short time.

experiment, sufficient energy needs to be mustered by a 7-nm motor to pull a 1,000-nm microsphere. The mass ratio exceeds 1:1,000,000, and the job is done within 20 ms. This is a formidable task—even for a driver not obliged to leap (Fig. 12.5).

PHASE-TRANSITIONS ALONG THE MICROTUBULE

An alternative way of approaching microtubule transport is to consider it as an evolved version of actin-filament transport. Similarities of structure abound: microtubules and actin filaments are both multi-chained polymers built of small globular elements; both bind numerous ligands; both can form parallel arrays; both incorporate bound nucleotide (ATP or GTP); and both polymerize with a similar nucleotide-splitting-based dynamic. And like actin filaments, microtubules can easily form gels (Weisenberg and Cianci, 1984). Thus, the potential for parallel function seems worth exploring.

To produce anything like the peristaltic wave that sweeps along the actin filament, three essential features are required: Tubulin must undergo some kind of structural change; the change must propagate along the microtubule; and the propagating change must include a concomitant change of water structure.

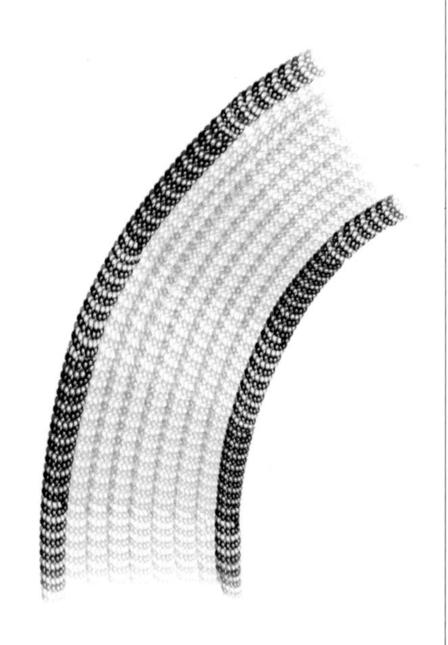

Figure 12.6. As the axostyle bends, microtubles on the inner edge shorten.

Tubulin structural change is confirmed. Depending on whether kinesin is or is not bound, negatively stained electron-micrographic reconstructions show striking differences of tubulin conformation, the difference residing chiefly in tubulin's beta subunit (Hoenger and Milligan, 1997). That such conformational change could lead to microtubule shortening is suggested in studies of the axostyle. The axostyle is a bending organelle consisting of several thousand microtubules cross-linked along their length. During the bend (Fig. 12.6), the axial spacing of tubulin molecules on inner-edge microtubules is observed to become smaller than on outer edge microtubules. Although the diminished axial repeat was not detectable in all electron micrographic fields, when present it was of substantial magnitude (McIntosh, 1973). Thus, the capacity for conformational change and shortening that occurs in the actin filament is also present in the microtubule.

The second requirement is for the structural change to propagate. In many cells, microtubules form radial tracks from the nucleus to the cell periphery. These tracks are thought to make up a kind of "intracellular nervous system" for communicating information in both directions (Albrecht-Buehler, 1985; Dayhoff *et al.*, 1995). For example, cells are often attracted to light. When exposed to a source of infrared light, cultured CV1 cells respond by sending out pseudopods, which enable them to crawl toward the light source. The cell's infrared sensor, the centrosome, is situated on the nuclear membrane, far from the incipient pseudopod. Communication between these sites is evidently necessary. As the cell receives the stimulus, the microtubules that interconnect these sites undergo physicochemical change (Albrecht-Buehler, 1998). Hence, some information apparently propagates along the microtubule, enabling the cell to crawl in the proper direction.

Propagation is usually less obvious in microtubules studied in isolation, possibly because such preparations are routinely treated with the drug

Taxol to promote "stabilization." But the following experimental result nevertheless implies propagation (Fig. 12.7). To a bath containing microtubules with translating kinesins, additional kinesins were added. The added kinesins did not settle at random sites along the microtubule; they bound preferentially in the vicinity of the translating kinesins (Muto and Yanagida, 1997). Binding would not have been preferentially localized

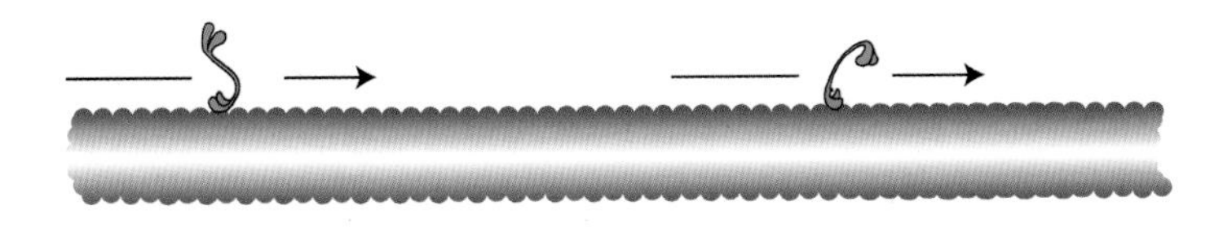

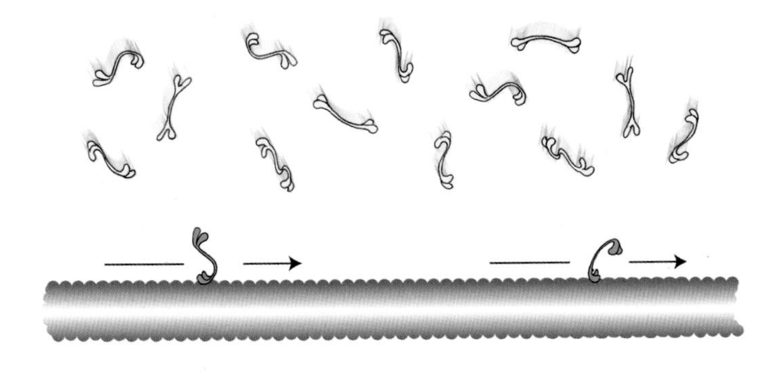

Figure 12.7. *Added kinesins bind preferentially in the region of already translating kinesins.*

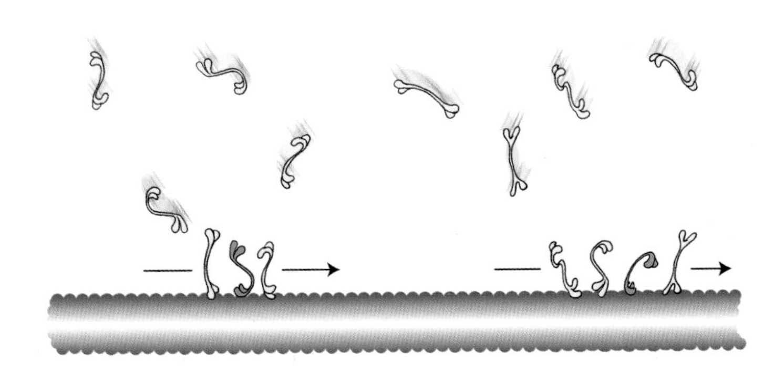

unless translating kinesins had induced some local change that propagates along with the translating kinesins.

Propagation is most obvious when the microtubules are organized into characteristic parallel arrays known as axonemes. The axoneme is the backbone of the cilium and the flagellum (see below). These organelles exhibit a familiar bending action, which can be seen in the microscope to propagate from one end to the other as the organelle functions. Along the axoneme, then, there is little question of propagation; and the studies of Figure 12.7 imply that some structural change propagates along constituent microtubules as well.

The third requirement is that the propagating signal involve water destructuring. For destructuring to take place, it is evidently necessary to begin with structured water, and this is implied by tubulin's abundant negative surface charge. It is also implied by the microtubules' ability to gel with protein concentrations of less than 1% (Buxbaum *et al.*, 1987). High capacity for water structuring is also implicit in electron micrographic images, which show microtubules surrounded by annular "clear zones" apparently devoid of solutes. Such zones are documented both in freeze-etched specimens (Stebbings and Willison, 1973) and in conventional transverse thin sections (Stebbings and Hunt, 1982). They extend ~20 nm from the microtubule surface, and they persist even when 75% of the cell's water has been osmotically removed (Albrecht-Buehler and Bushnell, 1982). Hence, these zones appear to reflect water that is tightly structured around the microtubular surface.

That such water destructures during propagation is implied by the added-kinesin experiments (Fig. 12.7, above). Added molecules of any kind will experience limited accesss to the microtubule if the microtubule is surrounded by structured water. Ready access near the transition zone implies locally high solvency—which in turn implies local destructuring. The scenario is therefore comparable to that of the actin system, the local structural transition encompassing both the protein and the surrounding water.

ROLE OF KINESIN

Following the microtubule-actin parallelism, kinesin's role is anticipated to be similar to that of myosin—a facilitator. This is consistent with kinesin's ability to induce a tubulin transition (above). On the other hand, the parallelism between kinesin and myosin can extend only so far; at some stage something must differ because the microtubular system ultimately diverges from the actin system in its functional features, especially in cargo specificity.

A possible divergence point lies in the way that kinesin facilitates, and some suggestion of this comes from the kinesin-translation experiments (Fig. 12.3). Such experiments reveal a stepwise advance of 8 nm (sometimes 16 or 24 nm). In between these advances, kinesin remains stuck on the microtubule for appreciable periods, during which the tubulin-conformational change is presumably induced. If kinesin-induced conformational change is required for propagation (and transport), and kinesin binds the specified cargo, then cargo transport will be tied directly to the propagating transition, as required. Garden-variety solutes could tag along for the ride, but the energy will be focused on designated cargo transport.

To envision more concretely how such a system might work, consider the nerve axon. Figure 12.8 schematizes the arrangement of microtubules. With limited alignment and loose packing, propagation can oc-

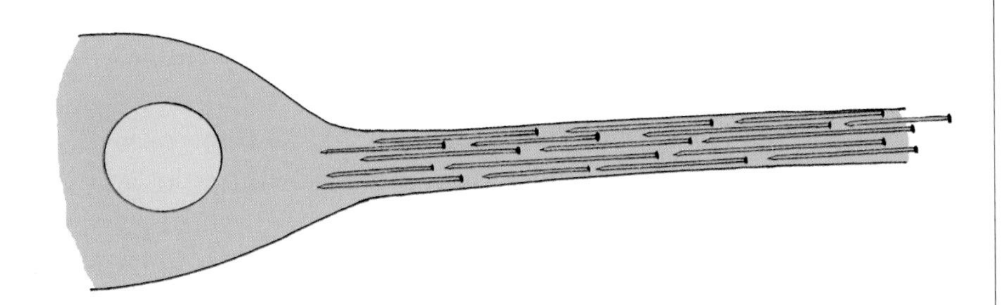

Figure 12.8. *Arrangement of microtubules in the axon.*

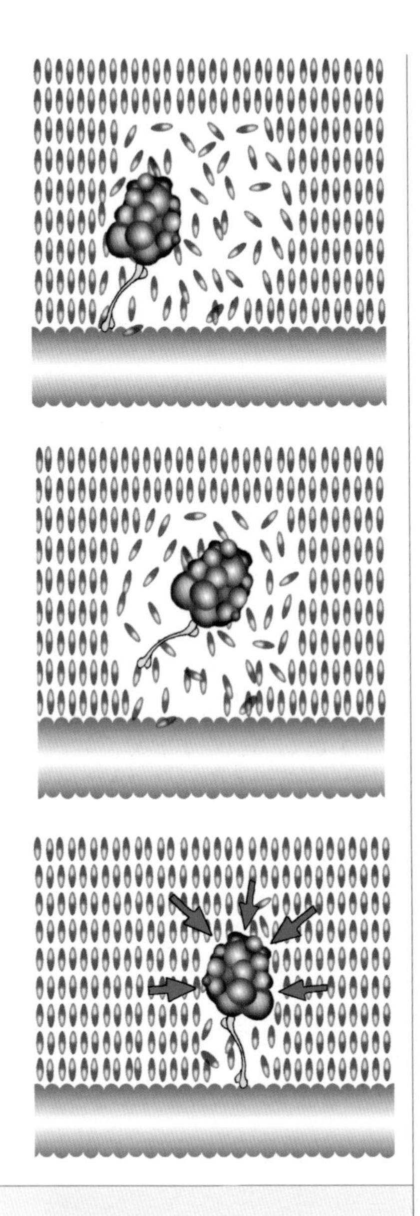

Figure 12.9. *Mechanism of kinesin advance along the microtubule.*

cur along one microtubule without necessarily impacting all others. Microtubules can operate independently. On the other hand, the ample presence of nearby microtubules sets up a sea of structured water surrounding the operating microtubule—an environment that ensures that any transiently destructured zone will quickly restructure.

Now consider a local phase-transition along the microtubule, induced by kinesin binding (Fig. 12.9, top panel). To confer directionality, the induced change must have some polarity—*i.e.*, it must extend farther in one axial direction than the other. Asymmetry of this sort seems all but inevitable, as the tubulin molecules are themselves polarized along the microtubule, with beta subunits consistently lying on the same side of alpha subunits (Fig. 12.1). With the kinesin-induced transition, vicinal water will destructure. The high solvency of destructured water allows bound kinesin to unbind, releasing itself from spatial constraint and readying itself for directed movement (middle panel).

The driving force for translation comes ultimately from the advancing front of re-structuring (lower panel). This feature operates as in the actin system, except that the destructured zone does not encompass parallel units because neighboring microtubules are too remote; the zone is therefore restricted to the vicinity of the transition. Surrounded by an ambient sea of structure, this destructured pocket will be immediately forced to restructure. Margins close in from all directions (arrows), nudging the freed kinesin along the microtubule, and forcing it to settle at some new site—either at the next tubulin, or one or even two beyond. Re-binding initiates a new transition, which sets in motion the same sequence of events at the next position along the microtubule. In this manner the kinesin-load complex advances by a step at a time.

In a sense, the kinesin molecule does "leap" as it advances (Fig. 12.4). But the limited molecular length issue is moot because the leap is not generated by the kinesin molecule, but by the microtubular phase-transition. The "motor" lies in the microtubule, not in the kinesin. Also answered is the question of kinesin's return to the microtubule surface:

this is driven by the radial force of the water-restructuring front. And, so long as the destructured zone is restricted enough to preclude excessive lateral diffusion, the freed kinesin molecule should return consistently, as observed, to the same protofilament track.

Reverse Gear

An intriguing feature of microtubule transport is that it is a two-way street. Kinesin transports in one direction, whereas dynein and a kinesin-like molecule called *ncd* transport in the other. Simultaneous bidirectional transport along the same microtubule is a feature well recognized.

Figure 12.10 shows that in addition to the linear tracks, tubulin-tubulin contacts also define helical tracks. Such tracks might be functionally inert, or, if activated by dynein or *ncd*, they could support retrograde transport. Any such scheme would require that the kind of conformational change induced by dynein or *ncd* be different from the conformational change induced by kinesin, and this is confirmed (Hoenger and Milligan, 1997). Thus, kinesin activation could define one track while dynein and *ncd* activation could define others, at least theoretically. The tracks could operate with little interference so long as transition zones were small enough that one did not interfere with another.

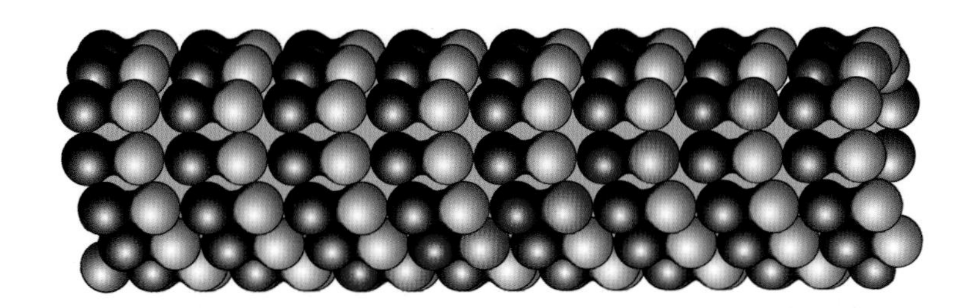

Figure 12.10. *Example of a helical track along the microtubule.*

That the proposal of separate tracks is more than speculation is demonstrated by direct visual observation (Fig. 12.11). Microtubules rotate about their long axes as they translate over a dynein-studded surface (Vale and Toyoshima, 1988; Mimori and Miki-Nonomura, 1995). Similar rotation is observed in *ncd*-based translation, with the helical pitch differing from that of the dynein-induced translation (Walker *et al.*, 1990). Thus, different activators may define different helical tracks for retrograde transport, while kinesin defines a linear track for antegrade transport.

Figure 12.11. *Microtubule rotation observed during translation past dynein or* ncd.

***

In sum, there is reasonable support for the proposal that microtubule- and actin-based transport systems function by a similar mechanism. In both cases the essentials are accounted for by a propagating structural transition. The microtubule system is more selective in that the energy is focused on transport of a distinct cargo rather than on all solutes. And, it is more sophisticated in that bi-directional traffic can be supported on the same microtubule at the same time. Yet, its operation is hardly more complex than the more primitive actin system.

Microtubule Arrays

Apart from intracellular transport tasks such as those considered above, transport requirements also exist outside the cell. Ova are propelled toward the womb; inhaled dust particles are expelled from the respiratory system; sperm cells are driven through the vaginal mucus; *etc.* The presence of extracellular transport tasks need not imply fresh principles, for the relevant machines are built of standard parts—all are based on microtubules.

Such extracellular transport systems rest largely on cilia and flagella, long, rod-like entities whose main structural unit is the axoneme. The axoneme is a rather exotic structure that contains a characteristic 9+2 grouping of parallel microtubules—two singlets at the center and nine doublets distributed around the rim (Fig. 12.12). The rim doublets are connected to the central microtubules by radial spokes, robust entities that terminate on the central split-ring structure called the inner sheath. Doublets are interconnected to one another by a thin link called nexin, and curiously, also by a molecule we have just encountered—dynein.

Figure 12.12. *Cross-section of the ciliary or flagellar axoneme.*

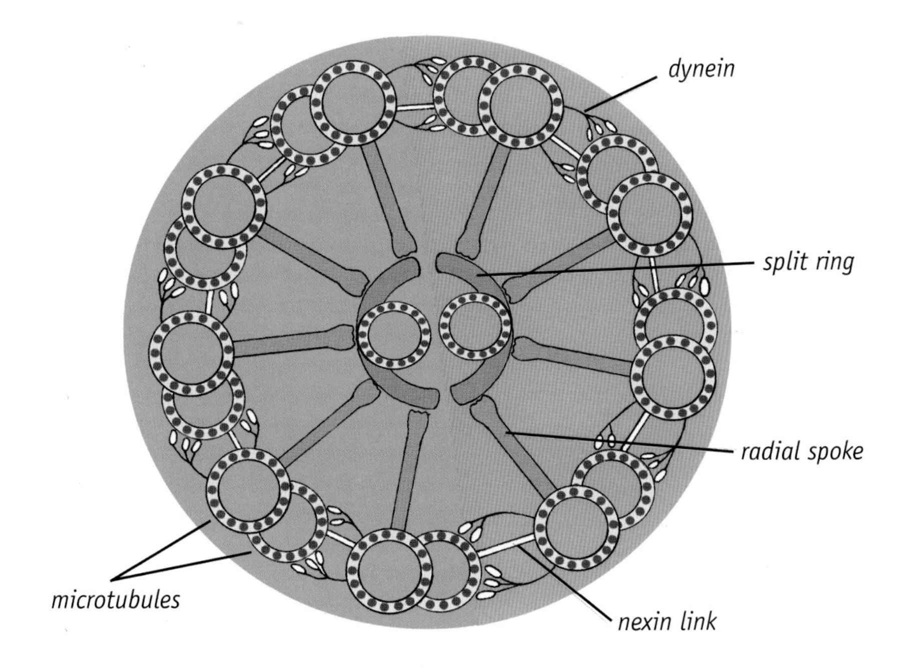

The axoneme performs two seemingly reciprocal functions. In cilia it transports fluid past the cell, and in flagella it transports the cell past the fluid. These reciprocal actions are presumably manifestations of the same working principle—just as propellers can move water, but they can also move a boat through water by using the same working principle. For the axoneme, the question is what that principle might be.

Since the axoneme is essentially a collection of microtubules, a question is whether the working principle could be the same as that of the microtubule. If the microtubule generates transport, could the functioning axoneme generate transport as well? Could transport underlie function?

Figure 12.13. *Microspheres are transported along the length of the activated flagellum.*

In now-classic experiments, Bloodgood and colleagues (1977, 1979) infused microspheres into a bath containing *Chlamydomonas* flagella (Fig. 12.13). Those microspheres that adhered were promptly transported along the flagellar surface at ~2 μm per second in the axial direction. They were observed to follow along straight lines, as though "operating on tracks." Transport was not a feature of any particular marker type because microspheres of different size and composition could be transported, as could bacterial cells. Nor was it a secondary effect of beating because mutant flagella that could not bend as a result of a focal defect were still able to transport. Transport of similar nature is also reported in cilia, as well as in needle-like cellular extensions called axopods, which are also built of arrays of cross-linked microtubules (Troyer, 1975; Bloodgood, 1978).

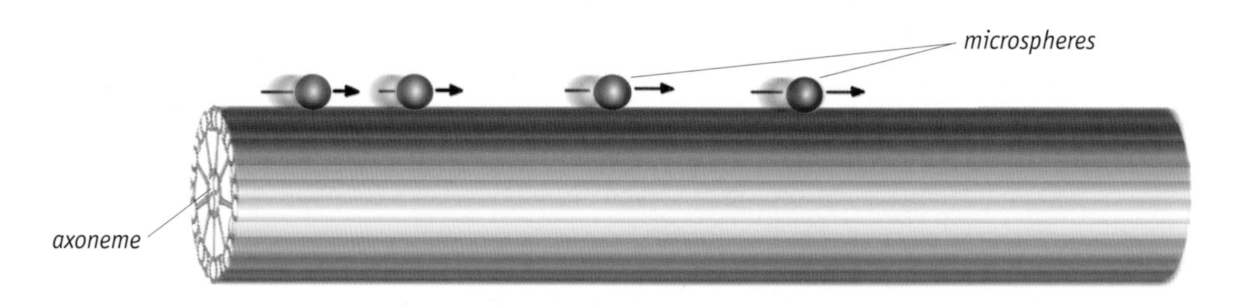

Thus, transport is as much in evidence along the flagellar and ciliary axoneme as it is along the isolated microtubule. Yet, a role for transport along the axoneme is not immediately obvious, for the axoneme's most prominent feature is bending, not transport. On the other hand, if transport and bending alike arose from some common mechanistic principle, then this concern might vanish—and we might also be able to figure out how the axoneme does its job.

Basis of Axonemal Bending

Of the conceivable mechanisms to create a bend, the most natural options are sliding and shortening. Sliding of one microtubule doublet past its neighbor is the accepted model of bending and will be considered shortly. Microtubule shortening is another option. In bimetal rods, for example, shortening of one of the two laterally bonded metal strips more than the other (because of different temperature coefficients) generates a bend. In the inchworm-crawling gel (Fig. 8.9), a structural transition that shortens elements along one edge relative to the other creates a curl. Similarly in the axostyle, bending may arise as microtubules along one edge shorten relative to those on the other edge (Fig. 12.6 above). Differential shortening is a natural way of creating a bend.

An option to consider for the axoneme, then, is a bend that arises from shortening of some microtubule doublets relative to others. As this event propagates axially, the bend likewise propagates. If the underlying vehicle is a microtubular phase-transition, then transport will be an inevitable byproduct. The flow-generation requirement will be satisfied.

In considering whether any such bending mechanism could be grounded in reality, it is helpful to check for compatibility with structure, and several structural features seem relevant. The first is the split-ring. This bisects the center of the axoneme like a banana sliced lengthwise—in effect creating two half-axonemes that run axially (Fig. 12.12 above). A phase-transition in one half-axoneme but not the other could generate the bend (Fig. 12.14). The fact that the bend is observed to be consistently perpendicular to the bisection plane, even as that plane twists gently along the axoneme, lends support to this possibility.

A second compatible structural feature is the layout of dynein. Unlike the isolated microtubule where dyneins lumber along to facilitate propagation, axonemal dyneins are spatially fixed (Fig. 12.12). The cross-sectional pattern repeats at 24-nm axial intervals—frequent enough, perhaps, to ensure seamless propagation. Dyneins at each axial interval appear in clusters of 4 – 5 heads (Fig. 12.15). Clustering implies simultaneous transitions in multiple protofilaments at the same axial level, a feature that may be necessary for triggering the localized shortening. Thus, dyneins appear to be juxtaposed to ensure that bending and propagation proceed effectively.

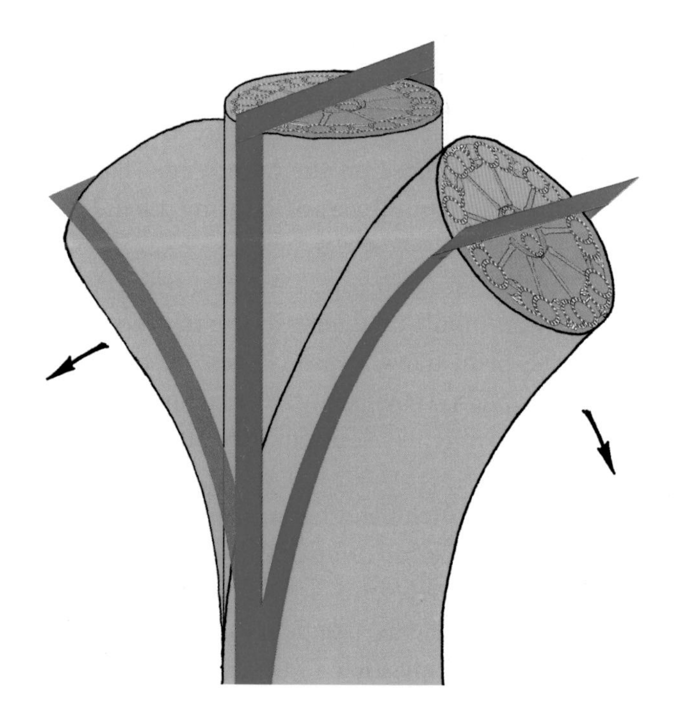

Figure 12.14. *Axonemal bending occurs normal to the split-ring bisection plane.*

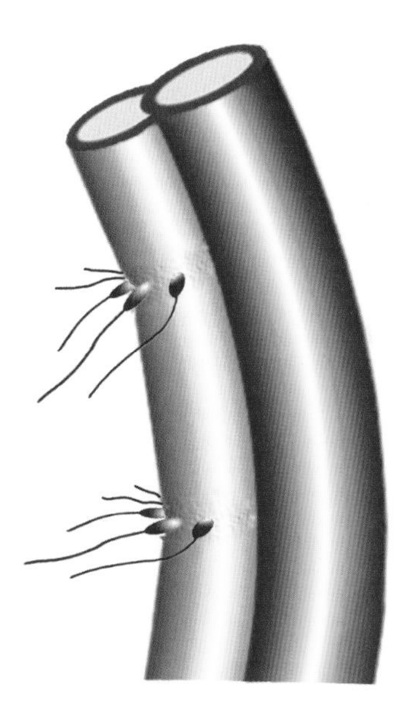

Figure 12.15. *Transition regions shorten when dyneins interact, creating local bends.*

A third noteworthy structural feature is the rim microtubules' peculiar doublet configuraton. This idiosyncrasy can make sense here. Dynein heads bind only to the doublet's incomplete microtuble, leaving the complete partner functionally inert (Fig. 12.12, above). The inert microtubule may then serve as a stiff backbone to facilitate easy bend reversal. Without such backbone, the bend would be sharp rather than gentle, and the return to straightness would be mechanistically less simple.

It appears then, that all three of these odd structural features fall into place naturally within the framework of the proposed mechanism. Structural exotica serve distinct purposes.

The proposed bending mechanism differs qualitatively from textbook views. Current wisdom has it that dynein-based motors propel one rim doublet alongside the neighboring doublet. This is presumed to induce bending—although bend propagation is somewhat less obvious. Evidence for sliding comes largely from classical observations on axonemes treated with trypsin: devoid of cross-links, the loosened microtubules telescope past one another in the presence of ATP, extending the axoneme up to nine times its initial length (Fig. 12.16). For many, this observation is taken as compelling evidence for a sliding mechanism.

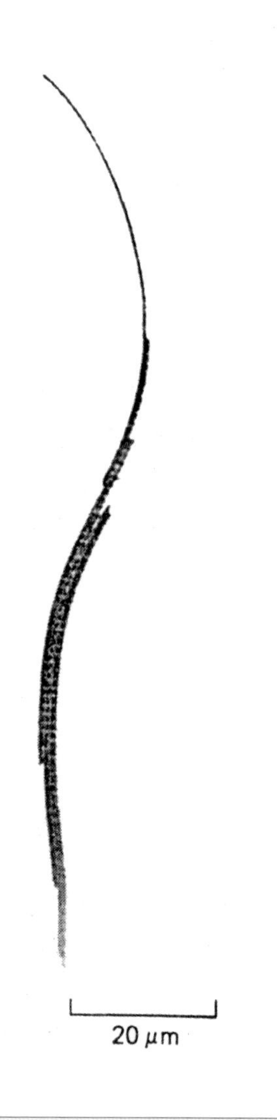

Figure 12.16. *When disrupted by trypsin, the axoneme's microtubules telescope as shown. From Warner and Mitchell (1981).*

The proposed model also anticipates telescoping. When doublets are disconnected from one another, dynein heads can translate actively along the microtubule by the propagating phase-transition mechanism described above. Since the translating heads originate on a neighboring doublet, that doublet must translate along with the heads. Hence, the axoneme telescopes (Fig. 12.17, left). When microtubules are not loosened, this same action will produce bending (Fig. 12.17, right). Bending may also be accompanied by some twisting if the cross-sectional dynein pattern repeats along a gentle helix. Thus, while telescoping is predicted by a sliding mechanism, it is not out of accord with the propagating phase-transition mechanism.

CILIARY AND FLAGELLAR FUNCTION

How could the propagating phase-transition translate into useful work? Classically, the significant feature is presumed to be the bending action. In cilia, bending is thought to propel fluid along the surface from which the cilia emanate. In the flagellum, a propagated bend is presumed to push on the surrounding fluid, thereby propelling the attached cell. These explanations seem intuitively reasonable. On the other hand, one needs to deal with the fact that apart from bending, these organelles also generate flow (Fig. 12.13). Since flow is generated and flow is the product, it seems prudent to ask whether one flow might be connected with the other. Flow could be an irrelevant byproduct of axonemal action; or it could contribute to—or even create—functional flow.

Let us first suppose that the generated flow is irrelevant, and functional flow is created solely by ciliary bending. Several design features then seem curious. In order to get maximum bang for the buck, pushing implements such as oars and fins are typically paddle shaped—but ciliary cross-section is distinctly round. Also curious is the stroking dynamic. Cili-

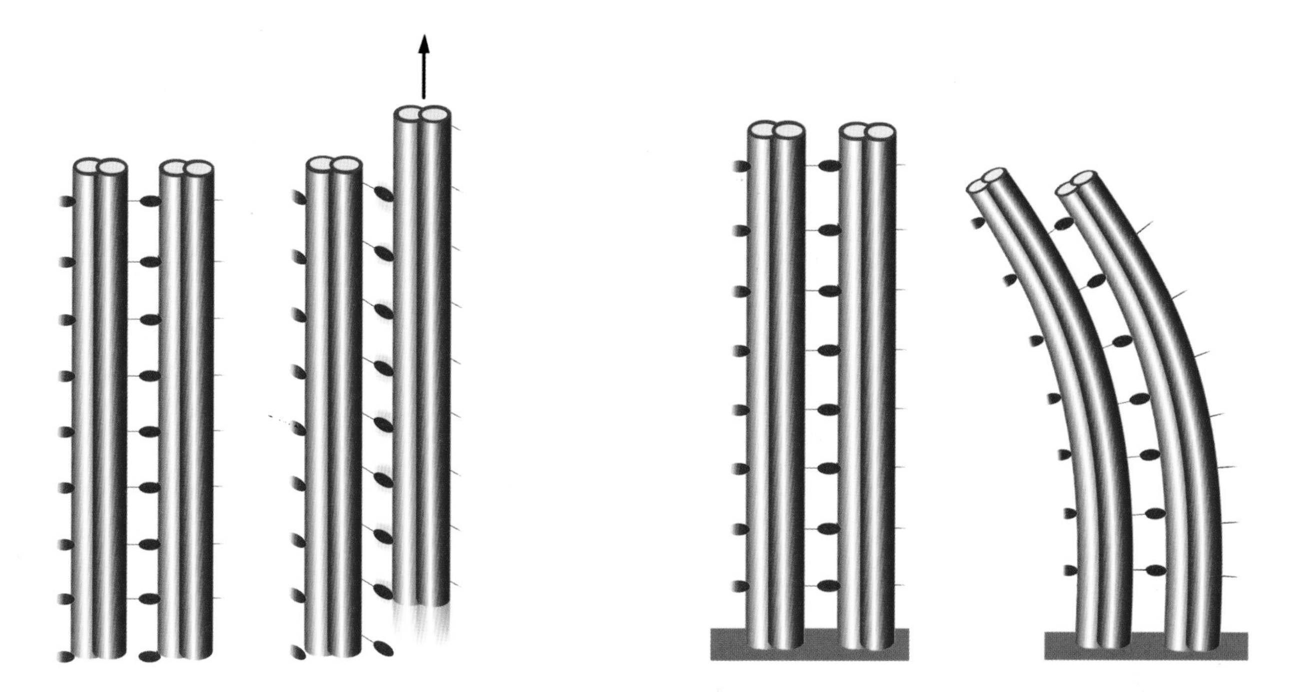

ary populations stroke in a wave-like progression resembling a field of wheat blown by a gentle wind. The effectiveness of any such "metachronal" dynamic for shifting a bed of fluid is unclear—rowing a scull to victory, for example, involves synchronous stroking, and not stroking in a wave-like progression. Design concerns also arise in flagella. Calculation based on screw-like bending action shows that 99% of the energy is used to push water molecules laterally; only 1% of the energy remains for useful propulsion of forward motion (Purcell, 1977).

Figure 12.17. *Left: Loosened microtubules telescope as a result of dynein transport along the microtubule. Right: the same mechanism results in bending in the intact axoneme.*

The alternative option is that axonemally generated flow does play a role in creating functional flow. In cilia, axonemal flow would do the job directly. In flagella, the attached cell would be propelled by the reactive force of the generated flow, similar to the way the squid, or the boat, propels itself.

These proposals may seem wickedly radical, but they are logical sequelae of the experimentally observed flow along the axoneme. The main question is whether such flow could be quantitatively adequate, and the microsphere-surface-translation experiments cited above imply that it may be. Those experiments were carried out two ways: markers translating along the spatially fixed axoneme; and axoneme translating over spatially fixed markers. Translation velocity was the same. The fact that the axoneme could translate at the same velocity even with its bulky cell body attached shows that the propulsive force capacity must be considerable.

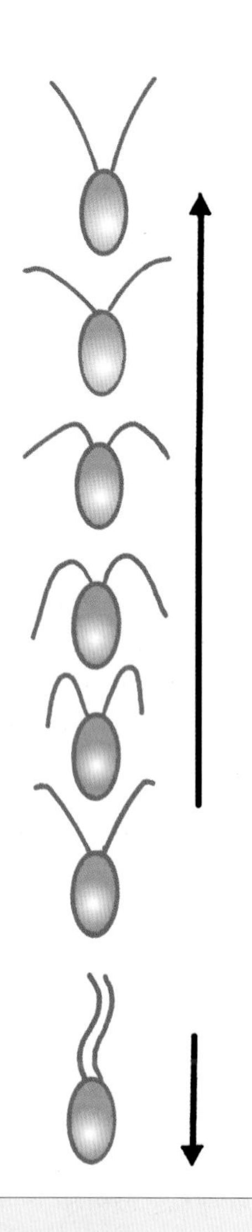

Figure 12.18. *In* Chlamydomonas, *flagellar tips point chiefly in the direction opposite movement. After Bray (1992).*

To illustrate how this mechanism might operate in a real situation, consider the propulsive action in the common unicellular green alga *Chlamydomonas*. *Chlamydomonas* has two antennae-like flagella that propel the cell through water. The cell ordinarily swims forward, antennae first; but during an avoidance reaction it can swim backward. Differences of respective flagellar orientation are telling (Fig. 12.18). During forward swimming (top sequence) the flagella undergo a kind of breast-stroke. By the time the propagated wave draws anywhere near the axoneme's tip, the distal section of the axoneme is pointing rearward, so exiting flow can propel the cell forward. When swimming in reverse (bottom) the flagellae remain forward-directed—propelling the cell backward. Depending on the desired swimming direction, then, bending may confer the orientation required for appropriately directed fluid flow.

Operation of cilia would follow along similar lines. Cilia are typically shorter structures than flagella, designed to propel fluid in a direction perpendicular to their nominal orientation. To generate the flow in the required direction, the distal section of the cilium would need to reorient. And that indeed is what the bending stroke achieves. The distal sections spend most of their time oriented parallel to the cell surface (Fig. 12.19). Hence, generated flow is appropriately directed.

For a flow-based mechanism of this sort to be effective, many cilia would need to operate collectively; a solo act could not transport a bed of fluid. On the other hand, synchronous action would not work well because fluid drawn simultaneously from the base of each cilium would leave a vacuum, which would impede the process. More natural is a scenario in which the transported fluid is passed along progressively from one element to the next. Thus, the observed metachronal progression makes sense within the proposed framework.

Conclusion

Like the diesel engine, the microtubule engine can produce wondrous results with few moving parts. In principle, the necessary structural elements are the microtubules, accessory proteins needed to trigger the microtubular phase-transition, and fuel such as ATP. This simple system can transport both solutes and fluid. It can accomplish this task within the confines of the cell—or outside the cell when embodied in a structural framework such as the axoneme.

A point to emphasize is the thematic consistency between the proposed microtubule- and actin-transport mechanisms. Both involve a propagated phase-transition. The microtubule system is an evolved version of the actin system, affording higher efficiency and versatility. Collectively, these systems offer an array of options for the cells' diverse intracellular and extracellular transport tasks.

Among the most consequential of such transport tasks is the one considered in the next chapter—transport of chromosomes. Chromosome transport takes place during cell division, when microtubule-based and actin-based systems come together in an orchestrated scenario for apportioning chromosomes into incipient daughter cells and dividing the cytoplasm into two. The plot that unfolds is thick and fascinating. And it is replete with familiar players that carry with them ample mechanistic clues.

Figure 12.19. *Time course of ciliary bending stroke (left to right). The distal section spends much of the time in the direction parallel to the surface, particularly during the "power stroke." After Bray (1992).*

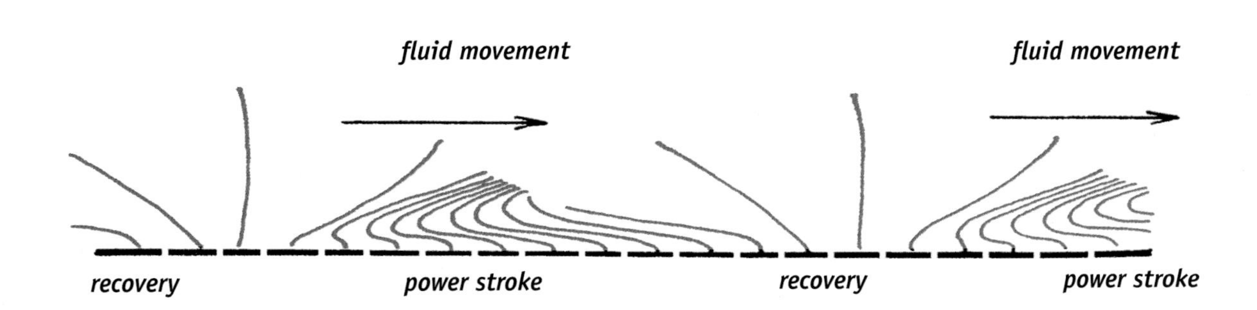

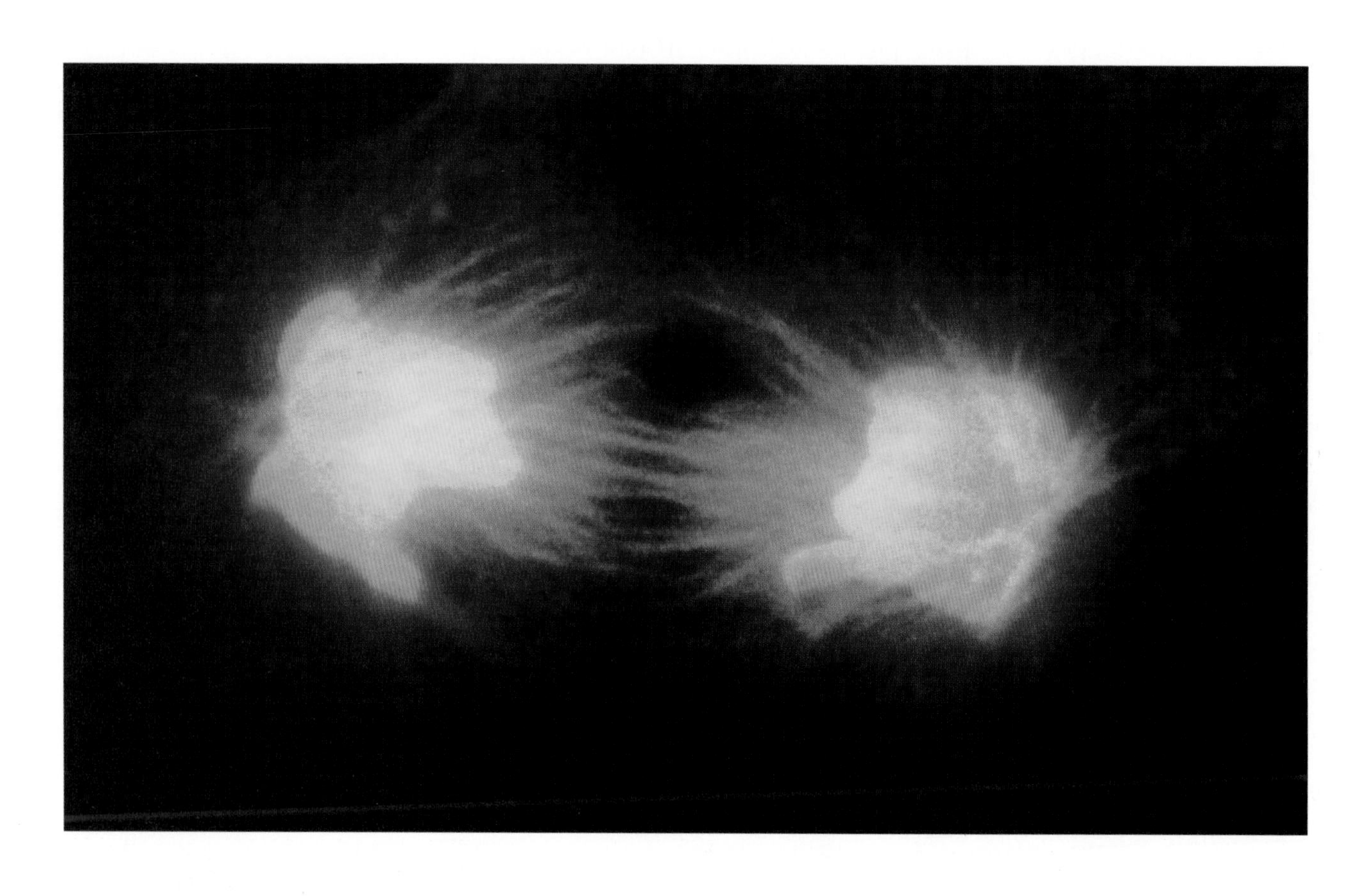

Immunofluorescence image of mammalian cell mitosis, courtesy Dr. J. R. McIntosh

13

CHAPTER 13

CELL DIVISION

Cell division is a familiar process driven by familiar players (Fig. 13.1). Microtubules frame the primary event. They create the mitotic spindle, a scaffold within which chromosomes can be deftly segregated into two incipient daughter cells. The cell then turns to actin for the subsequent step—cytokinesis. A ring-like gel of actin filaments forms around the spindle's equator, and divides the cytoplasm by contracting like a sphincter. One cell becomes two.

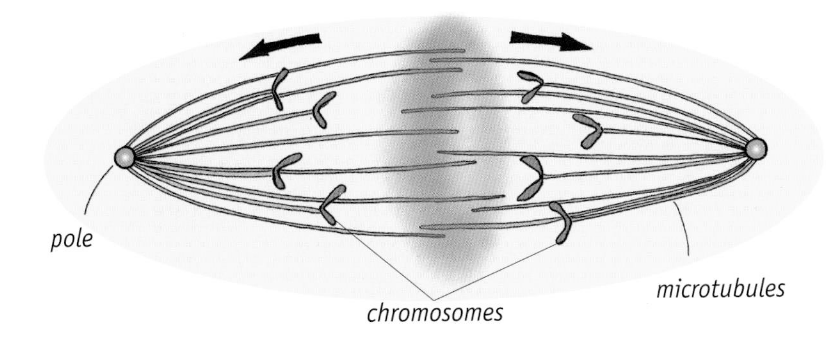

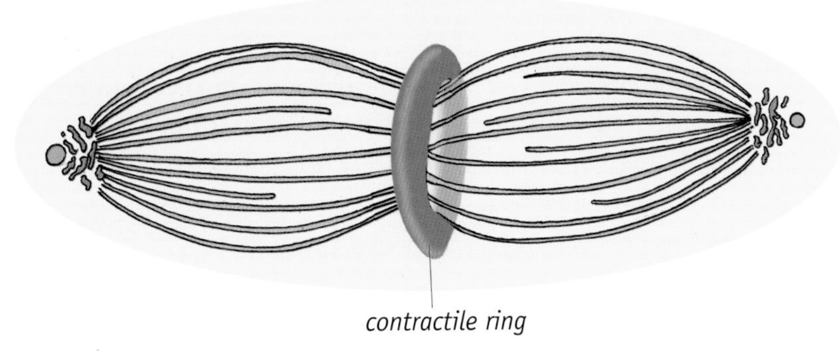

Figure 13.1. *Cell division process consists of mitosis (top) and cytokinesis (bottom).*

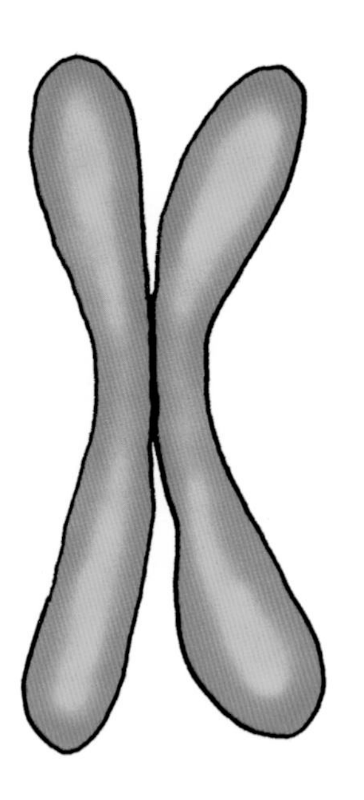

Figure 13.2. The chromosome consists of a pair of sister chromatids joined at their waist.

Although the dynamics of mitosis and cytokinesis are known to a gnat's eyelash, the underlying mechanisms remain vague. The principal players, however, are the same as those of the last two chapters and the goal is to determine the extent to which the mechanisms of those chapters might suffice. Could chromosome transport operate by the same principle as organelle and solute transport?

To keep the presentation streamlined I focus on the main events only, and thereby avoid trekking through the terminological tangle of mitotic stages (prophase, prometaphase, *etc.*), which implicitly emphasize morphology. The focus here is on mechanisms. For those interested in additional detail I recommend the excellent textbook by Bray (1992), as well as focused reviews by Inoue and Salmon (1995), Hyman and Karsenti (1996), and Nicklas (1997).

THE CAST

Apart from the spindle and the contractile ring (Fig. 13.1), four characters play leading roles in the drama of cell division:

(i) The chromosome (Fig. 13.2). This structure contains the genetic information in duplicate, one set in each of its two sister-chromatids. Sister-chromatids are joined at their waist by a belt-like constriction, and will eventually split apart and migrate in opposite directions to the two incipient daughter cells.

(ii) The kinetochore (Fig. 13.3). Kinetochores are specialized structures oppositely oriented on either side of the chromosome's waist. Kinetochore morphology is curious. Closely spaced bristle-like projections emanate from a densely packed trilaminar base, creating a structure sufficiently distinctive to suggest a special role.

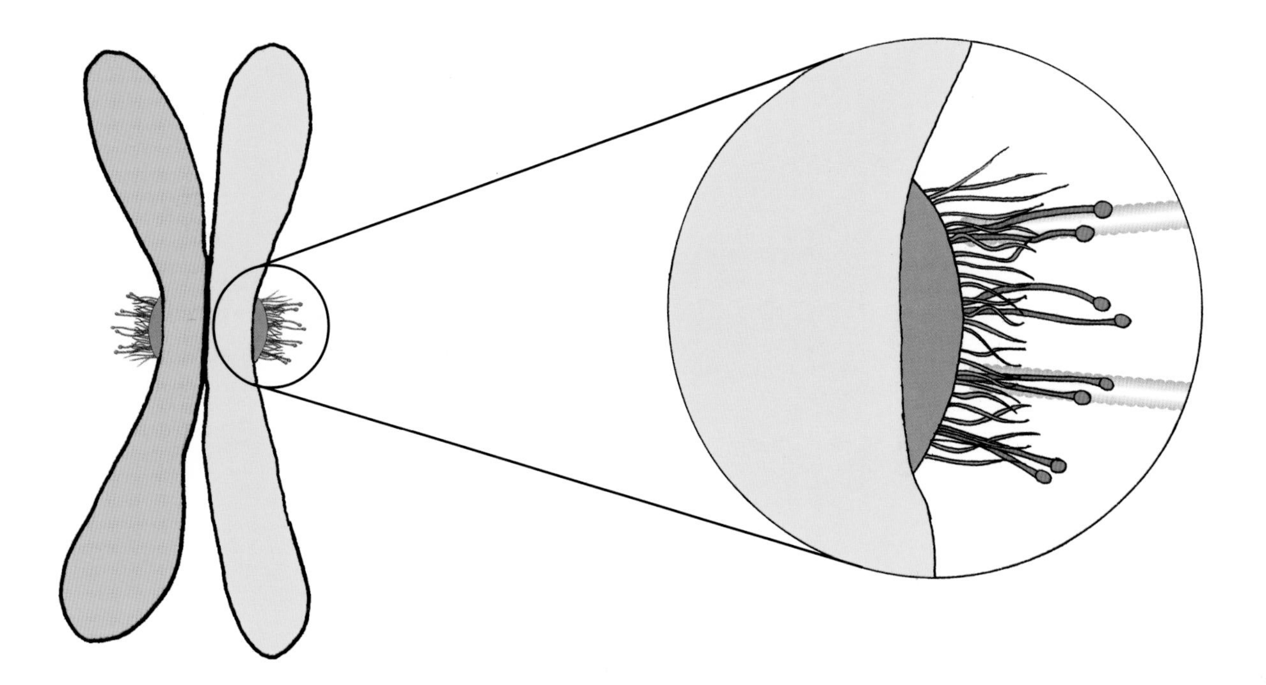

(iii) Microtubule-activating proteins. Stuck onto the kinetochore, these proteins project beyond the bristle-like projections (Fig. 13.3). They include dynein, as well as kinesin-like proteins MCAK and CENP-E (Yen *et al.*, 1992; Wordeman and Mitchison, 1995; Walczak *et al.*, 1996). These proteins are widely presumed to facilitate transport along the spindle's microtubules.

Figure 13.3. *Kinetochores lie on either side of the sister-chromatid pair (left). Expanded view (right) shows bristle-like projections and activating proteins that interact with microtubules.*

(iv) Centrosomes (Fig. 13.4). The centrosomes will become the mitotic poles—the nucleating sites from which microtubular buds will eventually grow to form the spindle.

With this cast, the play proceeds. The play is dominated by a theme that recurs again and again in cell division and also in other cellular processes: polymerization. Polymerization had been considered briefly in the context of the ameoba, where growth of actin filaments was fundamental to pseudopod extension (Chapter 11). Polymerization is central to mitosis, and we therefore treat the subject here in more detail.

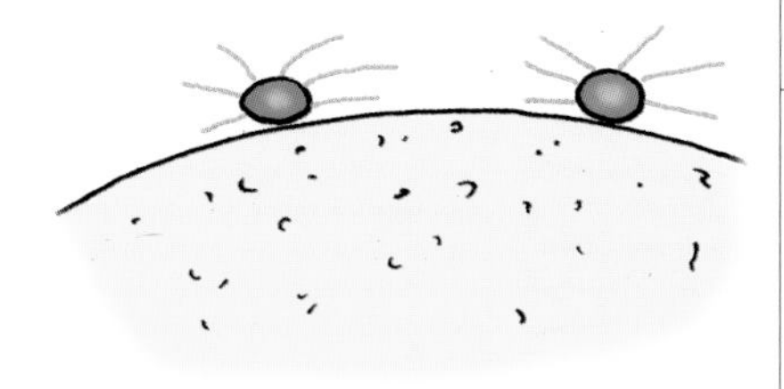

Figure 13.4. Centrosomes lying on the cell's nuclear membrane will eventually grow the spindle.

Polymerization

In order to polymerize, monomers must associate. Factors that promote association will therefore promote polymerization, and one of the most pervasive of these is environment. Environments dominated by structured water exclude solutes, which must then be forced toward nearby surfaces, where association can take place. If association occurs at the end of a filament, the filament will polymerize, or lengthen. In nonstructured environments, by contrast, any such pressure to polymerize will be considerably reduced because solutes are more soluble and less excluded. Thus, the higher the water structure, the higher the proclivity toward polymerization.

These same principles should apply not only for polymerization-depolymerization equilibria but more generally for any kind of association or binding. Structured environments promote binding, whereas unstructured environments promote unbinding. A transition toward structure could, for example, promote association of an enzyme with its substrate; if destructuring were a byproduct of enzyme action, the enzyme would subsequently dissociate, thereby completing the enzymatic cycle.

With this backdrop, consider the process of microtubule growth, which is required for spindle construction. Recall the previous chapter. Envision a pocket of unstructured water drawing along the microtubule (Figure 13.5). As the pocket moves to the end of the microtubule and collapses into the surrounding sea of structure, contained solutes will be forced toward the microtubule. If such solutes include tubulin, which can bind to the end of the microtubule, the microtubule may grow. Conversely, the microtubule could depolymerize if the pocket were somehow to linger.

How might such principles be used in spindle construction?

Spindle Growth

The process of spindle construction commences with two upheavals initiated by the cell-cycling program. In the nucleus, dispersed strands of chromatin condense into pairs of sister-chromatids. In the cytoplasm, the cytoskeletal network dissolves: Through a water-destructuring gel-sol transition, microtubules, actin filaments and other skeletal elements (except intermediate filaments) suffer complete depolymerization and disperse throughout the cytoplasm. These transformed players await their call to action.

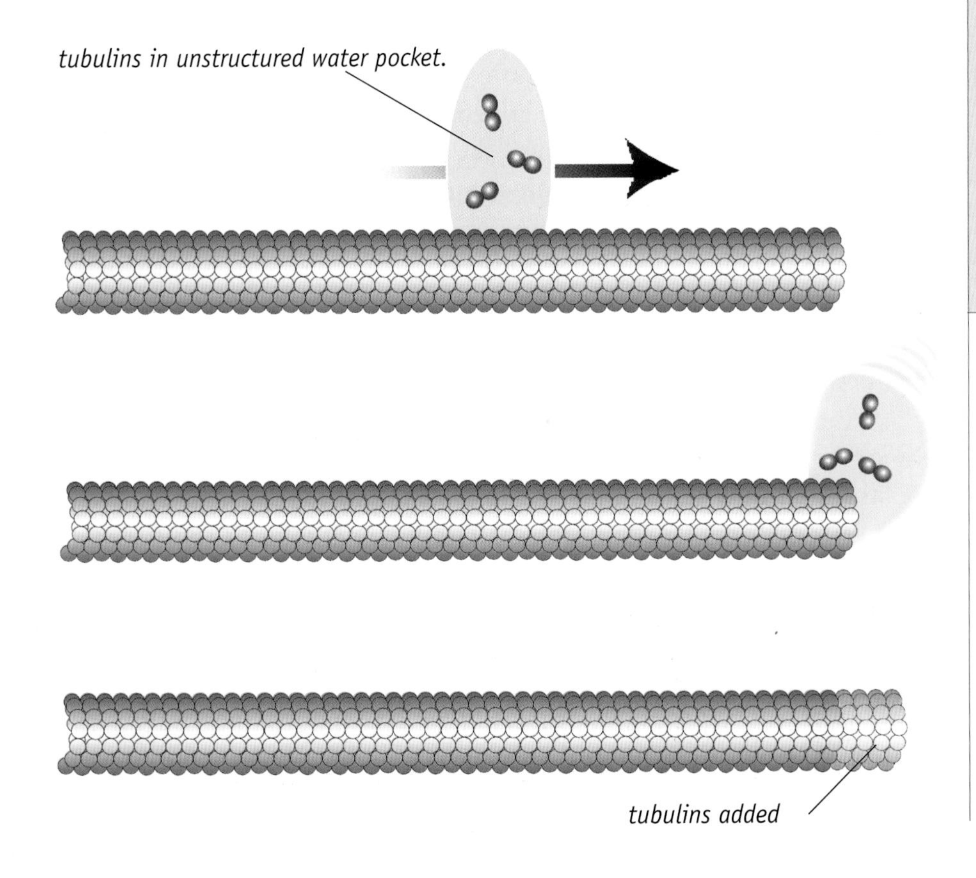

Figure 13.5. *Tubulins contained within a moving pocket of unstructured water can lengthen the microtubule as the pocket collapses at the tubule's plus end.*

The first call is for spindle construction. Construction begins at the microtubular buds (Fig. 13.4), which will eventually evolve into the spindle. The buds grow as dissolved tubulins bind and polymerize. Growth will remain slow, however, because random diffusion is an undirected process and there is little pressure to polymerize.

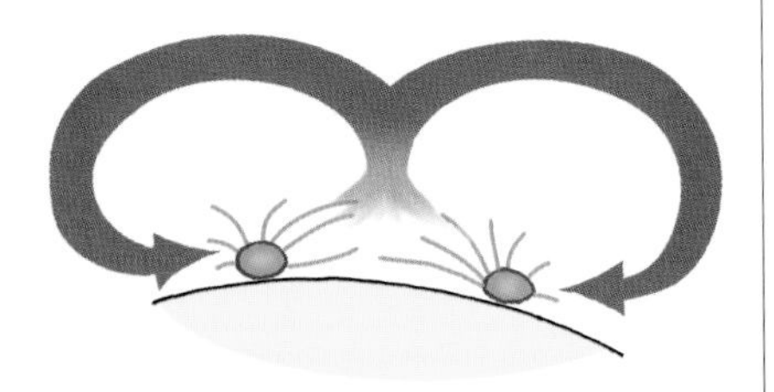

Figure 13.6. *Directed flow brings solubilized tubulins toward the centrosomes, which enhances polymerization of microtubules.*

To effect polymerization with reasonable speed, the cell requires a more directed system. Such a system lies in abeyance, for among the molecules dissolved in the cytoplasm is the transport-activating protein kinesin. Kinesin action can induce actively directed transport along microtubules (Chapter 12), and the budding microtubules of the polar (centrosomal) region should be no exception. The induced streaming will draw solutes from the cytoplasm (Fig. 13.6). Drawn tubulin molecules can enhance microtubule growth, while drawn kinesins perpetuate flow—which in turn draws more tubulins. Positive feedback drives rapid microtubule growth. The growth direction will be mainly parallel to the nuclear membrane because the perpendicular direction, toward the open cytoplasm, is less conducive to growth: the cytoplasm is a sol whose unstructured character does not readily support polymerization. Thus, the spindle grows, and matures toward its characteristic shape.

The central element of this mechanism is flow, and the logical question is whether there is any evidence for flow originating in the polar region.

Since the 1930s it has been clear that particles inserted into the spindle's polar region are ejected. The so-called polar-ejection force has remained of mysterious origin, and because of its diffuse nature it has been termed the "polar wind." Particles confirmed to be blown by this wind include chromosome-arm fragments, small membranous particles, and nucleoli. These particles flow from the pole toward the spindle's equator or periphery. The force does not arise as some kind of tug of war between complementary half-spindles because it can be observed in half-spindles in isolation—particles move away from the pole and stop moving as they exit the microtubule array. The magnitude of the ejection force depends on local density of microtubules and on the particles' surface area projected normal to the microtubule axis (for review, *see* Rieder and Salmon, 1994).

All of this evidence is consistent with an ejection force based on flow—hardly a surprise since flow is generated by constituent microtubules as well as by microtubule arrays (Chapter 12). The flow stream carries particles irrespective of their nature, and ejects them from the microtubule array. Dependence of ejection force on particle-surface area is anticipated: larger particles are pushed by a larger number of streaming water molecules. Indeed, the full force of flow can be strong enough to drive the pole itself: When drugs such as nocodazole or colcemid are applied to dissolve those microtubules immediately adjacent to the pole, the residual spindle generates a large enough force to suck in the entire pole assembly (Ault *et al.*, 1991). Thus, spindle microtubules appear to generate substantial amounts of flow.

A consequence of flow generation must be a reactive force. In the same way that a canoe is propelled in the direction opposite the paddle stroke, the half-spindles will move opposite the flow (Figure 13.7). It is a simple matter of action and reaction. With little resistance, the two comple-

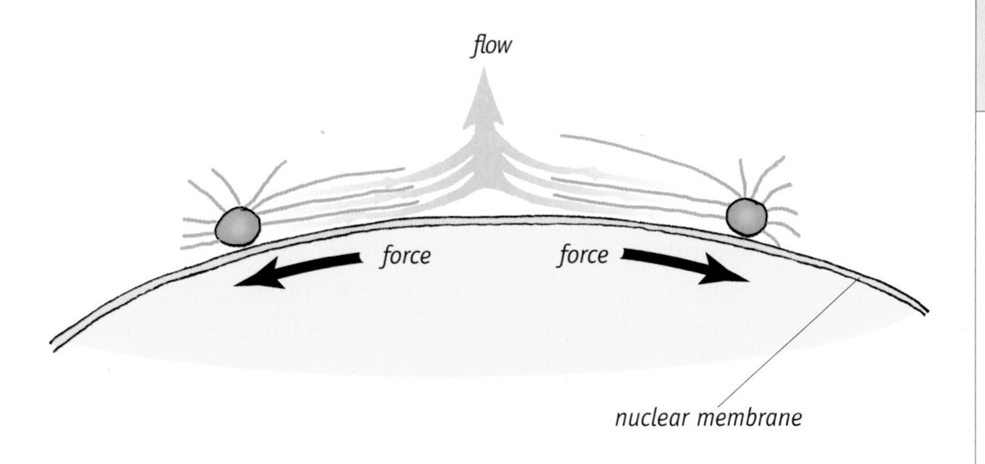

Figure 13.7. *Generation of flow produces a pole-directed force.*

mentary microtubule assemblies separate. Thus, pole separation—an essential element of mitosis—is accomplished as a natural byproduct of spindle construction.

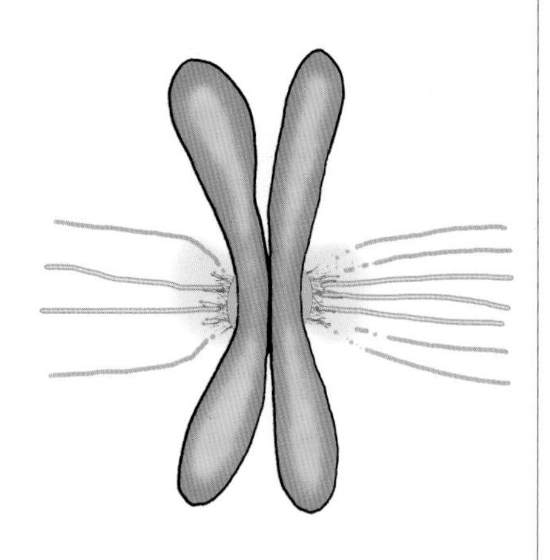

Figure 13.8. Growing microtubules are drawn toward the kinetochores because of their highly structured environment.

Construction proceeds also in the lateral direction. Once the cell-cycling program triggers nuclear membrane dissolution, spindle growth can proceed laterally. Microtubules invade the former nuclear region and progressively envelop the suspended chromosomes. Once enveloped, the chromosomes should follow the flow stream in the same way as the tubulins and kinesins, heading toward the middle of the spindle. Thus, the flow stream explains in a simple way how the chromosomes migrate to the equator, where they need to be. At the equator, the chromosomes are caught between two oppositely directed streams, pounded ceaselessly by thrusts from both directions and thereby suffering the characteristic to-and-fro agitation commonly known as the "prometaphase dance."

Meanwhile, microtubule polymerization continues as a result of this flow. But there is a problem—the equator is now laden with chromosomes, which obstruct the path of the growing microtubules. The chromosomes, however, are more than passive obstructors: As condensed polymers, they create structured microenvironments. Because structure promotes polymerization (above), microtubule growth will be directed toward the chromosomes. Such drawing potential ought to be especially strong in the kinetochore region because of the tight structuring anticipated from the densely packed bristle-like projections. Thus, growing microtubules should—and do—condense preferentially onto the kinetochores (Fig. 13.8).

Kinetochore-microtubule association brings several consequences. Most conspicuous is the restraint of chromosomes, whose jitterbug action fades progressively into quiescence with increasing association. Association also links the two half-spindles. This linkage arrests the process of pole separation—at least for the moment. Finally, microtubules that bypass the chromosomes and escape the clutches of the kinetochore may continue to grow unimpeded and eventually overlap those microtubules grow-

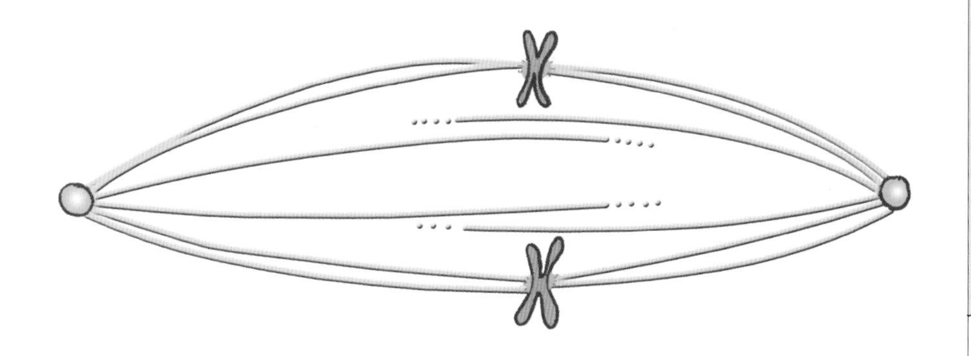

Figure 13.9. *Once the chromosomes link the two half-spindles, only those microtubules escaping the clutches of the chromosomes can continue to grow.*

ing from the opposite half-spindle (Fig. 13.9). Overlapping microtubules are poised to mediate further pole separation.

Transport toward the Poles

All actions up to now set the stage for the main event: chromatid separation. Separation allows a copy of the genetic material to migrate into each of the two incipient daughter cells. This process begins when the linkage between sister-chromatids ruptures, allowing each chromatid to begin its poleward journey.

In principle, chromatid transport could follow along the lines outlined in the previous chapter. Microtubule-activating proteins on the kinetochore could facilitate poleward transport—but there is a curious problem. The poleward track (in mammals) comprises not one but some 40 – 60 densely packed microtubules. How could the bulky chromosomal cargo work its way through so dense a thicket? And why should there be more than one microtubule per chromosome anyway?

Figure 13.10. *As the chromatid moves along the microtubules, the array depolymerizes.*

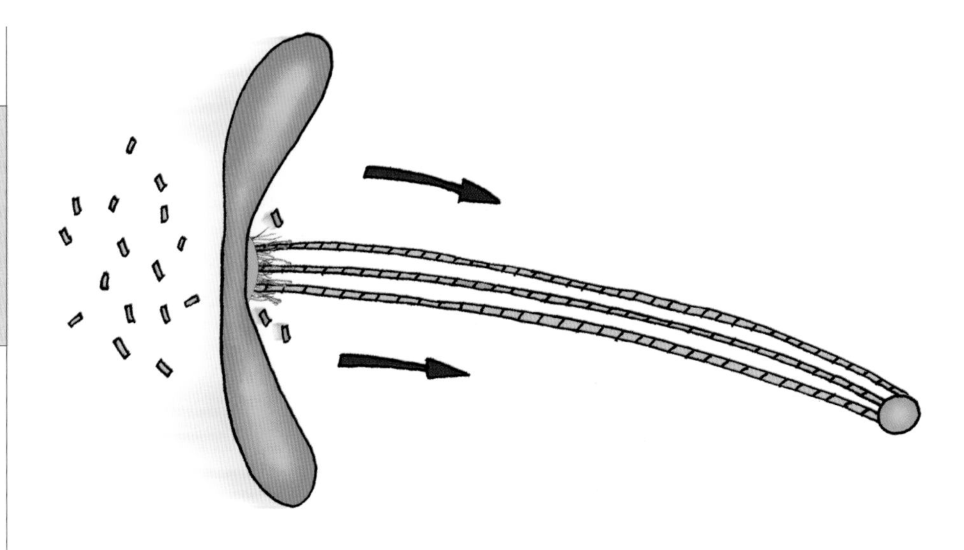

The problem resolves through the expedient of depolymerization: the chromatid "eats" its way through the thicket (Fig 13.10). This recognized "Pac-man" like phenomenon explains how the chromatid can penetrate through this spatial obstacle—the transport process gaining strength and certainty by the presence of multiple microtubules. It also throws something of a monkey wrench into our development because it implies either that the phase-transition mechanism is not the answer to all movement processes after all, or that depolymerization can be an implicit feature of phase-transitions, surfacing when circumstances are appropriate.

To test for the latter, we look toward the transport system's unique element, the kinetochore. The kinetochore contains the microtubule-activating proteins. These proteins are curious in that they contain long stalks, typically >100 nm, which are anticipated to be fully extended in the kinetochore's structured environment (Chapter 8). The heads of these proteins will therefore project well beyond the dense coronal brush (Fig. 13.3). This arrangement implies some special function.

One possibility is that this idiosyncratic arrangement provides a vehicle to couple transport with depolymerization, and a highly tentative scheme is given in Figure 13.11. The initial condition is shown in panel *i*. In panel *ii*, a microtubule phase-transition creates local destructuring, which predisposes monomers to dissolve and the stalk to fold. Stalk folding translates the chromosome-kinetochore complex rightward, toward the pole. The system recovers as the tightly packed coronal fibers nucleate restructuring (panel *iii*). While the chromatid is held in place by microtubule - corona-fiber association, the restructuring front straightens the stalk and advances the heads to the next position along the microtubule. Once the heads rebind at the advanced position, the system is set for the next poleward-translation step (panel *i*).

Details of this inchworm-like scheme must be taken as highly provisional inasmuch as the complement and disposition of microtubule-activating proteins is just now beginning to emerge. Evidence is scant. A good test would be to check whether transport occurs as implied, in stepwise fashion. In principle, this transport mechanism is little different from the one of the previous chapter: Transport is driven by a restructuring front that drives the activating protein forward. The difference arises from the idiosyncratic nature of the kinetochore, which may be purposefully designed to link transport with depolymerization. Otherwise, the mechanism is similar to the one we have been dealing with.

Pole separation can resume once the final sister-chromatid bond has been broken—by which time most chromatids will have already progressed substantially poleward. Microtubules of each half-spindle drive

Figure 13.11. *Possible mechanism coupling transport with depolymerization. See text for details.*

i.

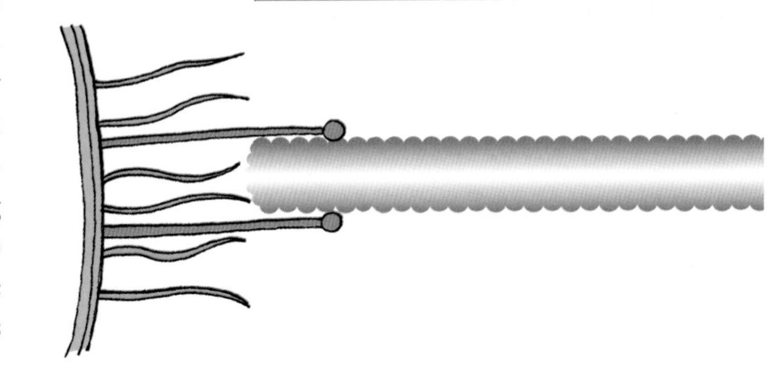

ii.

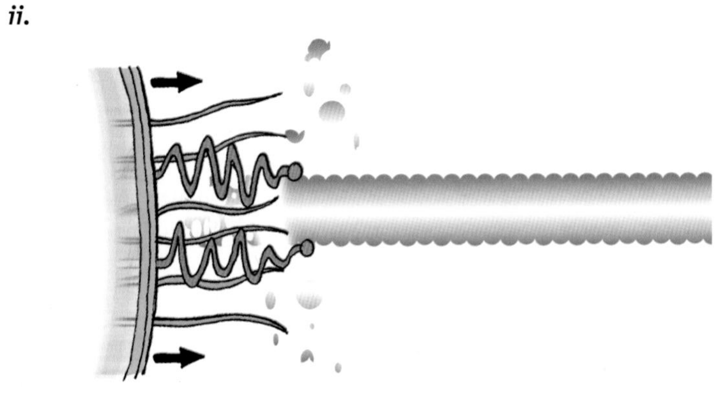

iii.

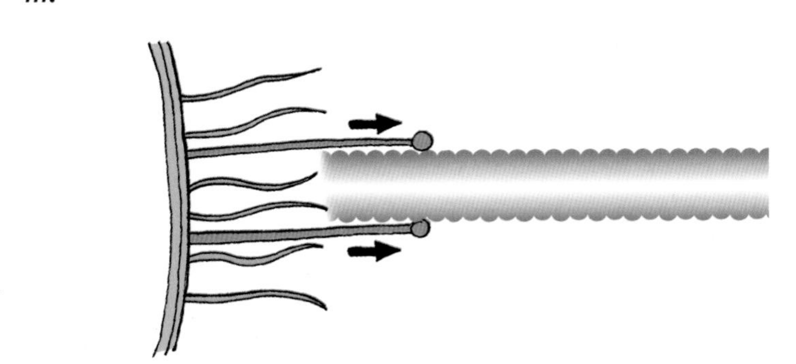

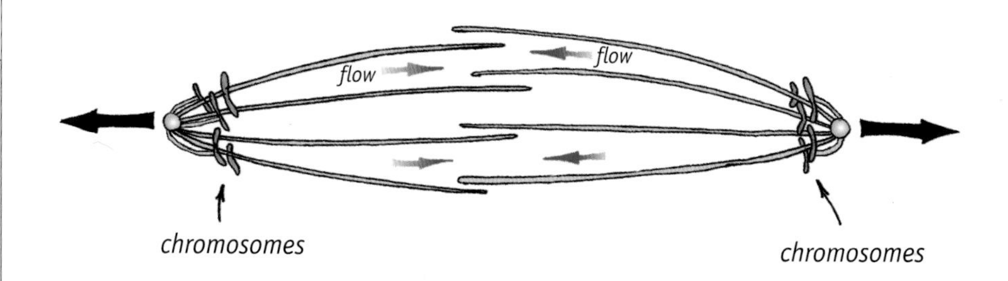

Figure 13.12. Once chromatids are released, the reactive force resumes pushing the poles apart.

flow in opposite directions (Fig. 13.12), pushing the two half-spindles apart. The separation process is now particularly effective because the chromatids are tethered to the half-spindles, and as the half-spindles separate, so do the chromatids. Separation continues until microtubules at the equator no longer overlap. Then it stops.

Cytoplasmic Division

Although the above-described events are orchestrated to segregate chromosomes, they also set the stage for the subsequent event—cytoplasmic division. The stage begins to be set when the sister-chromatids are aligned at the equator. The dense equatorial plate deflects axial flow, which must therefore bend radially and exit the spindle on either side of the plate (Fig. 13.13). Because the solutes exit into a zone of high structure conferred by nearby microtubules, conditions for polymerization are apt—high solute concentration and high structure. Thus, abundant solutes including actin and myosin can eventually gel into a ring around the spindle's equator (Fig. 13.1).

Contraction of this ring divides the cell physically into two. This process requires appreciable shortening, and Figure 11.8 shows that the actin gel is quite up to the task. The trigger may well be the thinning of microtubules that occurs as microtubules depolymerize (Fig. 13.10): Thinning of microtubules diminishes water structure within the ring's aperture, which shifts the ring's equilibrium toward transition (Chapter

8). This mechanism does not preclude other likely factors such as myosin phosphorylation from playing a role, but the destructure-based trigger can provide an appropriately seamless link between mitosis and cytokinesis.

The cycle draws to an end as the contractile gel dissolves. Cytoplasm and genetic material are now properly apportioned into incipient daughter cells. The processes that took place at the beginning of the mitotic cycle may now reverse, allowing the daughter cells to re-establish their working configurations and go about their business. One cell has become two.

PATHOLOGY

When mitosis and cytokinesis go rampantly out of control, the result, unfortunately, is cancer. The subject of cancer follows so naturally from any discussion of cell division that a few speculations on etiology seem irresistible.

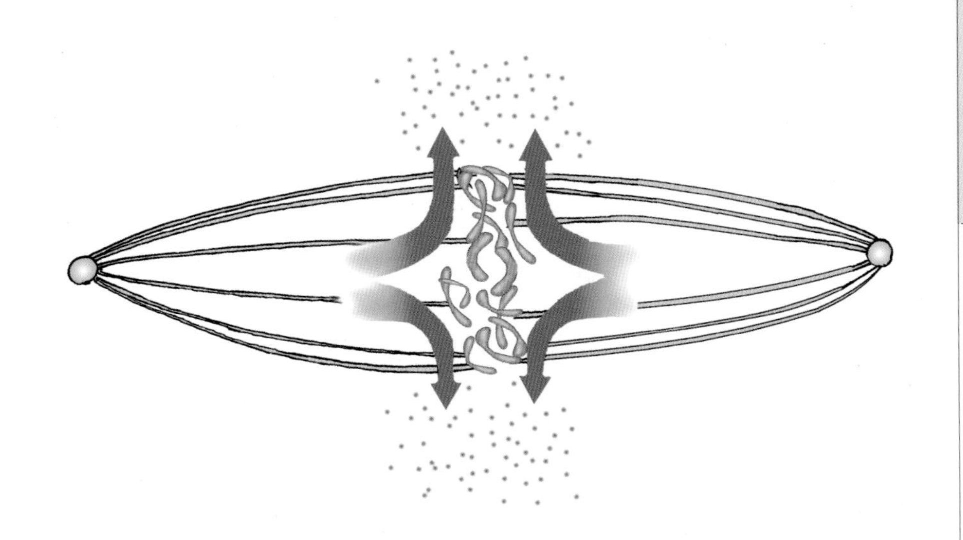

Figure 13.13. *Solutes exit the spindle in narrow zones on either side of the equator, thereby concentrating the solutes.*

Within the considered paradigm, prime attention should be accorded elements that affect not only protein but also water structuring. The story of cancer-cell water dates back several decades when a physician, Raymond Damadian, came to the laboratory of Gilbert Ling (whom you may remember from earlier chapters) for help. Damadian had been wondering whether a difference of water structure might underlie organ pathologies, and began exploring the potential of nuclear magnetic resonance as a tool to test for such differences. The collaboration proved fruitful. Its evolution led to a breakthrough technology called magnetic resonance imaging (MRI), for which Damadian holds the patent. This powerful technology arose directly out of the concept of structured cell water.

One of the most far-reaching consequences of MRI was that it could be used to distinguish tumor cells from non-tumor cells. The distinction rested on an apparent difference of cell water, for differences of water-protons' relaxation times underlie the MRI image. Water appeared less structured in tumor cells, and this result was amply documented (*e.g.*, Damadian, 1971; Beall *et al.*, 1982). It now appears that much of the difference may arise from differences of paramagnetic ion content between cancer and non-cancer cells (Ling *et al.*, 1990). Nevertheless, it is clear from experimental studies that cells do undergo a change of water structuring during mitosis (Cameron *et al.*, 1987), a change anticipated from the associated sequence of phase-transitions. Thus, tissues undergoing rampant division might well show at least some difference of water structuring relative to those that are not.

The rationale, then, is to look for agents that may impact water structuring—for such agents should promote mitosis and could promote rampant mitosis. An interesting clue is that tumor cells express "signature" proteins, which arise as a consequence of mutation. Among different cancers, different signature proteins proliferate. These proteins are unwanted guests in the cell's abode, and cannot easily be evicted because they escape recognition by the immune system. Their destructive power is evidenced by the success of antibody therapies directed specifically at their destruction: eliminating these proteins effectively eradicates the tumor.

In this context, a possibility to consider is a linkage between mutation and water-structuring capacity. Mutant proteins might not be able to structure water as effectively as normal, wild-type proteins. This speculation arises because the capacity to structure is an implicit design feature in the proposed paradigm—it is responsible for setting the cell's basic state. At the same time, the protein must have the capacity to act, which commonly translates into a capacity to destructure. Protein design must therefore reflect a balance between these two competing strategies, and in the mutant, neither one of the strategies may survive unscathed. The signature protein is likely to be functionally inept and inept in its capacity to structure water.

Water disorder has predictable consequences for mitosis. Since many of the mitotic events involve a transition between ordered and disordered water, factors preserving water's order will inhibit those transitions, whereas factors promoting disorder will enhance them. A disordered aqueous environment may thus facilitate mitosis—the cell will be biased toward replication.

The hypothesis, then, is that cancer may have a largely unrecognized etiology. The first step is a mutation triggered by any of a variety of environmentally or genetically based agents. If the expressed protein is not destroyed by the immune system, it will accumulate inside the cell. The protein wreaks its havoc by compromising the ordering of water and thereby promoting mitosis. The cell begins proliferating, and because daughter cells contain both the mutated gene and its gene product, they continue to proliferate and eventually deplete the body of its limited resources.

A prediction of this hypothesis is that two therapeutic courses could prove effective. The first is a direct attack on the protein, which already appears to be stunningly effective. The second involves the water. Agents that promote water ordering are predicted to inhibit tumor proliferation. To my knowledge, this approach has not yet been tested. Nor is it established whether the reputed efficacy of some current anti-cancer agents might lie in such a mechanism.

Conclusion

The intricate chain of processes collectively known as mitosis seems largely explainable by a common mechanism—the phase-transition. The propagating phase-transition achieves the essentials: it transports building materials along microtubules for constructing the spindle; it propels chromatids toward the poles for segregation; and it pushes the two half-spindles apart in anticipation of cell division. Mitosis, then, can be viewed as a specialized application of the principles of transport (Chapter 12), the transported solutes being the materials for spindle construction, and later, the chromosomes.

A second and more subtle point is that these transitions take place in functional workhorses that get recycled. Microtubules of working cells dissolve in preparation for mitosis—and then reassemble to frame the spindle. Elements are re-used. Actin filaments of the cytoskeleton likewise dissolve, and subsequently reassemble to form the contractile ring. Even accessory proteins such as kinesin, dynein, and myosin are recycled for use in the mitotic process. For the drama of cell division, then, few new players are needed.

Cell division is thus a fairly conservative process. Although the players may be differently costumed for this performance, they are the same players as those employed by the cell to carry out its routine activities. As a result, their roles are well-rehearsed. With familiar players in familiar roles, the cell directs the intricate script that transforms one cell into two.

Having spent three chapters dealing with transport of one kind or another, we turn our attention to another one of the cell's major tasks: contraction. The contractile process also employs familiar players, including actin and myosin, as well as several new players. The question, once again, is whether an unexpectedly central role may be played by the phase-transition.

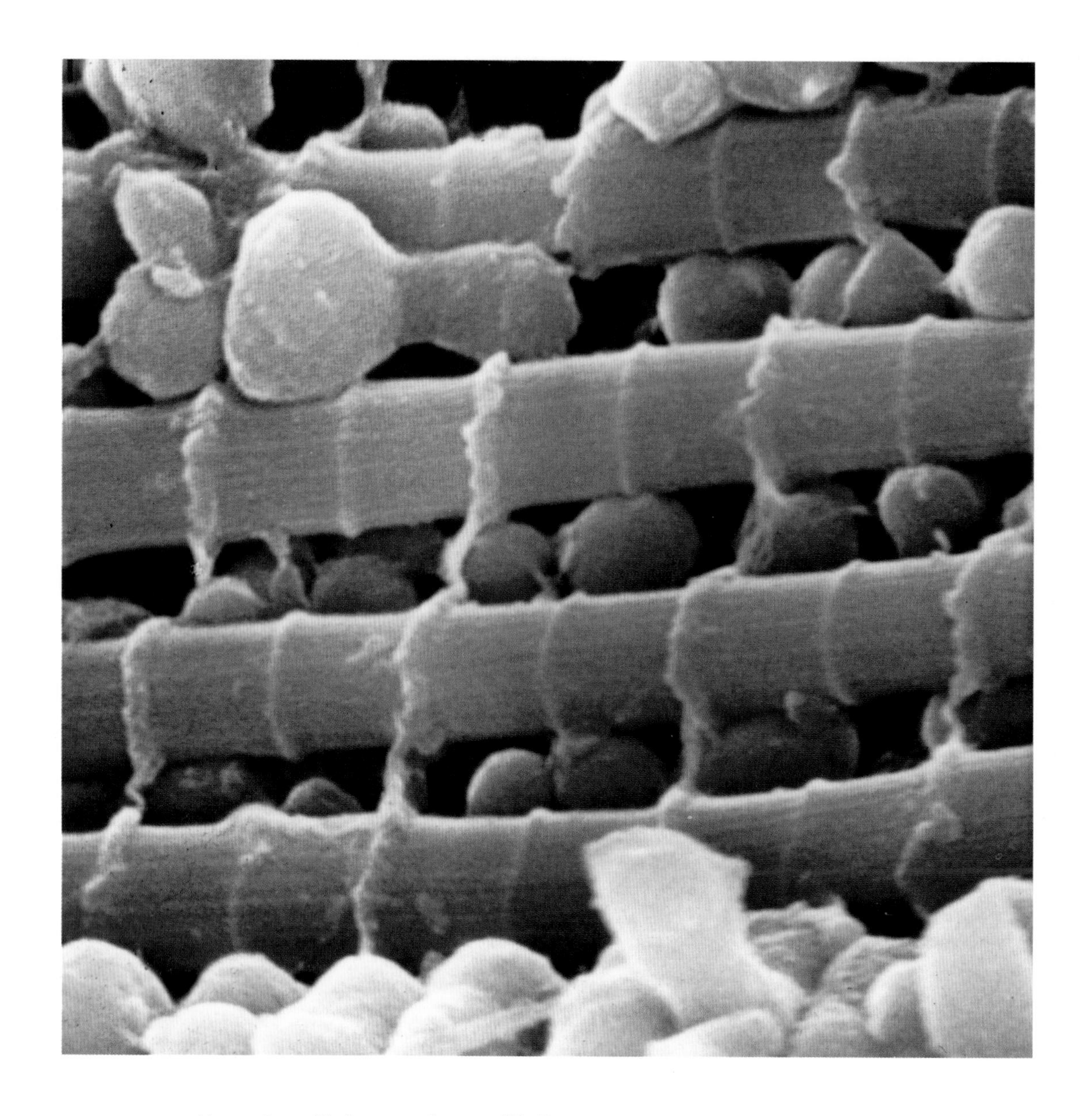

Honeybee flight muscle myofibrils, courtesy Dr. Károly Trombitás.

CHAPTER 14

MUSCLE CONTRACTION

Muscles are built of micron-sized contractile units called sarcomeres, which contain three filament types: thick, thin and connecting (Fig. 14.1). Thick and thin filaments are recognized to play a central role in contraction. Connecting filaments, identified fairly recently by groundbreaking contributions from Koscak Maruyama and Kuan Wang, link the thick filament to the ends of the sarcomere and serve as the sarcomere's molecular spring.

All three filaments are polymers: Thin filaments are built largely of repeats of monomeric actin; thick filaments are built around repeats of myosin; and connecting filaments of vertebrates are built of titin, a huge protein consisting mainly of repeating immunoglobulin-like domains. Together with water—which is held in the sarcomere so firmly that even centrifugation cannot remove it (Ling and Walton, 1976)—this array of polymers forms a gel-like lattice.

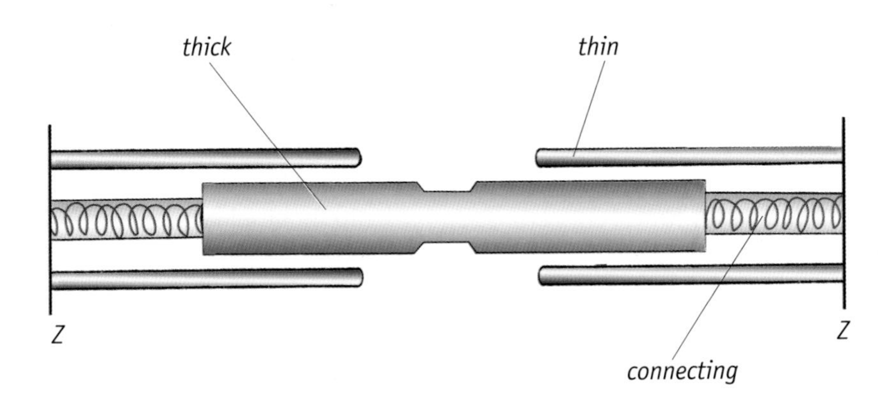

Figure 14.1. *The sarcomere is built largely of thick, thin, and connecting filaments. Each thick filament is typically surrounded by six thin filaments.*

Whether this gel contracts by a phase-transition is once again the question of interest. Before treading too far into such seemingly uncharted territory, it is prudent to consider the prevailing view, which is unusually well established. Is there reason to question its adequacy?

THE SWINGING CROSS-BRIDGE THEORY

Until the mid-1950s, muscle contraction was held to occur by a mechanism similar enough to the one that will be suggested here that it must be considered a progenitor. The dominant feature was protein folding.

With the discovery of interdigitating filaments in the mid-1950s, it was tempting to dump the notion of protein folding, and suppose instead that contraction arose out of pure filament sliding. This supposition led Hugh Huxley and Sir Andrew Huxley (unrelated) to examine independently whether filaments remained at constant length during contraction. Their optical microscopic studies were published back-to-back in *Nature* (Huxley and Hanson, 1954; Huxley and Niedergerke, 1954). Together with more detailed follow-up (Huxley and Niedergerke, 1958), these studies became the field's pivotal works.

The studies explored relaxed and activated specimens alike. Relaxed specimens were manually stretched and released. During these maneuvers, the width of the A-band remained absolutely invariant. Since A-band width corresponds to thick filament length (Fig. 14.2), the result implied that thick filaments did not change length. This strengthened the emerging notion of filament sliding.

Less conclusive, however, were the results obtained when muscle length changed actively, for here sarcomere shortening was sometimes accompanied by A-band narrowing. The narrowing was dismissed as quantitatively inconsequential in the Huxley – Hanson study, and in the Huxley – Niedergerke study it was relegated to the microscope's limited resolution. With these dismissals, all results could remain interpretable within the framework of constant-length filaments.

These back-to-back papers by the Huxleys, one of whom shared the Nobel prize for his earlier work on membrane currents, dispatched the protein-folding paradigm to the basement of misbegotten notions. The constant filament-length paradigm took hold, and has held remarkably firm notwithstanding findings of thick filament shortening or A-band shortening in more than 30 reports published subsequently (for review *see* Pollack, 1983; Pollack, 1990).

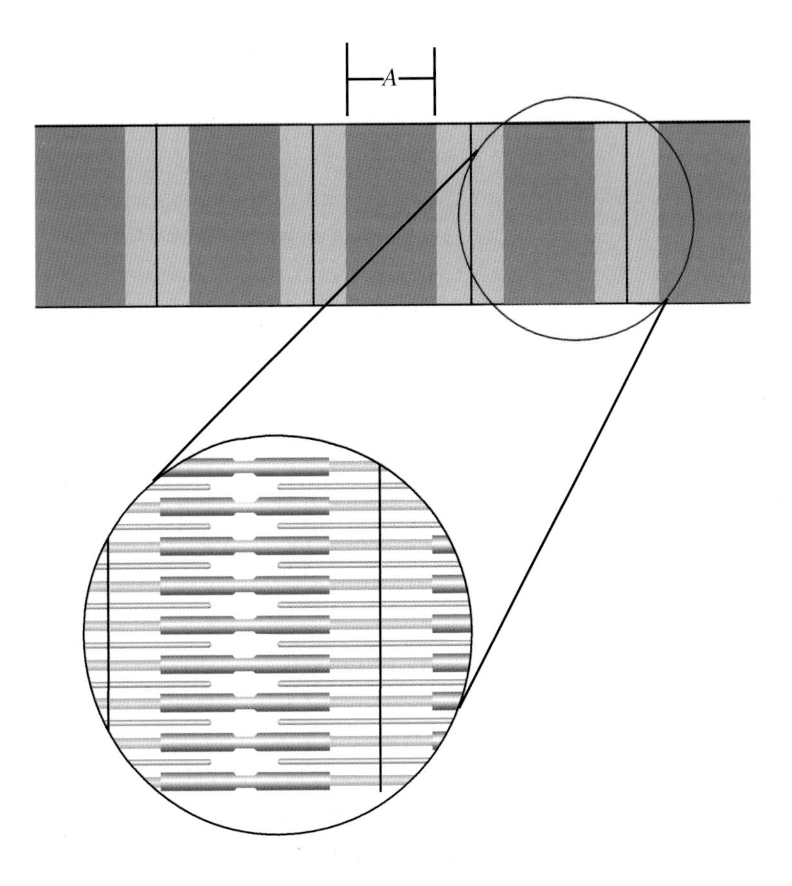

Figure 14.2. *Optical microscopic schematic of muscle striations (above), with corresponding molecular structure (below). A-band width corresponds to length of the thick filament.*

With the emerging notion of sliding filaments, the central issue became the nature of the driving force. What could propel sliding? I cannot resist drawing a parallel here with the field of cell physiology described in the book's opening chapters. To approach ground truth, it was necessary to peel back layers of foundational assumption, whose flimsiness had misdirected the field toward inappropriate conclusions. The situation here is analogous. Filaments are assumed to be passive elements of essentially fixed-length, driven past one another by extraneous drivers. Only if the fixed-length assumption is unambiguously validated can meaningful inferences be drawn on the nature of the driving force.

The concept of sliding filaments nevertheless caused great excitement in the muscle field and impressive numbers of researchers were inspired to search for the driver. These researchers soon settled on the mechanism schematized in Figure 14.3. In this mechanism the cross-bridges attach to the thin filament, swing, and then detach in the presence of ATP, readying themselves for the next cycle. Bridge swinging drives the thin filament toward the center of the sarcomere, thereby shortening the sarcomere and the muscle. The mechanism explains many known features of contraction and has therefore become broadly accepted (for review, *cf.* Spudich, 1994; H. E. Huxley, 1996; Block, 1996; Howard, 1997; Cooke, 1997).

So what is the problem? Apart from the concern of building on an uncertain foundation of constant-length filaments, are there good reasons for considering this mechanism inadequate? One central issue is the absence of evidence for cross-bridge swinging. To test for bridge-angle changes, numerous molecular scale approaches have been applied (for review, *see* Thomas, 1987). Electron-spin resonance, X-ray diffraction, and fluorescence polarization have produced largely negative results, as has high-resolution electron microscopy (Katayama, 1998). The most supportive of these results has been an angle change of 3° measured on a myosin light chain (Irving *et al.*, 1995). This change is far short of the anticipated 45°.

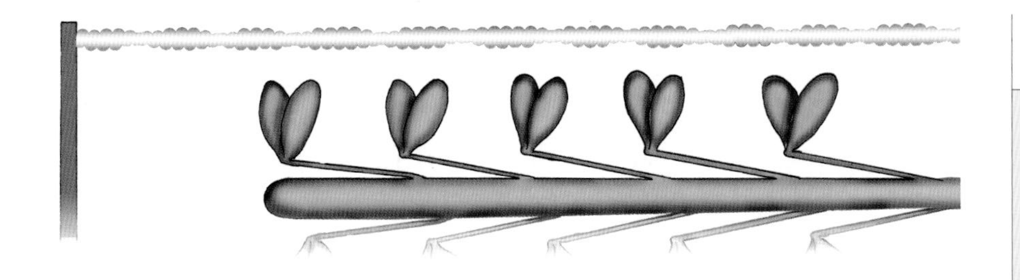

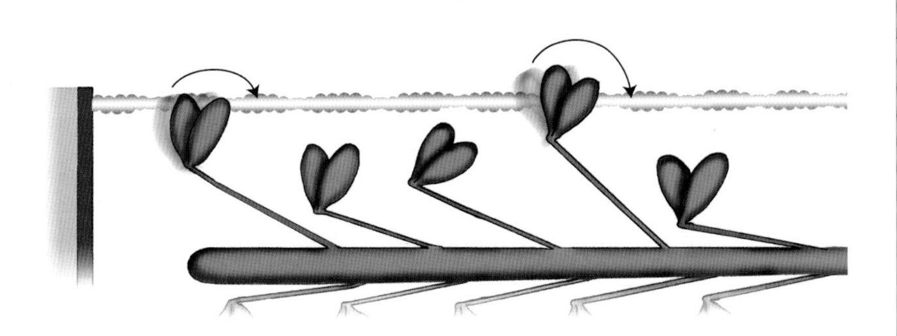

Figure 14.3. *Textbook mechanism of muscle contraction. Cross-bridge rotation drives thin filament past thick.*

Absence of swinging is only one of several areas of concern. Concerns run the gamut from instability (Iwazumi, 1970), to mechanics (Pollack, 1983), structure (Schutt and Lindberg, 1993, 1998), and chemistry (Oplatka, 1996, 1997). Many of these concerns are described in my 1990 book. But the full flavor is best gleaned by reading the original reviews, particularly those by Schutt and Lindberg and by Oplatka, whose spicy, no-holds-barred bluntness entertains as it educates. I also commend to you an insightful work by the late Graham Hoyle (1983), in which some understanding of the field's continued focus on seemingly unproductive paradigms is offered in a chapter labeled "Why do Muscle Scientists 'Lose' Knowledge?"

Here we pick up where the pioneers of a half-century ago left off. Whatever model evolves will be required to fit not only this older evidence but also the evidence of intervening years that has come to be considered supportive of the current sliding paradigm. Thus, evidence on mechanical transients, tension *vs.* filament overlap, ATP turnover, *etc.*, must be explained, and I refer the reader to the above-mentioned book (Pollack, 1990), where a detailed treatment is offered. The focus here is on evidence bearing most directly on the validity of the emerging paradigm—namely, structural changes in each of the three polymeric filaments.

STRUCTURAL CONSIDERATIONS

The functional unit of contraction is the myofibril, which comprises several hundred protein filaments (Fig. 14.4). Many parallel myofibrils make up the cell or fiber, and many fibers make up the muscle. The arrangement is thus hierarchical. The structure of the myofibril is striking in the almost crystalline precision with which parallel filaments align with one another. Such alignment is realized through extensive cross-linking throughout the sarcomere, both in the A-band and in the I-band.

I-band cross-links are illustrated in Figure 14.5. The top panel is a freeze-fracture image, which reveals the cross-links in life-like form. The bottom panel is an ultrathin section, which shows the links' ~40 nm axial repeat spacing. This spacing is a clue that the cross-linker may be troponin (Pollack, 1990), an abundant sarcomeric protein known to bundle actin and known to repeat along the sarcomere at ~40 nm intervals. Cross-linking is a general rule among diverse actin filaments and therefore no surprise that it occurs in the sarcomere.

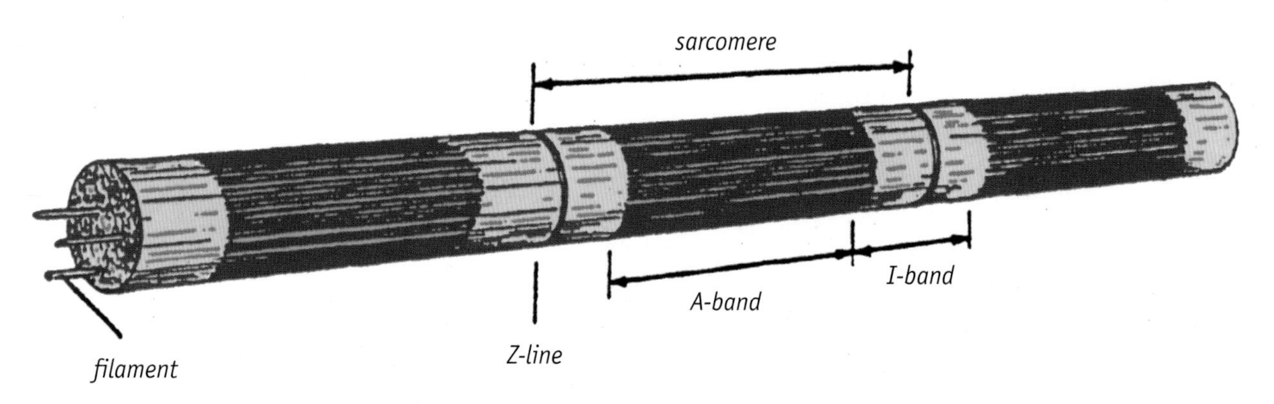

Figure 14.4. *The myofibril is a distinct bundle of regularly arranged filaments.*

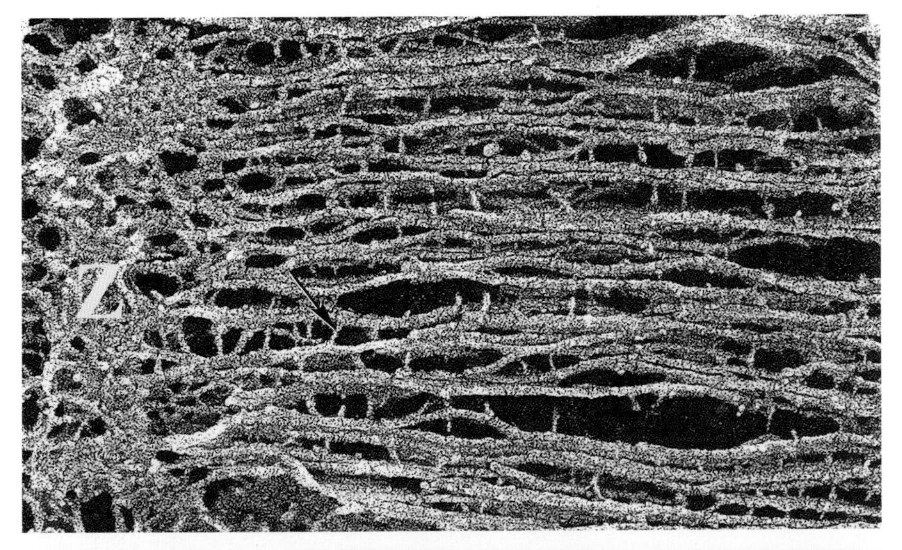

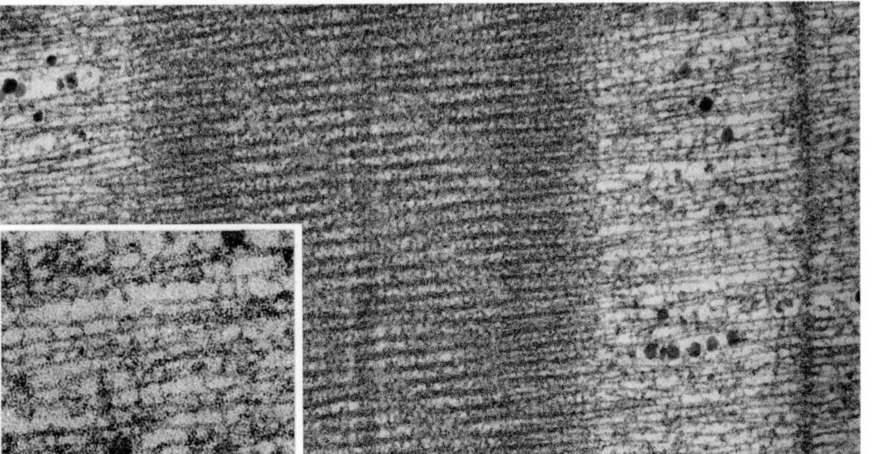

Figure 14.5. Top: *Freeze fracture image of I-band filaments. Vertically oriented mesh is the Z-line. Arrow indicates interconnection.* Bottom: *Ultrathin section showing regular repeat of I-band interconnections. View from shallow angle, below. Inset shows I-band at higher magnification. Micrographs courtesy of Peter Baatsen and Károly Trombitás.*

A-band interconnections are shown in Figure 14.6. The top panel shows a specimen that had been stretched to withdraw thin filaments from the lattice in order to reduce visual congestion. Stretch distorts the lattice somewhat but cross-links between thick filaments nevertheless remain clear. The bottom panel shows a cross-section. This sarcomere had been extended by about 30% from its resting length and the section grazed the tips of the thin filaments, some of which are seen as small dots amongst the more abundant and larger thick-filament dots. Interconnections between thick filaments are evident.

The thick-to-thick interconnections are almost certainly built of myosin, for myosin is the only A-band protein in sufficient quantity to account for structures so abundant. How the myosin interconnection might arise physically is shown in Figure 14.7. Links of appropriate length can be created if cross-bridge heads from adjacent filaments bind to one another at their tips (Pollack, 1990). This essentially static element may

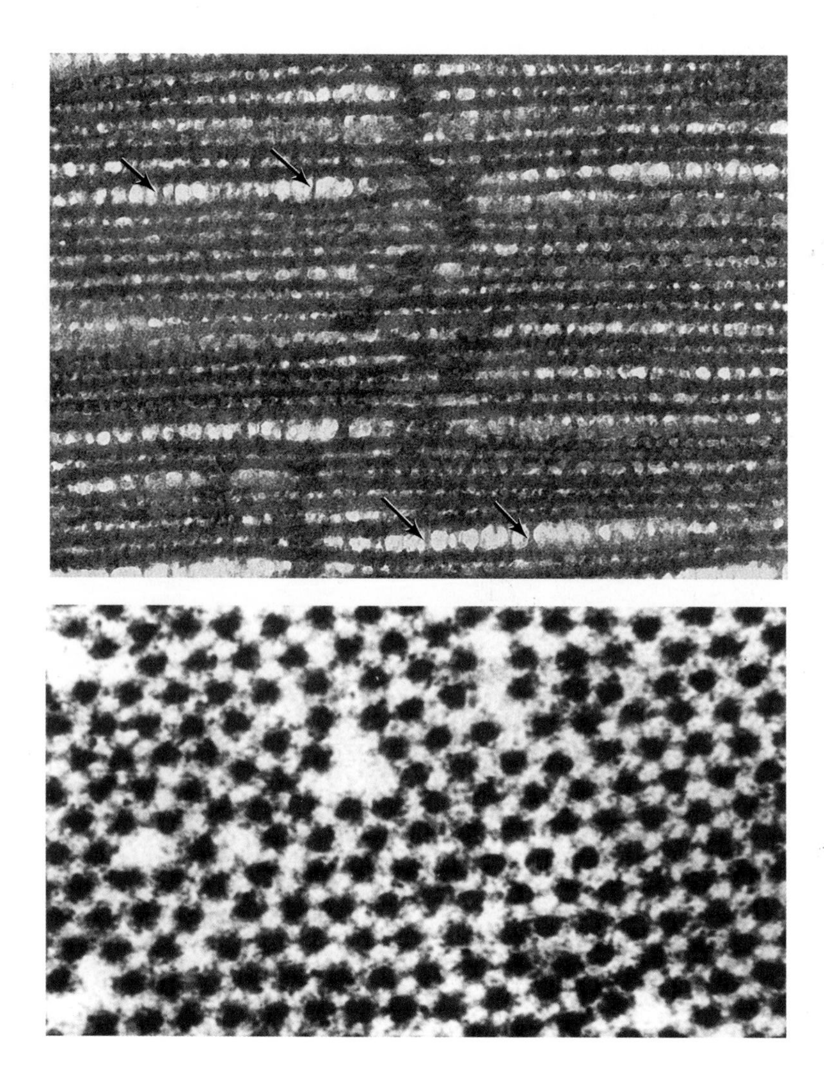

Figure 14.6. Top: *Thin section of bee muscle stretched to withdraw thin filaments from the lattice. Thick filaments remain. Arrows show interconnections.* Bottom: *Cross-section of moderately stretched rabbit sarcomere. Thick - thick interconnections are apparent. Micrographs courtesy of Károly Trombitás and Victor Popov.*

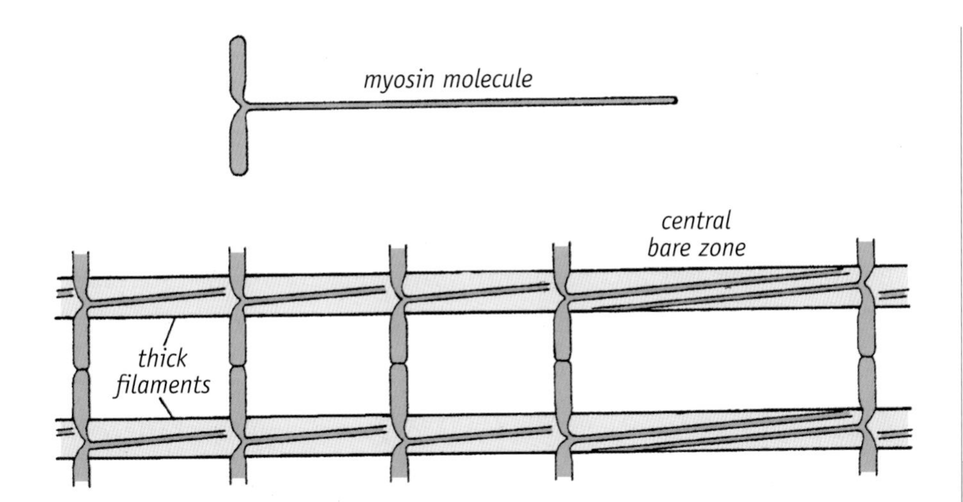

Figure 14.7. *Possible origin of the thick filament interconnection.*

explain why attempts to detect myosin-head tilting have failed; rungs cannot tilt.

Cross-links align the filaments and thereby confer regularity on the lattice. They also maintain lattice integrity by limiting swelling. Highly charged polymers such as those of the sarcomere can imbibe water up to thousands of times their volume (Osada and Gong, 1993), and may ultimately dissolve if not cross-linked. Sarcomeric cross-links restrain expansion in much the same way as they do in the peripheral cytoskeleton (Chapter 10). The muscle-filament lattice, then, is essentially a highly cross-linked, water-filled polymer gel. Because of its striking regularity, one might say it is a "supergel."

This supergel is evidently designed to contract, and certain features of its behavior imply yet another phase-transition. Triggers, for example, are the same as those for transitions in ordinary polymer-gels (Table 8.1). Contraction in demembranated cells can be initiated by: increased salt, pH change, temperature jump—even electrical current will trigger contraction of an isolated myofibril (Garamvolgyi, 1959). And like polymer-gel transitions, triggering is "razor-edge." Nothing happens until a threshold is crossed, whereupon contractile action is massive (Fig. 14.8).

With contraction, structured water becomes destructured, and is easily released from the lattice (Bratton, *et al.*, 1965; Hazlewood *et al.*, 1969; Ogata, 1996; Yamada, 1998).

Ironically, such critical behavior had been evident in model studies carried out more than a half-century ago, and that is perhaps the reason why contraction was presumed to involve something akin to a phase-transition. Vanguard experiments of the era focused on suspensions of actin and myosin. Such suspensions could form a gel, the condensation of which was considered a working model of muscle contraction. The gel remained stable until ambient conditions were edged just past a critical threshold; then it contracted massively. As it contracted, the matrix folded and water was released—much the same as it is released in the polymer-gel (Fig. 14.9).

These critical behaviors carry the phase-transition's unmistakable signature. The signature is evident in the random gel (Fig. 14.9), as well as the organized lattice (Fig. 14.8). The question is the exact site of the phase-transition—the sarcomere is built of three types of filament, and in theory any/all of these filaments could mediate the transition. We consider these three filaments one at a time, exploring whether each one might undergo a molecular phase-transition.

Figure 14.8. *Critical, all or none behavior of muscle. After Hoeve* et al. *(1963) and Brandt* et al. *(1984).*

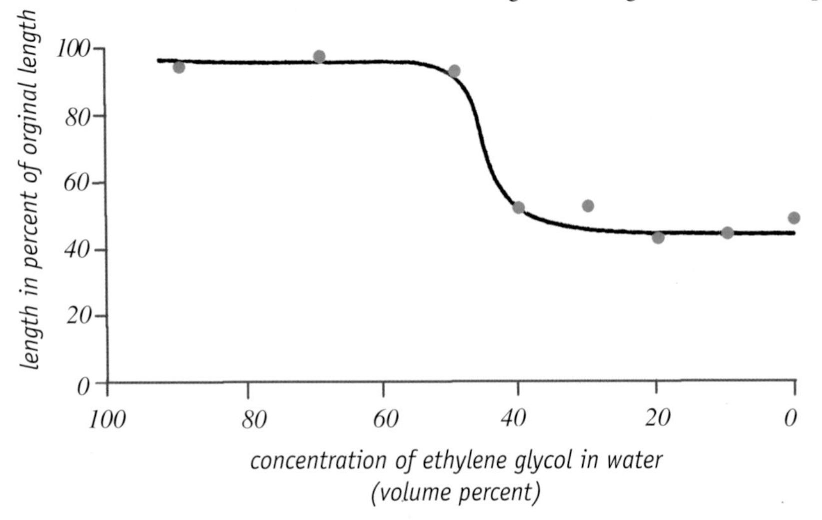

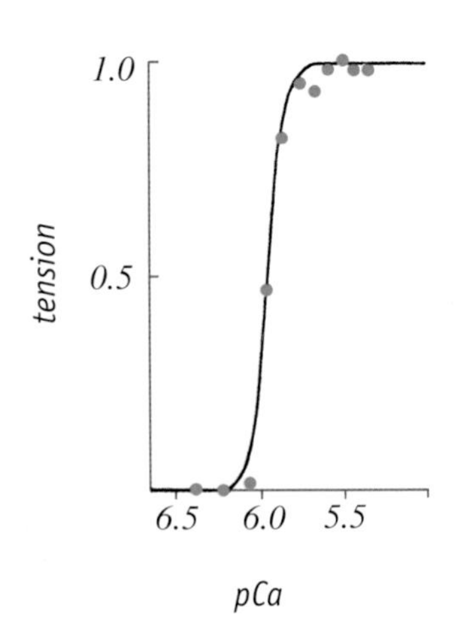

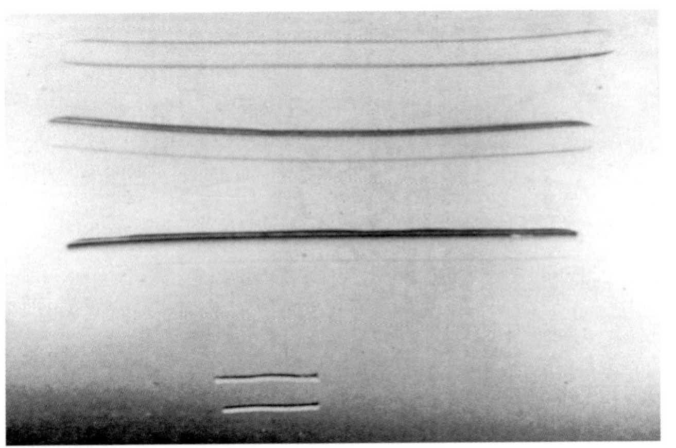

THIN FILAMENT PHASE-TRANSITION?

Thin filaments are built principally of actin. Evidence for a propagating transition along the actin filament is appreciable (*see* Chapter 11), and modern studies support a similar structural change in thin filaments (for review, *see* dos Remedios and Moens, 1995; Pollack, 1996; Schutt and Lindberg 1998). Crystallographic evidence, for example, shows that actin packing along the thin filament can occur in two configurations—a "long" configuration and a shorter one, the length difference being 10 – 15% (Schutt and Lindberg, 1992). Such length change could be driven by a calcium-induced condensation similar to the one seen in the actin-containing cytoskeleton (Chapter 8). Although the precise mechanism of this transition requires further study, the evidence above implies that whatever structural transition propagates along the actin filament propagates similarly along the thin filament.

Through such a mechanism, actin filaments can propel solutes along their length (Chapter 11). Various solutes are driven, including myosin—which is in fact required for propagation to continue. In the sarcomere, by contrast, myosin lies in a fixed framework. The question arises whether the relative motion could take place nevertheless—manifested as thin

Figure 14.9. *Contraction of the actomyosin gel triggered by an increasing concentration of ATP (left). Compare to contraction of the polyacrylamide gel (right) as the ratio of organic to non-organic solvent is edged progressively past a threshold (top to bottom). After Szent-Györgyi, 1951 (left) and Tanaka, 1981. (right).*

filament translation past this framework. Could a propagating actin transition drive the thin filament past myosins fixed on the thick filament framework?

A possible vehicle for such action is the inchworm mechanism (Chapter 8). Recall that a rod-like gel could advance itself incrementally along a saw-tooth rail by repeated cycles of folding and straightening. The thin filament could advance itself similarly. By a shortening transition that propagates, the filament could reptate over a latticework of myosin cross-links, each cycle advancing the filament incrementally toward the center of the sarcomere (Fig. 14.10).

Figure 14.10. *Reptation model in which actin filament snakes its way toward the center of the sarcomere.*

Among predictions of such a model is that function will depend critically on the integrity of actin's structure. If the thin filament's role is as central to contraction as implied, even subtle alterations in actin structure that produce little or no detectable morphological change could seriously impact contractile performance. Indeed, mechanics are profoundly affected by cleavage of a single molecular bond in actin, or by conservative substitution of a single residue (Schwyter *et al.*, 1990; Drummond

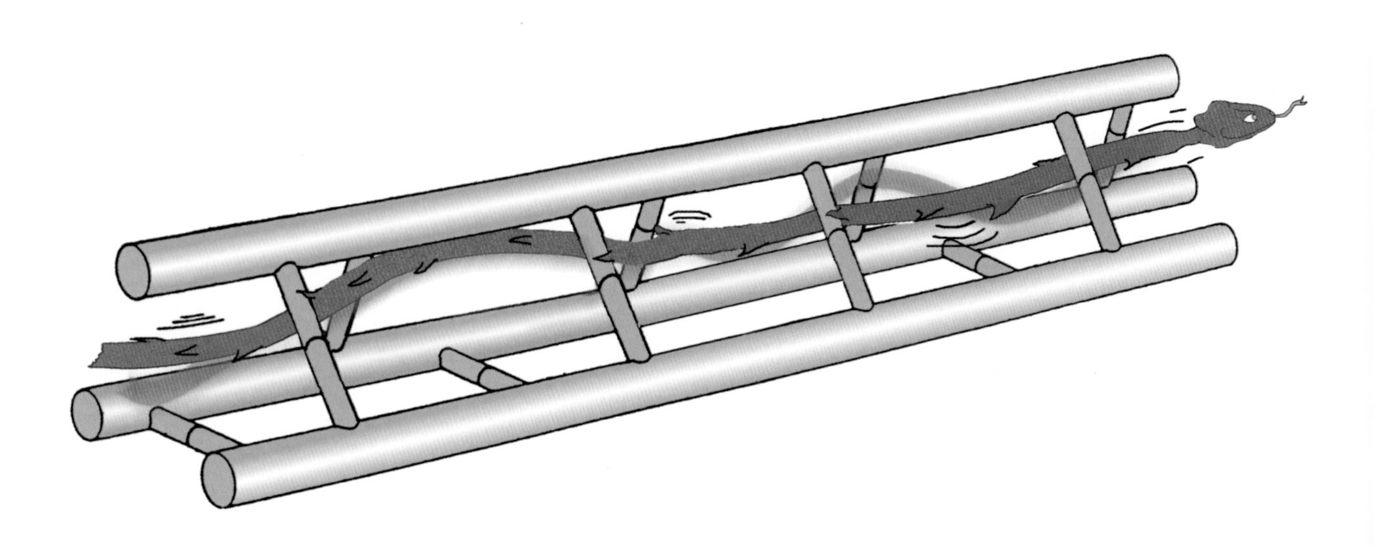

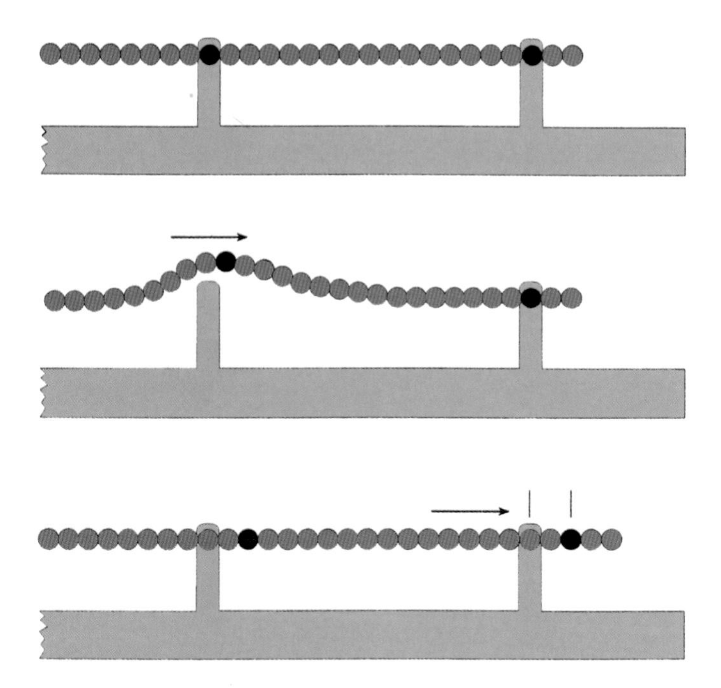

et al., 1990). And cross-linking of residues on contiguous actin monomers has a dramatic effect on motility and mechanics (Kim *et al.*, 1998). Thus, the filament does not behave as a simple tie-rod.

Figure 14.11. *Reptation model predicts that each advance of the thin filament will be an integer multiple of the actin-monomer repeat along the filament.*

Since the propagating transition mechanism is similar to the actin mechanism of Chapter 9, another prediction is sarcomeric streaming. Whether streaming occurs in the sarcomere is not yet established; however, *in vitro* systems derived from muscle actin do exhibit streaming (Tirosh and Oplatka, 1982). Streaming in the sarcomere could transport essential enzymes and substrates to their target sites. It could also mediate the shift of water that must take place as the sarcomere shortens and changes shape.

Perhaps the most critical prediction of all is the anticipated quantal advance of the thin filament (Fig. 14.11). With each propagation cycle, the filament advances by a step. The step ends as actin re-binds to myosin. Because binding to the interconnection must involve one or another actin monomer along the filament, the filament advance must be an integer multiple of the actin-repeat spacing. Translation magnitude should be quantal.

This prediction is confirmed (Fig. 14.12). Step size is indeed an integer multiple of the actin-monomer spacing. This result is seen both in the isolated molecular system (Kitamura *et al.*, 1999), and in the intact sarcomere (Blyakhman *et al.*, 1999). Agreement is striking enough to lend confidence that a mechanism closely similar to the one under consideration must play a role.

Figure 14.12. Top: *Stepwise translation of single myosin molecule along actin filaments (from Kitamura et al., 1999). Step size is ~5.4 nm, which is equal to the actin spacing along a single actin strand.* Bottom: *Sarcomere-shortening-step size measured in activated myofibrils (Yakovenko, Blyakhman and Pollack). Histogram peaks are separated by 1/2 of 5.4 nm. The half arises because the filament's two strands are axially displaced by half of 5.4 nm (see Fig. 11.1). The axial repeat along the filament is thus 1/2 of 5.4 nm, or 2.7 nm—the same as the quantal step.*

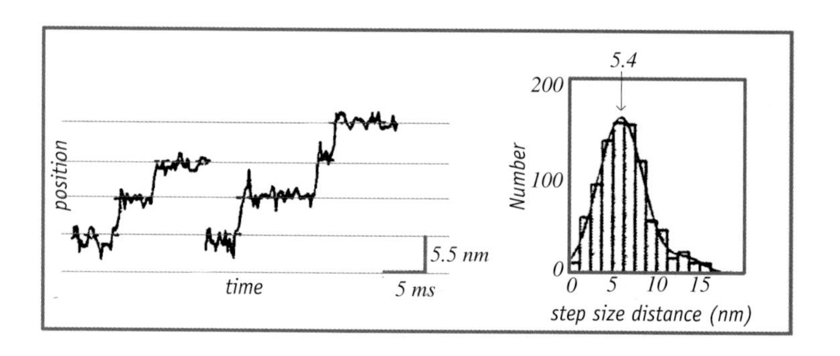

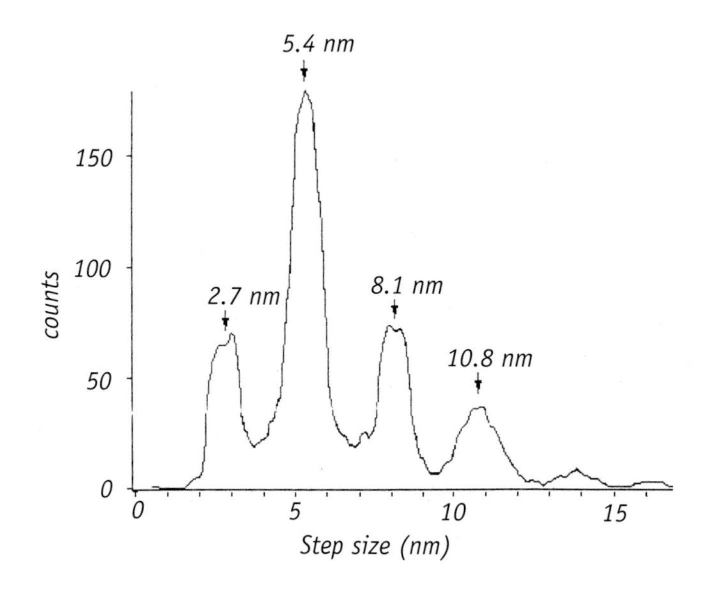

THICK FILAMENT PHASE-TRANSITION?

Although thick filaments are presumed to remain at constant length during contraction, in many instances the filaments can shorten. Evidence for shortening was alluded to earlier. The full array of evidence can be gleaned from two detailed reviews (Pollack, 1983, 1990), or better, from reading the original papers cited in those reviews. Representative examples of thick filament shortening are presented in the gallery of Figure 14.13. The extent of shortening varies with activation and loading conditions, but 20 – 30% of initial length is typical. Shortening is seen both in the optical microscope (A-bands) and in the electron microscope (thick filaments); it is seen in vertebrate and invertebrate specimens alike, and in a sizeable number of laboratories.

To appreciate how shortening might arise, one needs to consider the filament's structure. The thick filament is built of many repeats of myo-

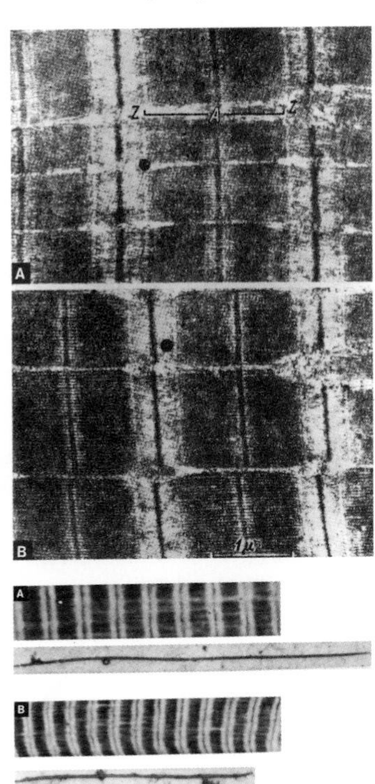

Figure 14.13. *Gallery of examples of thick filament shortening. Upper left: fibers side by side in same electron micrographic field—left not contracted, right contracted. Note difference of A-band width. All other panels: upper uncontracted, lower contracted. Lower right panel also shows thick filaments extracted from relaxed and contracted specimens. For details see Pollack (1990, p. 92 - 93).*

sin, a molecule that contains two globular heads and a rod-like tail (Fig. 14.14). The heads project out at right angles to form the radial cross-links (Fig. 14.7, above), while the rods coalesce to make up the filament backbone. If the backbone is to shorten, constituent rods must therefore play some role.

Figure 14.14. *Molecular structure of the myosin molecule.*

As it turns out, the rods themselves can shorten. The rod is an alpha-helix—a rod-like structure capable of undergoing transition to a random coil. Helix to coil is a classic phase-transition. It is triggered as hydrogen bonds that hold together the helix give way, allowing the helix to "melt" into a random structure whose equilibrium length is shorter than that of the helix. The transition is therefore tension-generating. The process resembles what happens if you put your fine wool sweater into the clothes dryer and turn the temperature too high. With the thick filament, however, the shortening event is more happily reversible.

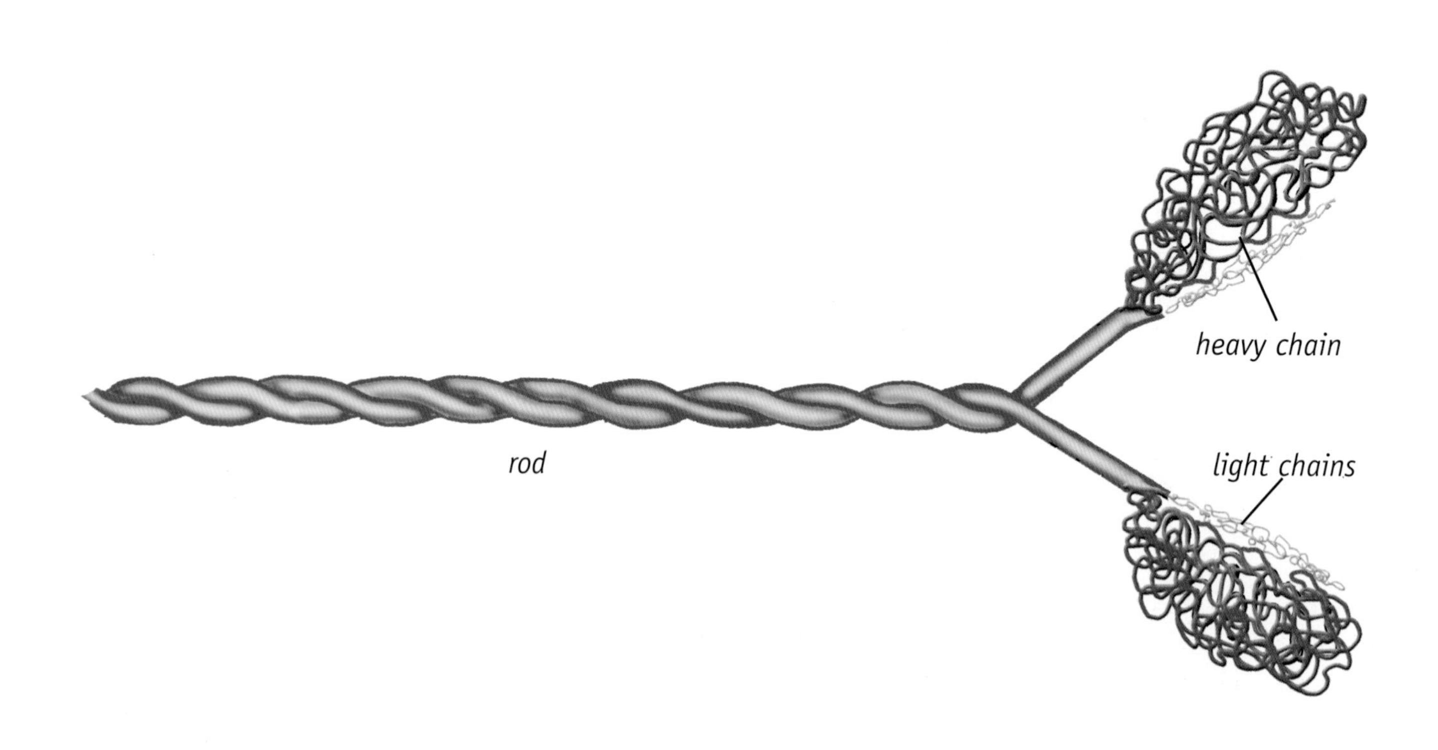

Figure 14.15. *Helix-coil transition in the myosin rod is a work-producing event.*

Details of the myosin-rod phase-transition are reasonably well known. Along the rod about 1/3 the way from head to tail is a region that is peculiarly different from the rest because of a high incidence of positively charged residues. It is this region that undergoes transition (Fig. 14.15). As it does, the rod shortens by 15 - 20 nm (Walzthöny *et al.*, 1986a,b; Walker and Trinick, 1986). The force generated by such transition is the correct magnitude to account for full muscle tension (Harrington, 1979). Thus, the thick filament shortens because constituent rods shorten.

Shortening of the filament could progress in either of two modes—all rods shorten simultaneously, or individual rods shorten in sequence, one at a time. The first option is incompatible with X-ray diffraction evidence because progressive diminution of myosin-repeat spacing is not seen during contraction. The one-molecule-at-a-time-mechanism preserves spacing constancy in filament regions that have not yet shortened, and thereby conforms to X-ray diffraction evidence.

In such a sequential model the transitions could start at both ends of the filament and progress inward; or, in a scheme that fits molecular architecture naturally (Pollack, 1990), they could begin at the filament's center-symmetry point and progress outward (Fig. 14.16). Either way, one transition enables the next, and the filament shortens stepwise. The size of the step, based on structural considerations, is anticipated to be multiples of 2.7 nm (Pollack, 1990). Therefore, both the myosin mechanism and the actin mechanism produce steps with the same size paradigm.

Figure 14.16. *Model of stepwise shortening of thick filament.*

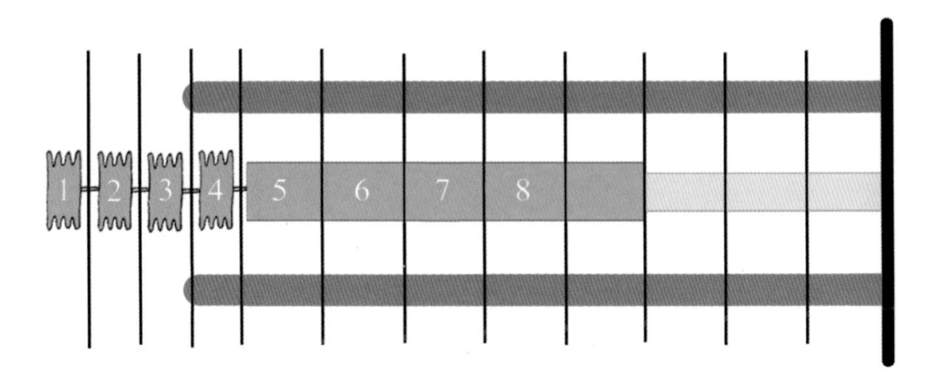

Connecting-Filament Phase-Transition?

The connecting (titin) filament is the third member of the filament set and yet another candidate for phase-transition. The filament is built of an axial string of repeating domains. Of these repeating domains, the most abundant by far is the immunoglobulin-like (Ig) domain, which has two states—folded and unfolded. The folded state is a short, compact beta-barrel, while the unfolded state is long (Fig. 14.17). Thus,

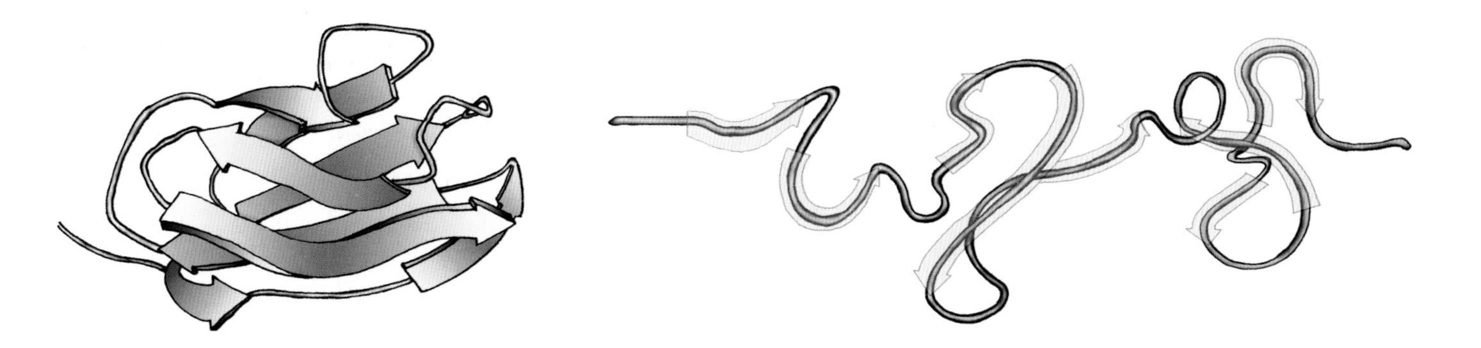

stretch imposed on the connecting filament may be realized through a series of domain unfoldings.

Figure 14.17. The Ig domain is shorter in the folded state (left) than in the unfolded state (right).

The unfolding paradigm has been confirmed by stretching sub-molecular constructs consisting of a string of Ig domains (Rief *et al.*, 1997). Smoothly imposed stretch produced a sawtooth tension response, the number of teeth never exceeding the number of domains (Fig. 14.18A); this implied a succession of domain unfoldings. The domain-unfolding model was confirmed in another single titin-molecule experiment (Tskhovrebova *et al.*, 1997). Here the molecule was stretched abruptly and held. Stretch increases tension, which shifts the equilibrium toward unfolded. If domains proceed to unfold at constant specimen length, tension should decay (stress-relaxation), and the expected decay was observed. However, the decay occurred in stepwise fashion, again implying a series of unfoldings (Fig. 14.18B).

With this model, the connecting filament stretches by a succession of unfoldings, and shortens by a succession of foldings. Foldings appear to take appreciable time, so an alternative, perhaps entropic, shortening mechanism may apply as well—but the capacity for domain by domain behavior is clearly shown. With each folding event a packet of potential energy is released, which can do work. Hence, the stretched connecting filament is much like an extended spring that can ultimately shorten domain by domain—in steps.

The connecting filament, however, may not consist of titin alone. Evidence that the connecting filament may also contain strands of tropomyosin is reviewed in Chapter 10 of my muscle book (Pollack, 1990), and I

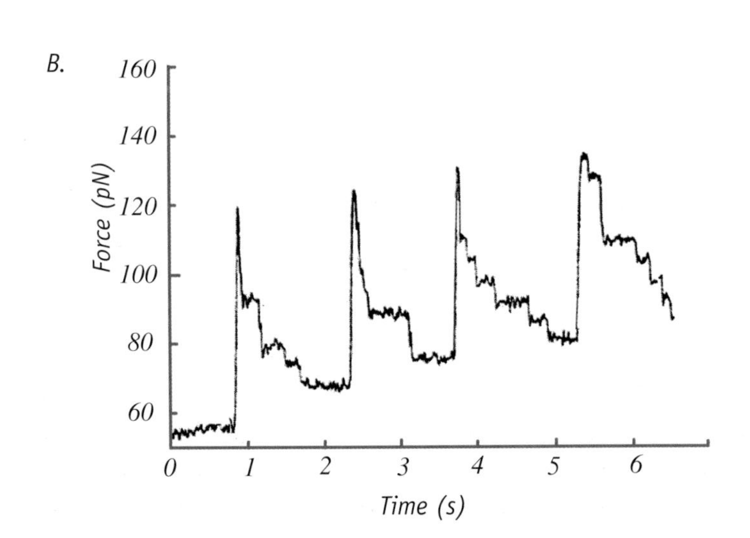

Figure 14.18. A: *Response to ramp stretch of single titin filament(Rief* et al., *1997).* B: *Response to a succession of lengthen-and-hold steps (Tskhovrebova* et al., *1997).*

believe this evidence is substantial. Tropomyosin molecules have surface charges that repeat axially every ~2.2 nm (Katayama and Nonomura, 1979), and it is relevant that when the connecting filament in single myofibrils is stretched or released, the step size is 2.3 nm (Blyakhman *et al.*, 1999). Thus, connecting-filament steps could be associated with sliding/melting of tropomyosin molecules, just as thick filament shortening steps seem to be associated with sliding/melting of myosin molecules (Pollack, 1990).

Fold-unfold events described in this section refer to specimens stretched or released in the unactivated state. On the other hand, factors associated with activation such as free calcium increase do commonly impact fold-unfold equilibria, and if they were to shift the equilibrium toward folding, the connecting filament would, in effect, contract. Connecting filaments might then contribute to active tension and shortening. At present there are scant data to indicate whether this could be the case—except perhaps for the classic study by Endo (1972), who showed that as calcium concentration was edged just above the resting level, resting tension shifted upward at all lengths. Whether indeed activatable, or merely inert, extended connecting filaments inevitably contribute a retractive force, so a modest contribution to sarcomere shortening is inevitable.

The Integrated Machine

It appears, then, that the contractile machine's vaunted strength is drawn from contributions from all three of its filaments—thick, thin and connecting. All filaments shorten. Shortening of filaments shortens the sarcomere.

This mechanism follows along the general lines advanced by Albert Szent-Györgyi, Aharon Katchalsky, Paul Flory and other pioneers of the pre-sliding-filament era. At that time it had been clear that muscle proteins shorten, but mechanistic details were lacking. It now seems that shortening is based in all three filaments, and that in each case the likely mechanism is a phase-transition: monomeric condensation, helix-coil transition, or fold-unfold transition. In the thin filament the transition propagates repeatedly, each cycle advancing the filament incrementally. In thick and connecting filaments local shortenings are cumulative, and translate directly into sarcomere shortening.

Because all three filaments shorten in discrete increments, the sarcomere is anticipated to contract in steps. This is confirmed. Stepwise shortening is observed in single sarcomeres of single myofibrils, as well as in specimens of variably larger scale up to whole cells (Pollack *et al.*, 1977; Delay *et al.*, 1981; Granzier *et al.*, 1987; Yang *et al.*, 1998; Blyakhman *et al.*, 1999). Such observations not only confirm the discreteness of the underlying process, but also the high cooperativity characteristic of phase-transitions in general.

The three filaments' contributions integrate in a complementary way. In the unactivated state, the sole mediator of sarcomere-length change is the connecting filament—other filaments do not change length. As activation crosses a threshold, active contraction may begin with the thin filament, which inches itself, step-by-step, toward the center of the sarcomere. But the thin filament mechanism has its limitations: In the same way that the inchworm's upward progress can be thwarted by an appended weight, thin filament translation is expected to terminate if the load becomes too high. From computations of the tension developed by a single activated thin filament (Ishijima *et al.*, 1991; Morel, 1991), the critical load appears to be about 1/3 maximal active tension. Beyond this load the thin filament cannot advance, and merely clings to the myosin cross-links.

The remaining operative agent at these high loads is the thick filament. Thick filament shortening is the sarcomere's heavy weapon—producing full active tension, and contracting against formidable loads. Such work-producing capacity is impressive and perhaps unique in biology.

Because the operational ranges of the three processes are roughly complementary, the domain of sarcomeric contraction is extensive. This confers versatility on the system. Where ranges do overlap, summation enhances contractile power. Thus, the integrated machine is both more versatile and more powerful than primitive contractile organelles such as the spasmoneme, which contain only a single shortening protein instead of three (Moriyama *et al.*, 1996). Muscle is an evolved organelle that masterfully exploits all practicable avenues of work production.

***

With this chapter we conclude the discussion of cellular processes—not because all processes have been exhausted but because this volume needs to draw to a close. For each of these process, we found evidence that one or another type of phase-transition plays a role. Recognition of the phase-transition's relevance is not entirely novel or original—in several instances phase-transitions have been pursued by other groups, who have gone on to gather substantial supporting evidence. But the collective force of this evidence has remained weak because advocacy has been organelle-specific and therefore scattered throughout the literature.

The chapters have shown, however, that the phase-transition may be a common mechanism among many organelles. It provides straightforward explanations for processes otherwise difficult to explain logically and coherently. It is simple in its decisiveness—on or off; and it is powerful in its effectiveness—a subtle environmental change triggers a massive response. To this author, it seems imbued with the stamp of understated elegance expected of Mother Nature's paradigms.

SECTION V

TYING LOOSE ENDS

This section considers issues common to the preceding chapters. It begins with cellular evolution and energetics, and then moves on to explore the underlying themes that integrate the material of the book.

Implosion of the Seattle Kingdome

CHAPTER 15

ENERGY

What powers the phase-transition?

A common feature among processes considered so far is the dump of potential energy. In the secretory vesicle, a compressed polymeric spring releases stored energy through uncoiling. In the muscle thick filament, an extended alpha-helical rod melts into a low-energy random-coil. In the cytokinetic contractile ring, an expanded, hydrated actin-filament matrix delivers stored energy through condensation. In each case, work is done by using energy that had somehow been stored in the organelle prior to action.

Storage of energy makes sense. Sailboats may be fun, but the likelihood of reaching your destination is higher if you circumvent the vagaries of wind by filling your boat's motor with fuel. Pre-mobilizing the energy is often the more effective route, and certainly more reliable. The challenge is to restore the potential energy surrendered during each working act, and the focus of this chapter is on the mechanism by which energy restoration is achieved.

To approach this question, we look for clues at the origin of life. Early life forms required energy much the same as contemporary life forms, but there were few frills. The underlying energetic strategies should therefore be more transparent than in modern-day life forms, where complexity can obscure fundamentals. If today's sophisticated energetic pathways evolved out of simple strategies used early on, then identifying these early strategies is a good place to begin.

Figure 15.1 (opposite). *The Seattle Kingdome, on the threshold of collapse in March 2000, triggered by small charges planted at critical sites throughout the structure. With permission, Seattle Times.*

ORDER

Before retreating a few billion years, it is worth recapitulating a basic energetic principle. According to the second law of thermodynamics, energy comes in two forms. One is chemical energy and the other is order. The first is familiar. The second is less familiar and therefore potentially intimidating. Ordering requires input energy, which can then be returned as the order gives way. Building the Seattle Kingdome required prodigious amounts of energy, much of which was returned in the form of mechanical energy and heat as the obsolete stadium's ordered structure was imploded into chaos (Fig. 15.1).

To illustrate the role of biological ordering consider the muscle cell (Chapter 14). Initially, water is ordered around thick and thin filaments, which lie in their extended, high potential energy states. Order then gives way to disorder: As filaments contract and surrounding water is released into disorder, potential energy is surrendered in the form of mechanical work and heat. Then the system must be reprimed—filaments need to be re-lengthened and water needs to be restructured. Re-ordering requires energy, much the same as rebuilding the stadium.

The energy associated with ordering and disordering in the cell can be substantial. Such effects depend on the number of participating molecules, and in the case of water the number is enormous—at least 10,000 times that of all other intracellular molecules. Hence, even if the energetic contribution of each molecule were modest, the summed energy deliverable through water disordering—not to mention protein disordering—could be substantial.

That ordering could be a feature of central significance to biology is not surprising since thermodynamics preceded life. The mere act of creating a pre-cellular entity out of a sea of dispersed solutes is an example of ordering. Ordering is also reflected in the cell's internal structure: monomers align into polymers, polymers coalesce into matrices, and matrices assemble into organelles. In each case, dispersed elements coalesce into larger-scale ordered structures. Structural order is, in fact, life's hallmark—as the cell withers and dies, so does the order. While all of this may seem obvious, it carries an implication of some significance: Order, the very hallmark of the living state, is in fact a cloaked mechanism for the storage of potential energy. At least some of the cell's energy may be stored in the form of order.

A challenge, then, is to determine how energy may be used to create structural order, and this question will be approached by looking at the roots of the tree of life, where the underlying energetic strategies originated. The origin-of-life field is a discipline unto itself and will be considered only briefly here. Details can be found in classical reviews (*e.g.*, Bernal, 1961; Oparin, 1964, 1971) and in monographs such the provocative 1995 book by de Duve entitled "Vital Dust." Our interest in this fascinating topic is restricted to ferreting out basic strategies by which energy is used to create order.

ORIGINS

Life began when the earth was relatively young—at least 3.8 billion years ago according to fossil records. A water origin is presumed, not only because water is critical for life, but because many primitive prokaryotes today can be found in water, near deep hydrothermal vents that belch hot gasses containing hydrogen, ammonia and methane.

Atmospheric conditions in the nascent earth differed from those of today. For starters, the atmosphere contained little or no oxygen. This is inferred from ancient iron deposits, which show no sign of oxidation. The exact atmospheric content is debated, but volcanic "reducing" gasses such as ammonia, hydrogen, and methane, as well as carbon dioxide appeared to be present. Thus, conditions on the earth's surface may have resembled those near the hydrothermal

Figure 15.2. *Demonstration that amino acids can be created under conditions thought to resemble those on the primitive earth. The following procedure could do the trick: place inorganic salts in a flask; apply heat; collect the steam and infuse a mixture of hydrogen, ammonia and methane; discharge electrical sparks into this gas to simulate lightning; condense the gasses to lower temperature, and voila!—a brew of more than a dozen amino acids.*

vents, and it is therefore possible, as Darwin first suggested, that life arose in the exposed shallows. With belching volcanoes, incessant lightning, and extreme radiation through a thin, putrid atmosphere, the primitive earth was hardly a bucolic scene; it more closely resembled the medieval concept of hell.

There is general agreement that original life forms, the so-called protocells or protobionts, must have been gels—blobs of hydrated polymer that remained distinct from their aqueous environment. Evidence for such accumulative condensation was produced almost a century ago when Bungenberg de Jong (1932) showed that even dilute solutions of polymers, when shaken, coalesced into droplets—then called coascervates—in which the organic matter became highly concentrated. The polymer concentration in such droplets could exceed the concentration in the surrounding bath by as much as 10,000 times. When placed in certain dyes, the droplets became progressively more colored, the intensity often many times exceeding that of the surrounding solution. Thus, gel droplets had the capacity to concentrate certain solutes, as the cell concentrates potassium.

Those observations lent credence to the emerging view that the protobiont may have been a simple polymer gel. But they did not address the polymers' nature—whether for example the polymers might have been built of amino acids. This question was considered in classical experiments by Stanley Miller and Harold Urey in the 1950s, which showed that under environmental conditions thought to exist in the nascent earth, amino acids could be created abiotically (Fig. 15.2); no cells were required. Thus, pre-cellular gels could well have contained protein-like polymers.

Whether any such amino acids and/or other organic solutes might polymerize and gel depends on concentration and environment. On both counts, estuaries are particularly opportune. Estuarine clay surfaces are known to drive polymerization—their surfaces are often replete with prokaryotic fossils. Estuaries also concentrate organic matter. Gels and other organic matter frequently concentrate on the water's surface rather than in the bulk, as evidenced by the gel scums seen to cover salt marshes and estuarine shallows. Concentrations increase further as waves break upon the shores and leave their residue—witness the appallingly high concentrations of beach oil left by fractured tankers. Thus, estuaries should be ripe, even today, for the emergence of simple protobionts. Nothing magical is required.

The earliest of these protobionts could not have evolved in any directional manner until genetics entered the picture. Protobionts could certainly grow—they could do so by annealing with other protobionts or by absorbing organic matter. And they could likewise break up from environmental insults. But such dynamics were likely to remain helter-skelter until primitive genetics emerged, and according to some, this happened early on with simple nucleic acids dissolved in the sea. With primitive genetics, evolution could assume some memory, and hence directionality. Features most critical for survival could be perpetuated.

Among the features critical for survival, one of the most central must be energy supply. Securing adequate nourishment is a struggle that persists to this day, and the struggles of early life are not likely to have been very much different. What differed was the complexity of tasks. The chief—perhaps only—task of the protobiont was to endure. In the face of nature's unrelenting proclivity to disorder, a protobiont without an energy supply would sooner or later dissipate as assuredly as the Kingdome without upkeep would ultimately crumble. There is no way around it: entropy is time's arrow. Maintaining order requires energy. Even for protobionts, survival depended on a supply of energy.

The energy for ordering came (as it does today) from the sun. It was tapped through environmental sources such as heat, tides, waves, lightning, and evaporation. Sun-induced evaporation, for example, could concentrate solutes, shifting the equilibrium toward polymerization and thereby promoting order. But such energy sources were largely out of the protobiont's control. Survival rested on the whims of the environment. Those protobionts that could derive energy from more reliable sources would certainly have been at a survival advantage.

ATP

Enter ATP. Apart from its vaunted capacity to produce energy through hydrolysis, two additional features conspire to make ATP a particularly facile source of energy. One is high charge concentration. The close proximity of its three phosphates confers an unusually high concentration of negative charge on the molecule, creating a focal point of intense charge. The second is high affinity. ATP's affinity for myosin for example is 10^{11} to 10^{13} M^{-1} (Trentham *et al.*, 1976), which implies that the association is practically irreversible. ATP in solution would be quickly sopped up and held by protein surfaces, which would then have acquired foci of intense negative charge.

How exactly free energy is extracted from this molecule to do work has remained surprisingly unclear, but a potentially simple route is to induce order—to drive contiguous proteins into their extended state and vicinal water into its structured state. Charge-induced protein extension follows from the classical phenomenon of induction, set forth a century ago by G. N. Lewis, and extended to larger scale structures by Ling (1962). Ling's association-induction hypothesis posits that certain ligands can exert long range effects—similar to the allosteric effects well known to occur in proteins. Charged ligands such as ATP in particular shift the electron cloud of the adjacent atom, which in turn shifts the cloud of the subsequent atom, *etc.* Through such sequential action, binding of ATP induces long-range protein extension, which is then reinforced by water structuring. The protein-length change will be extensive in negatively charged proteins because ATP is also negatively charged, inducing locally strong repulsive forces that drive the extension.

An example of such long-range action is in hemoglobin, the major red blood cell protein that carries oxygen. Hemoglobin's affinity for oxygen is dramatically reduced by ATP (Chanutin and Hermann, 1969; Klinger *et al.*, 1971). This action apparently results from ATP binding alone, because hemoglobin does not hydrolyze ATP. However, the ATP-binding site is remote from the four oxygen-binding sites. Thus, the effect of ATP binding must propagate, by a mechanism more generally categorized as allosteric. ATP evidently induces some kind of cooperative phase-transition that encompasses much or all of the molecule.

Another example is in muscle. Without ATP, muscle is in a state of *rigor-mortis*, in which constituent proteins are cross-linked to one another and contracted. ATP breaks actin-myosin cross-links and restores the proteins to their extended state. The site of ATP binding is remote from the sites of cross-linking and contraction, so the influence of ATP must again be long-range. The cut-end experiment introduced in Chapter 6 implies that the ATP state is one of high potential energy: When the cut muscle is exposed for extended periods to salt solutions, only those regions containing unhydrolyzed ATP show selective ion partitioning; regions devoid of ATP do not (Fig. 15.3). Thus, ATP drives the system to its high potential energy state.

These long-range actions of ATP are treated in depth by Ling (1992), who emphasizes binding *per se* as the event of primary significance. Although hydrolysis has been the accepted energy-supply vehicle, there is increasing focus on binding (*e.g.*, Visscher *et al.* 1999). Because ATP binding is so tight, the energy released is substantial—about twice that obtained from hydrolysis.

The actions of ATP are to some extent mimicked by simpler organic phosphates. Binding of phosphate is known to induce many of the same functional processes as ATP binding. Phosphates dissolved in the primeval soup may therefore have preceded ATP as an energy source. On the other hand, ATP itself would have been a notable addition because of the substantially larger free energy that can be extracted as a result of its much higher protein affinity and charge concentration.

The initial source of ATP was presumably exogenous. Under the then-prevailing atmospheric conditions, experiments have shown that ATP can be created abiotically in much the same way that amino acids and nucleic acids could be created. But such abiotic sources were unlikely to have produced ATP in very high concentrations, and eventually the protobiont developed its own means of production. Even the new endogenous ve-

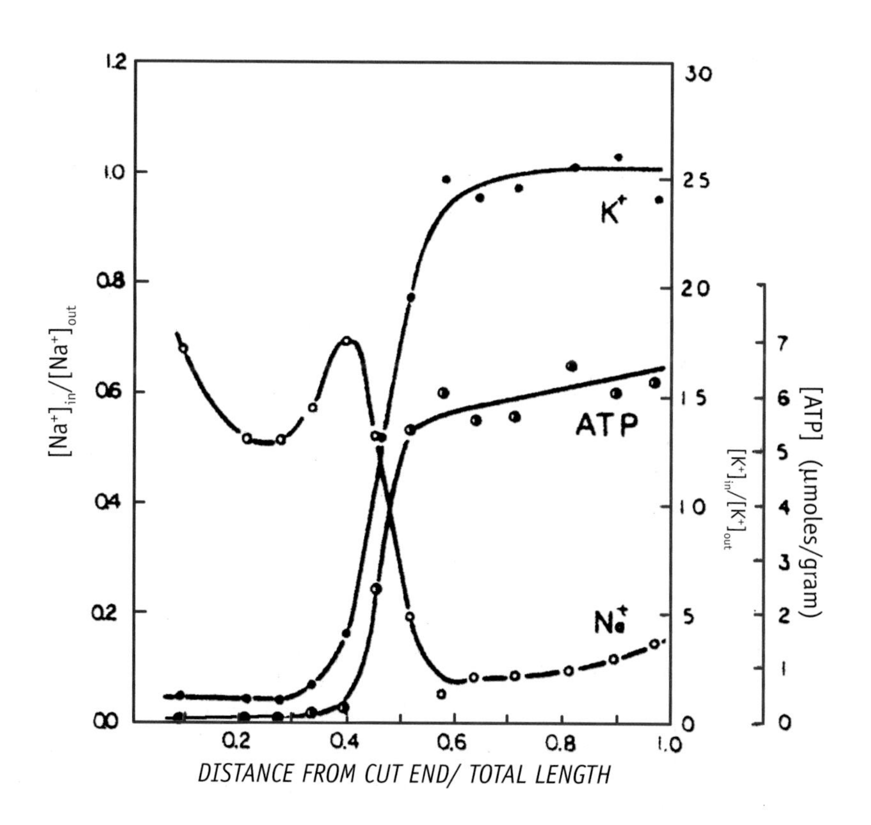

Figure 15.3. *Effect of ATP on ion concentrations in cut-end experiment (after Ling, 1992). Cut muscle was exposed to physiological salt solution at 25°C and examined after three days. ATP was depleted in damaged zone near the cut (left). In more remote regions (right), high ATP concentration correlates with potassium accumulation and sodium exclusion.*

hicle must have been of limited efficacy, however, for oxygen was not yet available in the atmosphere; cells had to rely on low-yield anaerobic processes that are the forerunners of today's glycolysis. Nevertheless, the emergence of endogenous ATP-producing machinery set the stage for the major evolutionary developments that would transform simple protobiont gels into prokaryotes and eventually into eukaryotes.

In sum, the supply of energy for the protobiont came ultimately from the sun. The sun's energy produced ATP—abiotically at first and later through biotic machinery. The energy of ATP in turn was used to confer potential energy, possibly through a protein extension / water structuring mechanism. Such energy helped keep the system stabilized in its ordered, high potential energy state. Essentially, ATP kept the system "alive."

Mitochondria and Aerobic Metabolism

The contour of this energetic landscape was dramatically altered by the advent of atmospheric oxygen. As primitive cells began releasing photosynthetic oxygen to the atmosphere, and as volcanic activity subsided, the atmosphere changed: reducing gasses were progressively replaced by oxygen. The response to this assault on ancient conventionality was a new metabolism that could capitalize on oxygen's presence. Although the anaerobic route has remained in reserve for exigencies of oxygen deprivation, the aerobic route sprung up in certain unicellular organisms including bacteria, and eventually surpassed the anaerobic route in efficacy. Aerobic prokaryotes could now flourish. Plentiful energy was available not only for maintaining order, but also in supporting the drive toward specialization.

But something happened along the way that was to make news. By some quirk, a primitive bacterial cell that had begun employing the new oxygen-based metabolic system invaded a larger cell that had not. The invader remained—and there began the story of the mitochondrion lodged within the host eukaryote cell. On the basis of genomic and other evidence, bacterial cells are known to have evolved into mitochondria.

Because ATP is produced on the bacterial surface, the invader must have been sheathed in ATP, and hence in negative charge. For the host cell, internalization must therefore have been a jarring experience, perhaps akin to your swallowing of a negatively charged watermelon. But the host turned this unpleasant misadventure to advantage: the invader's high surface charge could order host-cell water in the very same way that charged surfaces ordinarily order water (Chapter 4); and water ordering could induce extension of nearby host-cell proteins. So long as oxygen could diffuse through the host cell to fuel continual ATP production, the invader could be exploited to confer potential energy on the host (Fig. 15.4).

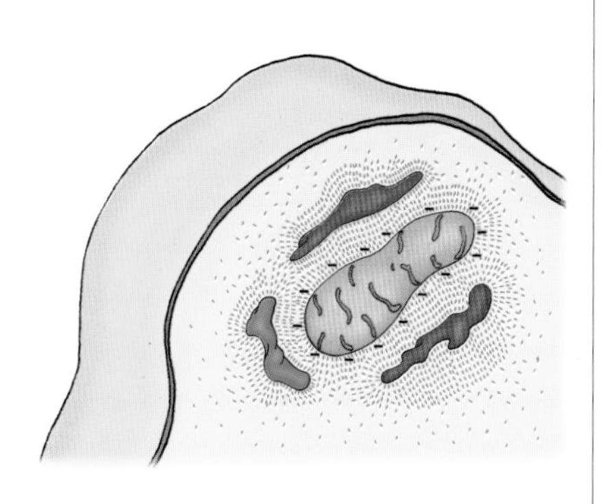

Figure 15.4. *Surface charge on the internalized bacterial cell organizes host cell water and extends host cell proteins.*

Such an ordering scenario is given support by fact that water in and around mitochondria appears to be structured. Water structuring inside the mitochondrion is confirmed by NMR studies (López-Beltrán *et al.*, 1996). And zones just outside the mitochondria often contain large clear zones (Fig. 15.5), implicative of solute-free water. Thus, mitochondrial surface charge, continually replenished by ATP synthesis, could structure vicinal water; and structured water could in turn drive nearby proteins to their extended, high potential energy state. This would promote and sustain order.

Surface-ordering mechanisms are not unique to biology. In the field of materials science for example, "surface energy" is a well-recognized vehicle for impacting nearby solvent molecules, which can be transformed to do work (*e.g.*, Fig. 11.12). The same applies in a capillary tube, where surface interaction can draw a column of fluid upward to considerable heights. The scenario proposed here is largely the same, except that a more fundamental energy source maintains the surface condition so that energy can continue to be supplied.

Shifting of the ordering site from the cell's surface to the mitochondrial surface must have changed the equation of life—for mitochondria could now evolve single-mindedly to enhance energy production. And indeed they did. The rate of energy production was enhanced through development of infoldings called cristae, which enlarged the productive surface area; and mitochondria also multiplied in number, and adroitly positioned themselves next to the most prodigious consumers of energy such as axonemes and myofibrils. With mitochondria increasingly effective at supplying energy, host cells could afford to turn their attention toward other developments such as specialization. The task of energizing was left to the mitochondrion—the guest that stayed.

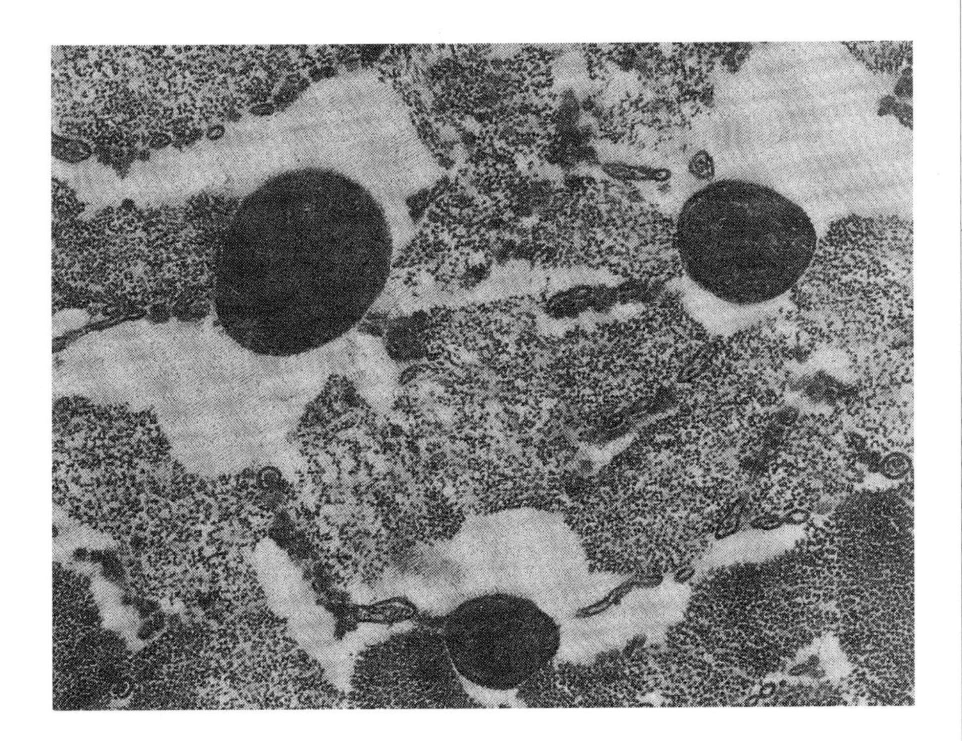

Figure 15.5. *Quickly frozen samples of frog muscle, illustrating extensive "clear" (light) zones around (oval) mitochondria. From Trombitás* et al. *(1993).*

CYCLING OF ATP AND CHARGE

The scenario described above implies an effect akin to a magnet's ordering of iron filings. For early host cells, such long-range action could well have been an energetic mechanism of some significance. But ATP did not remain locked on the invader's surface indefinitely—mitochondrial ATP is known to translocate to the cell's proteins. Once bound there, it is split into ADP and Pi, which translocate back to the mitochondria for recycling into ATP. This shuttling mechanism evidently supplanted, or at least augmented, the remote energy-supply mechanism.

Nevertheless, the energy-extraction principle could remain much the same as the one just described—ATP-induced extension of proteins, with consequent structuring of vicinal water. The energy released as ATP binds to the protein would drive the protein-water complex into to its ordered, high potential energy state, while ATP is hydrolyzed into ADP and Pi for recycling. The difference is that the source of ATP is now endogenous and more ample; and the action is local. Hence, the shuttling mechanism can be considered an advance over the more remote, indirect mechanism.

Functionality of any such protein-based ordering mechanism depends on the ability of ATP to translocate from mitochondria to proteins, and this should be possible during organelle action (Fig. 15.6). During this time, vicinal water destructures, which means that diffusion down con-

- ATP
- ADP

initial condition

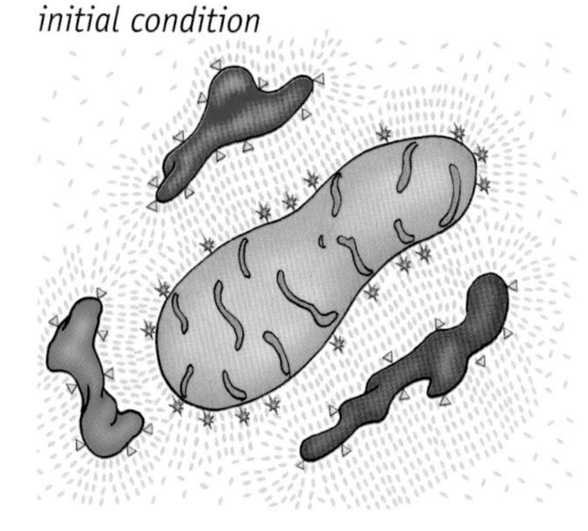

action

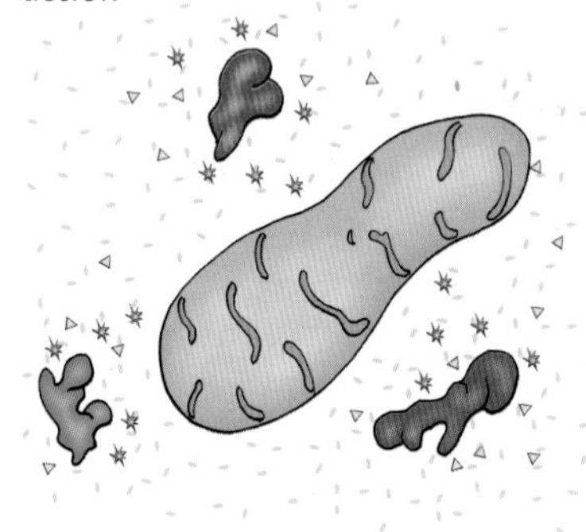

binding

splitting and initial recovery

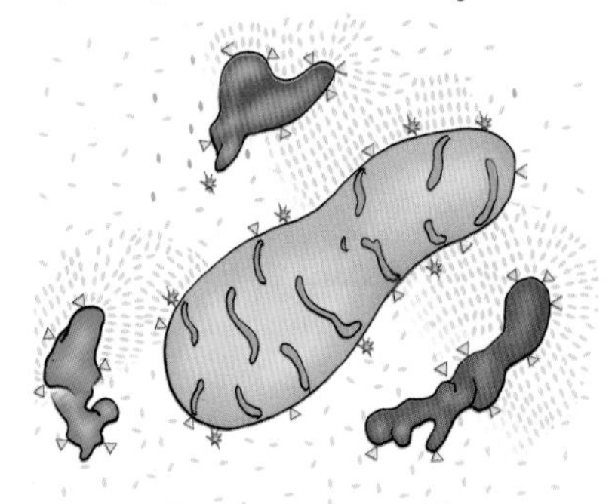

final condition

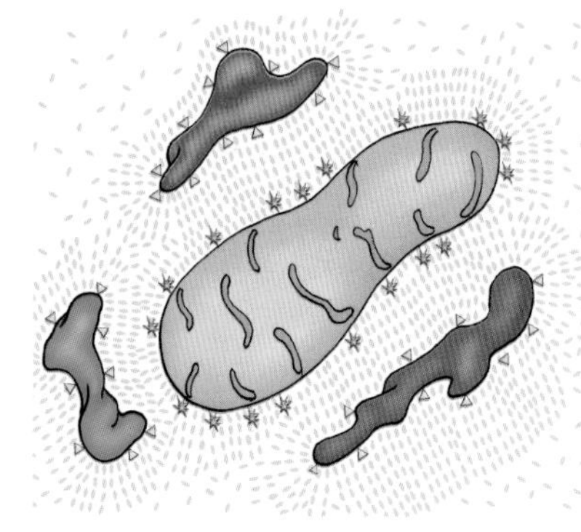

centration gradients is possible. Mitochondrially generated ATP will thus diffuse toward nearby organelles—where it easily replaces ADP because of its >100,000-times higher protein affinity (Lowey and Luck, 1969; Marsh *et al.*, 1977). The released ADP can diffuse down its concentration gradient and back toward the mitochondria, for recharging into ATP. All of these exchanges occur as the organelle acts. Recharge is therefore rapid, if not immediate.

The ATP energizing mechanism just described shares many features with phosphorylation—a universal mechanism for protein activation. In both scenarios, added phosphate mediates a change of protein conformation. The former is facilitated by enzymes (ATPase) that may be part of the protein itself, whereas the latter is facilitated by enzymes (kinase) that are often distinct entities. Otherwise, these two protein-activation vehicles appear to operate rather similarly.

A notable feature of the ATP-cycling mechanism is the concomitant cycling of charge. With each ATP that binds, the protein gains negative charge. Since charge cannot build indefinitely, the protein must eventually discharge, and this occurs as the hydrolytic products ADP and Pi dissociate, returning the protein to its discharged state. Protein is thus charged to impart potential energy, and discharged during action. Given the charge requirement, it is no surprise that an electron-transport chain is employed for ATP generation. Electrons are transferred along this chain in stages, ultimately for the synthesis of negatively charged ATP—whose charge is then transferred to the protein. The protein is literally "charged" with energy.

Figure 15.6. *Cycling of ATP and energy restoration by protein extension and water structuring.*

The continual flow of charge makes the energy-supply process seem very much like that of a battery. The mitochondrial battery (obtaining its energy through metabolism), supplies charge to the cytoplasmic matrix, which uses the charge to ready itself for work. The system operates much like your laptop computer or cordless drill, which require charging prior to each use. Again, the engineered system and the natural system operate by much the same principle.

CONCLUSION

Out of these considerations has come a central energetic hypothesis—the energy for cellular action arises ultimately from charge. The charge can reside on the mitochondrial surface or on the protein surface: charge on the mitochondrial surface orders vicinal water, which drives proteins to their extended state; charge on the protein surface extends the protein, which orders vicinal water. Either way, the protein-water matrix gains potential energy.

This potential energy can be used to perform work, which is delivered during the phase-transition. The system behaves much like a row of dominoes: Energy must first be invested to order the dominoes (water), before they can be toppled with the tap of a finger. The falling dominos can perform work as the system's entropy increases.

The detailed accounting of energy transfer in this two-stage process is complex. Among the many quantitative factors that need to be considered are: the decreasing separation of charge as ATP nears an oppositely charged site; the decrease in entropy as ATP is bound and immobilized; the change of protein-hydration energy; the hydrolysis of ATP; the efficiency of the process.... The list goes on.

If the process is to repeat, however, the principle of conservation of energy comes to the rescue. In the end, the only entries in the long list of energetic transactions that will not cancel out are: (a) the splitting of ATP (~ 7 kCal/mol); and (b) the heat liberated by the process (efficiency of < 100%). Thus, a cap can be specified on the amount of energy available per ATP molecule that can be converted into ordering, and subsequently to work.

This cyclic process can be tracked in Figure 15.6. With the triggering of action (second panel), ATP molecules from the mitochondria bind to protein (middle panel). The protein's main job is to transduce metabolic energy into water structure—the "setting up of dominoes." Thus, binding and splitting of ATP induce the extended protein conformation (last panel), which encourages the structuring of water to the tune of a few kCal per mol of ATP. The system is now primed (top panel). Then the dominoes are tapped into action, and the increase in entropy is used to do work.

Some features of the proposed energetic mechanism are shared with textbook mechanisms, while others are not. According to current views, the energy supply currency is the same, ATP. ATP is known to act directly on proteins, binding and hydrolyzing into ADP and Pi and thereby supplying energy. Here, the energy-supply mechanism of ATP is given specificity—ordering of water and extension of proteins. The mechanism emphasizes the setting up of potential energy ahead of time; and it emphasizes maintenance of potential energy through an intermediate step in which mitochondrial-surface ATP promotes water ordering to ensure that juxtaposed organelles stand poised for their next act.

In this sense, the cell is very much an entropy machine, preparing for action by building and maintaining order, and then releasing energy through disordering. This is not to suggest that enthalpy plays no role; free energy is obviously derived from some combination of enthalpic and entropic contributions. But entropy has been largely ignored in the past, and this is the reason it is emphasized here. The fact that cells are entities of exquisite order is not at all fortuitous. It reflects a huge capacity for energy storage—a feature that may be vital for life.

Available in local stores. Rehydrate to obtain pets.

CHAPTER 16

A NEW PARADIGM FOR CELL FUNCTION

As our journey of exploration draws toward a conclusion it seems opportune to step back and reflect, for the goal-oriented approach we have taken has left little time for considering the picture's broader aspects. It is much like life itself. As Mitch Albom relates in his classic book *Tuesdays with Morrie*, rarely do opportunities arise in our goal-oriented lives to reflect on what really matters—only as the end looms do the most meaningful perspectives often emerge.

So we now reflect on what this book's chapters have wrought. We consider themes that integrate, as well as themes that project beyond the functional edifice we have constructed.

FOUNDATIONS

Our attention was initially sparked by anomaly. Current wisdom anticipates that breaches of cell-membrane integrity ought to incapacitate the cell but a variety of observations implied otherwise: the cell could survive even major physical disruption, notwithstanding evidence in at least several cases that the breached membrane did not reseal. Barrier integrity seemed inexplicably unimportant.

We came to realize that the barrier concept had been founded on the view that the cytoplasm was an aqueous solution. A barrier was needed to keep cellular contents from spilling out. Cell scientists now recognize that the cytoplasm is not an aqueous solution after all but a crowded

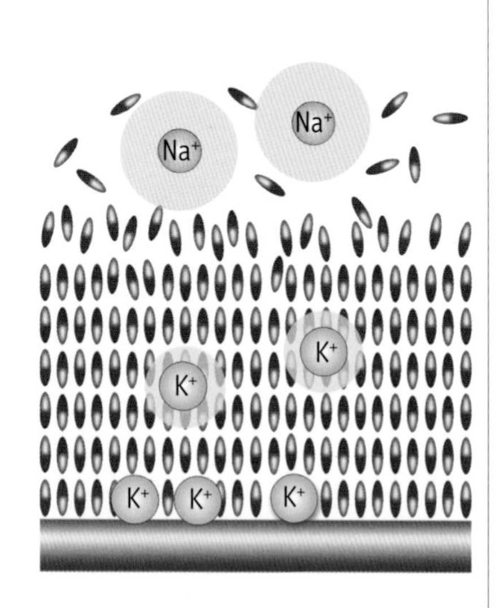

Figure 16.1. *Hydrated ion size is a critical determinant of solubility in structured water and of affinity for protein surfaces.*

polymeric matrix that can hold itself together even when the cell's membrane has been removed. But the 19th century view spawned mechanisms that endure to this day—free diffusion, for example, remains the presumed motor of solute displacements, in spite of evidence that solutes in resting cells are largely adsorbed. This lingering affinity for tradition comes at the price of unnecessary redundancy—an enveloping barrier for preventing mass dissipation, and channels and pumps for mediating flows across the barrier.

These elements are not required in a gel-like cytoplasm. Gel matrices retain solutes and water without an enveloping barrier; and without a barrier there is no need for ion-specific channels or pumps to permeate the barrier. Also, the issue of slicing the cell is moot: slicing the cell should be as innocuous as slicing a gel.

Treating the cell as a gel established new ground rules for solutes and water. Gel water is adsorbed onto polymeric surfaces. Adsorption imparts order, and ordered water is inhospitable to many solutes, which reside preferentially at water-polymer interfaces. In both gels and cells, we found considerable evidence that solutes were associated with polymer/protein surfaces, and largely excluded from the water.

These physical chemical features led to a direct explanation of physiological ion partitioning (Fig. 16.1). Partitioning of a solute depends on relative solubility in bulk *vs.* structured water, and on protein-surface affinity. Both depend on the size of the ion's hydration shell. The larger the shell the lower the solubility; the larger the shell the greater the energy needed to remove the shell for surface association. Sodium's relatively large shell implied substantial exclusion from the cytoplasm and limited affinity for proteins. Potassium's smaller shell implied higher solubility and higher affinity. Thus, partitioning of the cell's essential ions could be explained directly, with no need to invoke energy-consuming maintenance mechanisms.

Limited solvency of structured water also helped account for the cell's electrical potential. Because ions cannot enter the cell's structured water environment freely, their intracellular concentration is limited. Cations cannot contribute enough positive charge to balance the negative charge borne by intracellular proteins—particularly the highly negatively charged proteins of the peripheral cytoskeleton. The cell therefore remains negatively charged, and its potential is correspondingly negative.

It seemed at this stage that the gel concept was promising: it could circumvent the membrane-breach anomaly; it seemed to explain ion-partitioning; and it could account for cellular potentials—the latter two features being rudiments of cell physiology. On the other hand, these features are static. To pursue the gel paradigm further, it seemed important to determine whether a gel-based mechanism also had the potential to explain the cell's dynamic tasks.

To do this we looked at synthetic gels. We found that a common denominator of gel action was the phase-transition. Phase-transitions achieve wonders in polymer gels: they shift solutes, separate phases, change volumes, propel ions, *etc.*—actions suspiciously similar to those of the cell. Moreover, such actions are generated by relevant environmental triggers: they include changes of pH, temperature, pressure, ion content, *etc.*—stimuli to which the cell is regularly exposed. Hence, the common denominator of gel action seemed to hold potential as a common denominator of cell action.

The common denominator hypothesis was pursued in the book's second half as we explored the dynamics of various organelles. Notwithstanding inevitable biological intricacy—each of the ~100,000 human genome products must serve some function—the simple phase-transition could go a long way toward explaining the essential operation of each of the cell's major organelles. Explanations were not entirely new: in several instances the phase-transition's relevance had been recognized earlier and pursued doggedly. The new message is that the phase-transition may be generic: it could be the common thread that weaves the fabric of cell function.

Having recapitulated the book's central message, we now reflect on several of its emergent themes with an eye toward deeper understanding and broader relevance.

Structured Water

For traditionally trained biologists, the most jarring of the book's themes may be that of structured water. The presence of one or two layers of structured water around proteins is broadly recognized, but the potential for multi-layering is less appreciated even though the concept had at one time been a topic of vigorous debate among cell scientists. With the cytoplasm as crowded with macromolecules as it is now known to be, fewer than ten layers on average would suffice to encompass all cell water. Some estimates suggest only five or six water layers. Evidence for water structuring has been presented in both gels and cells, and it remains for the reader to judge whether or not this evidence is compelling.

If water inside the cell is structured, water outside the cell—and even unrelated to the cell—could be structured as well. After all, surfaces that organize water are not unique to the cytoplasm. Any system containing closely packed hydrophilic surfaces could contain structured water, and the functional significance of such water could be as profound as has been argued for the cell. Many such systems exist.

Consider the tree, for example. The tree's roots absorb ground water, which is transported upward to sustain growth and to replace moisture lost through evaporation. The transport conduit is a vertically oriented tube called the xylem. Water is drawn upward through the xylem by some kind of force thought to involve osmotic pull and/or capillary action. For trees of reasonable height these mechanisms suffice, but they run into difficulty explaining how water can rise to the height of a 100-m redwood tree, for such drawing forces must work against a formidable counter-force—the pull of gravity.

That capillary forces are at work seems beyond doubt. The xylem tube has appreciable affinity for water, and can draw water upward by the same surface-affinity force that draws blood up a narrow capillary tube. But long tubes present a problem. Even though the upward pulling force can be very high as the xylem tube narrows near its top, the weight of the water column presents a huge load that could easily lead to water-column breakage, or cavitation. This is commonly observed (Canny, 1998). With cavitation, even a herculean drawing force will be ineffective—as any child learns by sucking on a straw with fluid containing an air bubble in the middle.

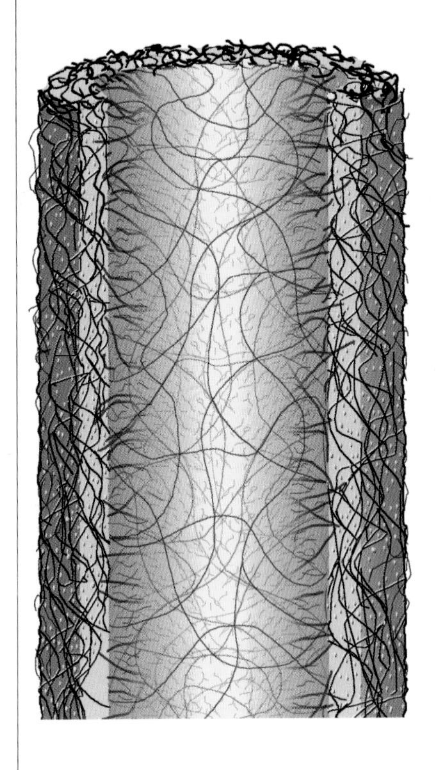

Figure 16.2. *Xylem tube consists of a fibrous matrix penetrating into the lumen. Clinging water can create a gel.*

Weight would be no problem, however, if the column of water clung tightly to the tube, and a suggestion along these lines was advanced some years ago (Plumb and Bridgman, 1972). The issue is whether all water in the xylem's cross-section could cling. To this end, it is now known that the xylem tube is built of a fibrous matrix extending into the lumen (Fig. 16.2). The matrix material is an extremely hydrophilic mucopolysaccharide, and the associated xylem fluid is reported to be viscous (Zimmermann *et al.*, 1994). Thus, the matrix apparently nucleates a gel that may well encompass the entire column cross-section. If the water clings to the matrix, column weight is irrelevant: So long as there is enough adhesive force, a sufficiently long tube could deliver water to the moon.

That the adhesive force is appreciable is substantiated by common experience. Ever try to burn a freshly cut log? In Seattle, the damp winter gloom coupled with a ready source of wood inspires frequent fireplace activity. Seasoned wood burns well, but everyone knows that wet wood merely sizzles. The surprise is that seasoning can require a year—even for split logs. Water clings tenaciously. Such tenacity is not seen in logs whose molecular structure has deteriorated from age; old wood dumped into a water tub until saturated will dry in short order and burn well. Thus, the water-adsorptive force lies in nature's surface design. The force is strong enough to resist all but fanatical attempts at dehydration, and

apparently strong enough to raise water easily to the top of a tall redwood tree (Fig. 16.3).

Structured water could play a role in geophysical realms as well, and at the risk of advancing speculation beyond this book's reasonable bounds I'd like to suggest two such realms: the earth and the sky. In the earth, the ratio of potassium to sodium is variably high, whereas in the sea it is only 1:50. Why such reversal? In part, the high potassium / low sodium paradigm of the earth arises from the particular mineral content of crustal rock. But water-adsorption features may also contribute. If water adsorbed onto the earth's hydrophilic clay surfaces is structured by surface charges, it will accumulate potassium and exclude sodium. In shale, for example, the surface potassium to sodium ratio is 3:1 to 4:1 (Krauskopf, 1967). Excluded sodium would then be washed into the sea, accumulating there in high concentration. The land may thus be a macrocosm of the cytoplasm—surrounded by a sea of excluded sodium.

Next consider the sky. The sky is the place where rising moisture gathers. But such moisture does not disperse uniformly; it condenses into puffy white clouds that float within a mass of otherwise dry air. Why it condenses into discrete units instead of remaining uniformly dispersed is one of those seemingly naïve questions that has to my knowledge yet to be seriously addressed by geophysicists. The structured water paradigm may again hold relevance. Grains of charged pollen or dust could nucleate arrays of oriented water, as do other macromolecules. If the surface of the water array retains any of the charge asymmetry present on the nucleator surface, then the array acts as a dipole. Dipole-dipole attraction will lead inevitably to clouds of moisture.

Although these ideas are speculative, it is not beyond reason that structured water could play an unsuspectedly pivotal role in realms well beyond those of cell biology. Structured water may lurk wherever hydrophilic surfaces exist—and may carry interpretive powers yet to be fathomed.

Pumps, Channels and Membranes

A second of the book's themes is the relegation of cherished functional entities to positions of secondary significance. These entities include pumps, channels and phospholipid-membrane barriers. In the proposed paradigm such elements are not central. On the other hand, they do appear to exist in some form, and their role needs to be dealt with.

The picture that has evolved looks like this (Fig. 16.4). The cytoplasm is enveloped by a multi-layered barrier. The innermost layer is the peripheral cytoskeleton—a thick, densely packed matrix of highly anionic filaments containing protruding globular elements. The next layer is the

Figure 16.3. Ground water absorbed by tree roots works its way to the top, even in 100-m redwoods.

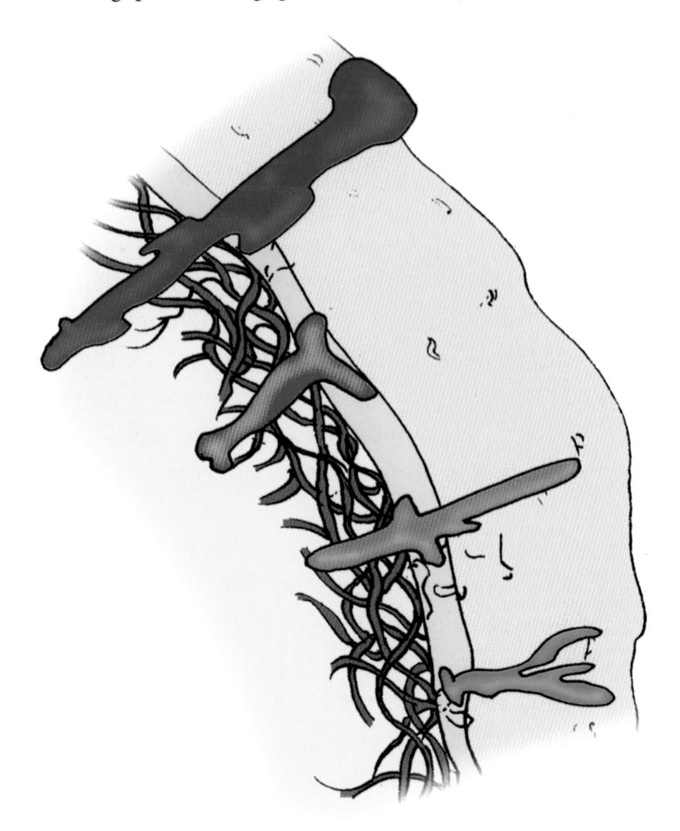

Figure 16.4. The cytoplasm is surrounded by a multi-layered boundary that is punctuated by protruding elements.

phospholipid membrane. This membrane is also punctuated by abundant elements including receptors, "pumps," "channels" and other molecules that may protrude from the cell. The membrane is therefore discontinuous. A tertiary barrier frequently found in eukaryotic cells is the glycocalyx—variously described as a mechanochemical buffer, an adhesion facilitator, and an agent of cell recognition. The glycocalyx is the cells' sugar coat, built of a thick layer of sugar-containing molecules such as glycoproteins, proteoglycans and glycolipids, some of which may penetrate back into the lipid-bilayer or even the cytoskeleton.

Clearly, then, the cytoplasm is enveloped by a multi-layered barrier. But the barrier is spanned by an array of molecules, whose crevices provide inside-outside flow passageways. Such pathways may be loosely thought of as "channels," gated open by the phase-transition. They could exhibit the kinds of pharmacological sensitivities that are observed, depending on the nature of the particular spanner molecule. Hence, the pathways behave in much the way that channels are currently conceived to behave, but are larger, more diffuse, and not necessarily exclusive for a particular solute.

Also in evidence in the new construct are pump-like elements. Pumps are defined pharmacologically: When a transmembrane ion gradient is compromised by a drug, the target molecule to which the drug binds is said to be a pump for that ion. For example, ouabain binds to a surface protein called Na/K-ATPase. Sodium- and potassium-ion gradients are compromised by this drug, so ouabain is interpreted as a Na/K-pump blocker.

Interpretations such as these, however, do not square with experimental observation. In the case of ouabain, ion gradients are compromised not only in intact cells, but also in cut cells (Ling, 1973, 1978), where pumps cannot be a major force. An alternative possibility is that ouabain binding induces a transition that allows sodium and potassium to flow down their respective concentration gradients. Clearly, the impact of binding to Na/K ATPase extends beyond just the membrane: Ouabain binding to permeabilized cells impacts processes as far away as the cytoskeletal microtubules (Alonso *et al.*, 1998) and the nucleus (Mentré and Debey, 1999). Hence, the impact is global. Binding does, nevertheless, ultimately influence ion gradients. While this is evidently not proof that Na/K ATPase necessarily "pumps" anything, conceptualizing the process within a framework of pumping could remain a convenient expedient so long as the underlying reality is borne in mind.

Finally, the membrane. In the previous chapter's discussion of evolutionary development, a notable feature was the absence of any specific membrane requirement. The cell began as a polymer gel and evolved from there. At no stage did survival depend on the presence of a membrane. Yet, membranes did eventually appear and one may rightly ask why. If they were not critical, why bother?

A plausible reason may be to restrict exchange (Fig. 16.5). Solutes can flow uniformly across an ordinary gel boundary. With partial blockage by a surface membrane, flow will be restricted to those regions devoid of membrane—regions ordinarily occupied by receptors and other proteins. If such proteins are activated by ions and/or ligands, then the most effective way to activate is to "channel" the required solutes to those sites.

Figure 16.5. *Relative to no barrier (left), a partial barrier (right) restricts solute exchange.*

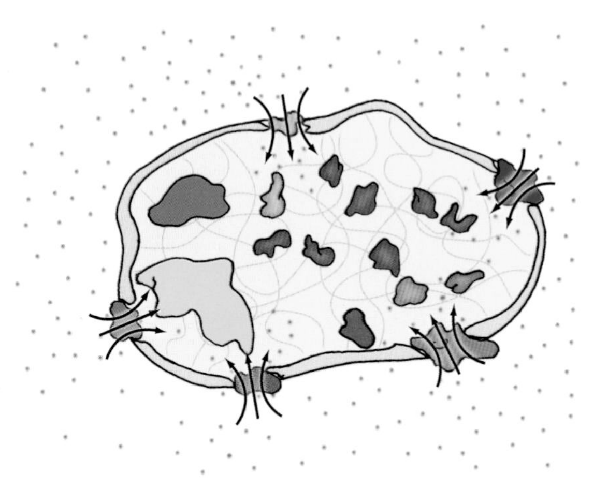

Hence, the membrane may have arisen as a matter of efficacy—to enhance activation sensitivity. At the same time, the presence of a membrane could help stem the inevitable loss of critical intracellular solutes such as ATP. In this scenario there is some sense to the progressive evolutionary increase of lipid-bilayer coverage (Chapter 2).

But a membrane that fully enveloped the cell was apparently never required, and because of this the cell never needed to make the abrupt transition from a system with no barrier, to one with a barrier. Any such transition could not occur simply. Mediating solute exchange across the new barrier would require the simultaneous appearance of channels and pumps (or their equivalents), as well as the machinery through which these elements could be manufactured. All of this would need to have appeared contemporaneously with the barrier, for there was no pressure to create them beforehand. The conundrum of sudden complexity is a formidable obstacle which, fortunately, does not need to be faced in the current paradigm.

In sum, the new paradigm contains elements loosely resembling the classical ones, and it is therefore no surprise that such elements can be isolated for study. On the other hand, the roles of these elements may be less specific and less central than generally envisioned.

PHASE-TRANSITIONS

A third theme of the paradigm is the pervasiveness of the phase-transition. The phase-transition is proposed as a central mediator of cell function, operating in various guises throughout the cell. Here we explore these transitions' common attributes.

For many biologists, phase-transitions lie within the realm of unfamiliar territory. Conformational changes, by contrast, are widely recognized, and the question arises whether the two really differ in a substantive way. Conformational change implies some modification in a protein's secondary or tertiary structure. Phase-change may involve similar modification, but typically involves an array of molecules instead of one—hence the prefix "cooperative" that commonly precedes "phase-transition." The difference, then, may be simply one of scale.

A good example of this parallel can be found in the dynamics of muscle, where translation steps are observed both in single molecules and molecular arrays. When the single myosin molecule translates along an actin filament, the displacement step is attributed to protein conformational change (Kitamura *et al.*, 1999). When an array of actin filaments translates along an array of myosin filaments, the displacement step is attributed to a phase-transition (Blyakhman *et al.*, 1999). It is almost certain that these steps of translation arise from the same source because in both instances their size is an integer multiple of the actin-monomer spacing, a feature too signature-like to arise fortuitously. The differ-

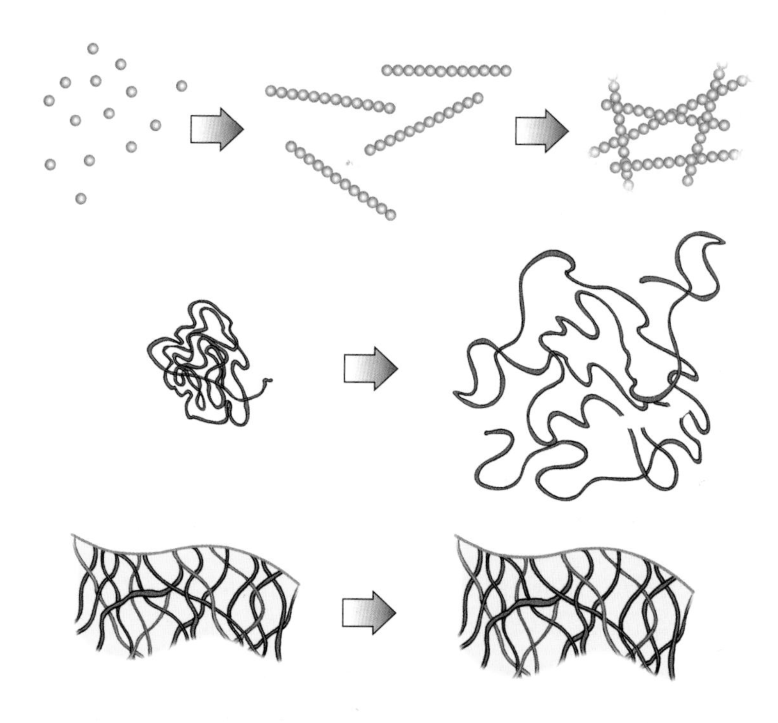

Figure 16.6. *Summary of expansive phase-transitions. Sol-gel transition (top); condensation-expansion transition without covalent cross-links (middle); and condensation-expansion transition with cross-links (bottom).*

ence, then, is largely one of scale.

Like conformational change, the phase-transition may come in many varieties. Threads of commonality are difficult to identify when the transitions are considered separately. When they are grouped according to the direction of change, *i.e.*, expansion or condensation, the common threads emerge more readily, and provide insight into function.

Among expansive transitions (Fig. 16.6), one of the most basic is the sol-gel transition (top panel). This transition is expansive in the sense that it embodies growth of monomers into polymers, and growth of polymers into gel matrices such as those of the mitotic spindle. Polymerization and gelation are critical phenomena: once begun, they continue toward completion.

Another class of expansive transition includes those that occur in the secretory granule and the peripheral cytoskeleton (Chapters 9 and 10). In the secretory granule (middle panel) a polymeric network is initially condensed into a tight, dry packet, with constituent strands bridged by divalent cations. As monovalent ions replace these divalents, the strands separate and adsorb water, allowing the divalents to diffuse out and accomplish their assigned secretory task. Likewise, the cytoskeletal matrix is ordinarily condensed (bottom panel). The condensed network creates a barrier that restricts solute exchange between the inside and outside of the cell. As monovalents replace calcium, constituent polymer strands separate and imbibe water. Because the strands are connected by covalent cross-links, however, expansion is limited. Elastic energy imparted to the network may then help return the matrix to its initial condition.

As for phase-transitions that are contractile (Fig. 16.7), a common example is the one that shortens the actin-filament bundle (left). Like the expansive transitions, this contractile transition appears to involve calcium. Calcium precipitates actin suspensions (Szent-Györgyi, 1951), condenses filamentous actin into bundles (Tang and Janmey, 1996), and contracts actin-filament gels into plugs (Bray, 1992). Calcium's role in the actin-filament bundle may thus be similar to that in the peripheral cytoskeleton and secretory granule: to condense through divalent bridging.

In condensation transitions such as these, covalent cross-links may also play a role—as they do in expansive transitions. In uncrosslinked gels, filaments can contract down to a small plug as water is expelled. Regularly cross-linked filaments cannot contract so dramatically because the links preserve filament separation; the lattice thereby retains water, whose structure in turn curtails the phase-transition's progress. Thus, striated muscle contracts modestly (except in pathological cases where rupture of cross-links results in a knot-like contracture). Smooth muscles on the other hand, as well as contractile rings, synthetic actin gels, and other organelles with few covalent cross-links and semi-random structures, can shorten massively. The extent of condensation depends on the extent of cross-linking.

Another contractile transition is the one that takes place in microtubules. Tubulin's homology with actin implies that the two transitions may operate by similar mechanistic principles—and indeed, propagation along microtubules shares many features in common with propagation along actin-filament bundles (Chapter 11, 12).

Figure 16.7. *Examples of contractile phase-transitions. Actin-condensation transition (left); helix-coil transition (middle); and extended coil to beta-sheet transition (right).*

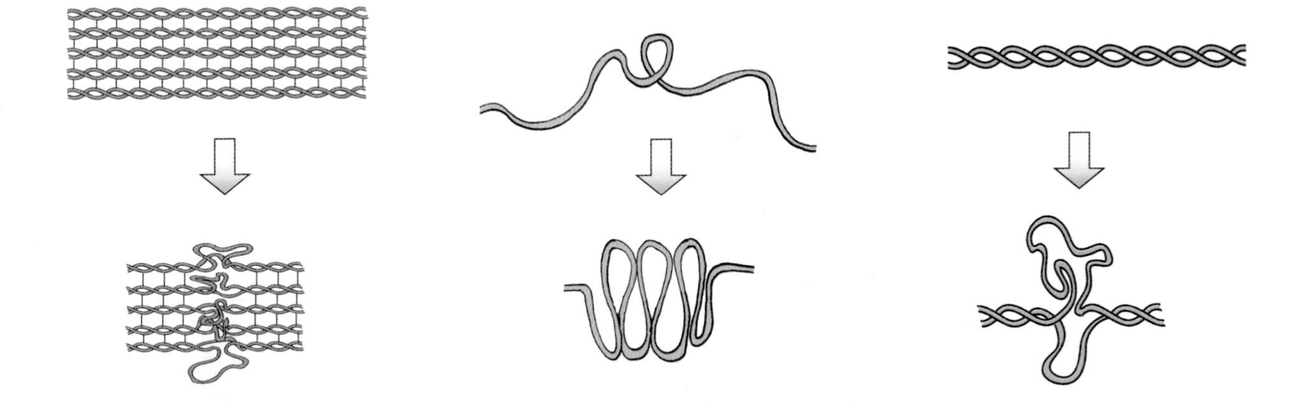

Finally, certain contractile transitions seem designed for specialized roles. In the helix-coil transition of myosin (Chapter 14), an alpha-helix contracts to a random coil to produce prodigious tension (Fig. 16.7, middle). In the titin molecule, extended coil domains condense into beta-sheets, producing equally prodigious shortening (right). Transitions of this ilk specialize in the extremes of mechanical behavior.

***

Out of this capsule review come a few observations. Perhaps the most obvious is the variety of transitions. Variety would seem to reflect the diversity of tasks—ranging from permeability change, ion transport and solute expulsion, to shape change, force generation, strand shortening, *etc.* Such tasks are disparate enough to warrant some degree of mechanistic specialization. Condensed spring-loaded vesicles are ideal for spewing out solutes; helix-coil transitions are just right for contractions against high loads; condensations from extended coils to beta-sheets are without peer in achieving large-scale length changes; *etc.* Diverse transitions meet diverse needs.

More illuminating than the differences, perhaps, are the common features. One common feature worthy of reiteration is the transitions' cooperative nature. Transitions are not unitary molecular events, but global actions. Once the critical threshold is crossed, the transition proceeds with the inevitability of a sneeze. In this way the system acts as an amplifier, converting a small environmental change into an inevitably magnificent response.

A second common feature is the involvement of divalent cations, particularly calcium (Fig. 16.8). This is no surprise. Calcium is involved in numerous aspects of cell function, and if such functions rest on phase-transitions then phase-transitions must in turn rest on calcium. The logic is clear. Calcium appears to operate consistently: By bridging anionic strands, it can condense the matrix and squeeze out the water. Con-

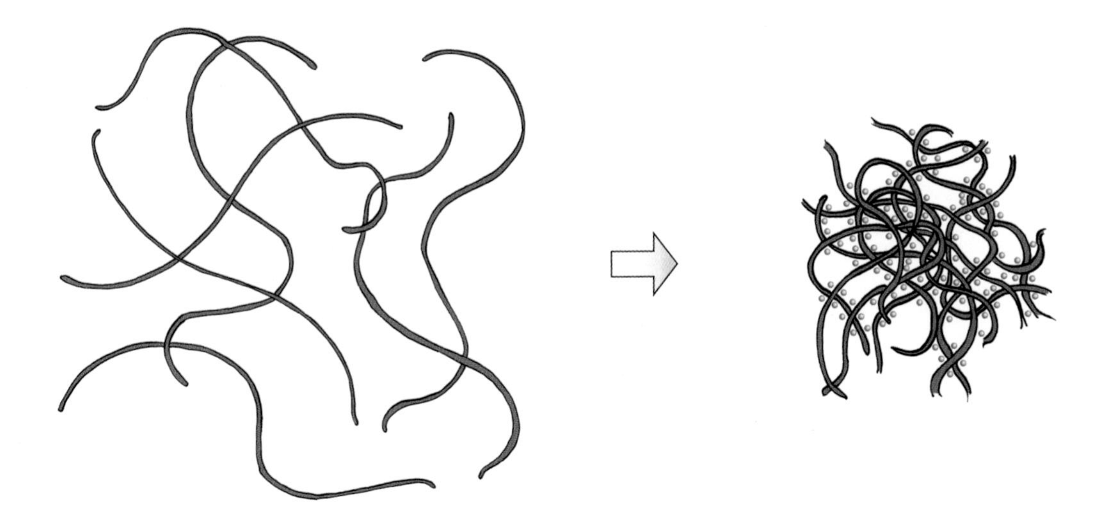

versely, when calcium is replaced through monovalent substitution, the network is loosened and the water is sucked back in. Shuttling of water into and out of the network is central to function, and at the core of all this is calcium.

Figure 16.8. Central role of divalent cations. By bridging neighboring anionic strands, calcium and other divalent cations precipitate condensation and water expulsion.

Calcium's condensing action depends on surface negativity, and this is perhaps an apt time to reflect on why nature has been so consistent in choosing this negative route. Structural proteins such as actin filaments, microtubules, intermediate filaments—even DNA—are all negatively charged. Cross-linkers, by contrast, are positively charged. These include bridging and bundling proteins, as well as divalents. All of this seems to make sense—but then again, why is it that the polarities are not reversed? Why couldn't the structural proteins be positively charged and cross-linkers negatively charged?

The negative route for structural proteins might have arisen out of an ATP requirement. According to the proposed hypothesis (Chapter 15), ATP's negative charge and the protein's negative charge act to induce protein extension, which is critical for energy transfer. The "choice" of surface negativity may therefore follow the dictates of this uniquely capable and peerless molecule. In theory, biological systems could be built of structural polymers with net positive charge. Divalent anions would then mediate condensation, and the energy supply system would require a positively charged counterpart of ATP. Such "reverse" systems could well be the basis of a life different from the one we know—perhaps lurking somewhere in another galaxy.

Lessons from Biology

Having dwelled on the past and present, I conclude with speculations about the future. If some projections were not to spring from this material it would seem out of character, for little reserve has been exercised up to now. Where might all this chutzpah take us?

Hopefully, to a productive way of thinking. The new paradigm charts a defined course. To approach an unknown cell-biological mechanism, the instruction manual states, "look first for some kind of phase-transition and proceed from there." This approach may seem single-mindedly narrow but the interpretive power exhibited by the phase-transition up to now implies that it may be a path worth taking.

Which brings me to a related point—the parallel between natural and artificial. The functional similarity between natural gels and artificial gels has been frequently reiterated. From solvency, to permeability, electrical behavior, volume changes, phase-separations and so on, natural and synthetic gels exhibit consistently parallel behaviors. Even dried animals can rehydrate to life just as dried polymers can rehydrate to function. Biological gels are therefore not particularly special—except to the extent that their operation has been optimized by long-term evolutionary forces. Nature has had four billion years to effect its mastery, whereas engineers and chemists working on synthetic gels have had but a few decades.

With that much of an edge, nature's product might be expected to have come a bit further than the human-engineered product. Indeed, natural phenomena are only just beginning to be exploited by engineers for practical use: *e.g.*, the beta-sheet to extended coil transition and the alpha-helix to random coil transition (Aggeli *et al.*, 1997; Petka *et al.*, 1998; Wang *et al.*, 1999); also, the packing of drugs into polyionic vesicles for secretion-like delivery to tissues (Verdugo *et al.*, 1995). What works for the cell should work for the gel. Conversely, achievements by engineers can provide clues as to the cell's function. Frequency modulation (FM) was used in electrical communication systems well before it became clear that the same principle was used in neural communication.

These expectations reach beyond the boundaries of biomimetics, a growing field that exploits biological principles for practical application. I would boldly suggest that once understood, every principle used by nature will eventually be used by engineers. This assertion cannot be proved or disproved, but I am reminded of computer-processing chips, where engineers have based artificial communication on silicon. For biological communication, nature has chosen carbon. Among all elements, the closest to carbon in terms of chemical properties is silicon. Coincidence or inevitability? As biologists ferret out nature's working principles, engineers may have much to learn as they gawk.

Finally, we return to the feature with which this book began: simplicity. Simplicity is a virtue to which engineers strive—it is practically hard-wired into their genome. Complicated mechanisms certainly have their charm but seasoned engineers understand that for reliable operation, mechanisms must be kept as simple as possible. The question is whether this principle should also hold for mechanisms that are engineered celestially. If cellular mechanisms are to work reliably, should they not be based on simple—even beautiful—principles?

Within such a framework lies the phase-transition, an amply powerful process triggered by a small environmental shift. Whether indeed this process is simple enough and powerful enough to reduce the currently bewildering array of intracellular epicycles to a few functional orbits remains to be seen. Please stay tuned.

ARTWORK CREDITS

Cover	HUVEC Cultures, F-actin – red, von Willebrand factor – green, nuclei – blue.	Jakob Zbaeron, Departement Klinische Forschung, Inselpital, Berne, Switzerland.
p. 24	"The Death of Marat Sade"	Jacques Louis David Musees rouyaux des Beaux-Arts de Belgique, Bruxelles
p. 38	"Woman IV", 1952-53	Willem de Kooning, American (1904-1997) The Nelson-Atkins Museum of Art, Kansas City, Missouri (Gift of William Inge)
p. 52	The Great Wave off Kanagawa, 1823-29 Japan Ukiyo-e Museum	Katsushika Hokusai
p. 98	"Dr. Megavolt" www.drmegavolt.com	Jeff Clark, photographer
p. 112	M.C. Escher's "Sky and Water I"	Copyright 2000 Cordon Art B.V. – Baarn – Holland. All rights reserved.
p. 184	"The Red Stiletto"	David Crow (photo by Harrod Blank), www.artcaragency.com
p. 206	Mammalian cells, strain PtK, in mitosis, as seen by immunofluorescence microscopy. Orange is propidium iodide, staining the DNA. Green is antitubulin.	J. R. McIntosh, MCD Biology, University of Colorado
p. 235	Figure 14.9, Right- Contraction of polyacrylamide gel	photo by Fritz Goro
p. 250	"The Kingdome Implosion - Dome's Final Roar"	Greg Gilbert and Jay Dotson/ The Seattle Times
p. 266	"Triassic Triops"	Triops, Inc. Eugene Hull, President

REFERENCES

A

Abbott, B.C., Hill, A.V., and Howarth, J. V. (1958). The positive and negative heat production associated with a single impulse. *Proc. Roy. Soc. B.* 148: 149-187.

Aggeli, A., Bell, M., Boden, N., Keen, J. N., Knowles, P. F., McLeish, T. C. B., Pitkeathly, M., and Radford, S. E. (1997). Responsive gels formed by the spontaneous self-assembly of peptides into polymeric beta-sheet tapes. *Nature* 386: 259-262.

Alberts, B., Bray, D., Lewis, J., Raff, M., Roberts, K., and Watson, J. D. (1994). *Molecular Biology of the Cell*, Third edition, Garland, N. Y.

Alberts, B., Bray, D., Lewis, J., Raff, M., Roberts, K., and Watson, J. D. (1989). *Molecular Biology of the Cell*, Second edition, Garland, N. Y.

Alberts, B., Bray, D., Lewis, J., Raff, M., Roberts, K., and Watson, J. D. (1983). *Molecular Biology of the Cell*, First edition, Garland, N. Y.

Albin, G., Horbett, T. A., and Ratner, B. D. (1985). Glucose sensitive membranes for controlled delivery of insulin: insulin transport studies. *J. Contr. Rel.* 2: 153-164.

Albrecht-Buehler, G. (1980). Autonomous movements of cytoplasmic fragments. *Proc. Nat'l. Acad. Sci.* 77(11): 6139-6643.

Albrecht-Buehler, G. (1985). Is the cytoplasm intelligent too? *Cell Motil. and Cytoskel.* 6: 1-21.

Albrecht-Buehler, G. (1998). Altered drug resistance of microtubules in cells exposed to infrared light pulses: Are microtubules the "nerves" of cells? *Cell Motil. and Cytoskel.* 40: 183-192.

Albrecht-Buehler, G., and Bushnell, A. (1982). Reversible compression of the cytoplasm. *Exp. Cell Res.* 140: 173-189.

Allen, R. D., Cooledge, J. W., and Hall, P. J. (1960). Streaming in cytoplasm dissociated from the giant amoeba, chaos chaos. *Nature* 187: 896-899.

Almdal, K., Dyre, J., Hvidt, S., and Kramer, O. (1993). Towards a phenomenological definition of the term 'gel.' *Polym. Gels and Ntwks.* 1: 5-17.

Alonso, A., Nuñez-Fernandez, M., Beltramo, D. M., Casale, C. H., and Barra, H. S. (1998). Na, K-ATPase was found to be the membrane component responsible for the hydrophobic behavior of the brain membrane tubulin. *Biochem. Biophys. Res. Commun.* 253: 824-827.

Anisimova, V. I., Deryagin, B. V., Ershova, I. G., Lychnikov, Y. I., Simonova, V. K., and Churaev, N. V. (1967). Preperation of structuarally modified water in quartz capillaries. *Russ. J. Phys. Chem.* 41: 1282-1284.

Annaka, M., and Tanaka, T. (1992). Multiple phases of polymer gels. *Nature* 355: 430-432.

Asakura, A., Taniguchi, M., and Oosawa, F. (1963). Mechano-chemical behavior of F-actin. *J. Mol. Biol.* 7: 55- 63.

Ault, J. G., Demarco, A. J., Salmon, E. D., and Rieder, C. L. (1991). Studies on the ejection properties of asters: astral microtubule turnover influences the oscillatory behavior and positioning of mono-oriented chromosomes. *J. Cell Sci.* 99: 701-710.

B

Bartels, E. M., and Elliott, G. F. (1985). Donnan potentials from the A- and I-bands of glycerinated and chemically skinned muscles, relaxed and in rigor. *Biophys. J.* 48: 61-76.

Bausch, A. R., Möller, A., and Sackmann, E. (1999). Measurement of local viscoelasticity and forces in living cells by magnetic tweezers. *Biophys. J.* 76: 573-579.

Beall, P. T. (1980). Water-macromelecular interactions during the cell cycle. *Nuclear-cytolasmic Interactions in the Cell Cycle, ed.* G. Whitson, Acad. Press, NY.

Beall, P. T., Brinkley, B. R., Chang, D. C., and Hazlewood, C. F. (1982). Microtubule complexes correlated with growth rate and water proton relaxation times in human breast cancer cells. *Cancer Res.* 42: 4124-4130.

Bernal, J. D. (1961). Origin of life on the shores of the ocean. *Oceanography, ed.* M. Sears. *AAAS.* 67: 95-118.

Berridge, M. J. (1994). The biology and medicine of calcium signaling. *Mol. Cell. Endicrinol.* 98: 119-124.

Bi, G.-Q., Alderton, J. M., and Steinhardt, R. A. (1995). Calcium-regulated exocytosis is required for cell membrane resealing. *J. Cell Biol.* 131: 1747-1758.

Bingley, M. S. (1966). Further investigations into membrane potentials in amoebae. *Exp. Cell Res.* 43: 1-12.

Block, S. M. (1996). Fifty ways to love your lever: myosin motors. *Cell* 87: 151-157.

Bloodgood, R. A. (1977). Motility occurring in association with the surface of the *Chlamydomonas* flagellum. *J. Cell Biol.* 75: 983-989.

Bloodgood, R. A. (1978). Unidirectional motility occurring in association with the axopodial membrane of *Echinosphaerium nucleophilum. Cell Biol. Int. Rep.* 2: 171-176.

Bloodgood, R. A., Leffler, E. M., and Bojczuk, A. T. (1979). Reversible inhibition of *Chlamydomonas* flagellar surface motility. *J. Cell Biol.* 82: 664-674.

Blyakhman, F., Shklyar, T., and Pollack, G. H. (1999). Quantal length changes in single contracting sarcomeres. *J. Muscle Res. Cell Mitol.* 20: 529-538.

Boyle, P. J., and Conway, E. J. (1941). Potassium accumulation in muscle and associated changes. *J. Physiol.* 100: 1-63.

Brandt, P. W., Diamond, M. S., and Schachat, F. H. (1984). The thin filament of vertebrate skeletal muscle co-operatively activates as a unit. *J. Mol. Biol.* 180: 379-384.

Bratton, C. B., Hopkins, A. L., and Weinberg, J. W. (1965). Nuclear magnetic resonance studies of living muscle. *Science* 147: 738- 739.

Bray, D. (1992). *Cell Movements*, Garland, NY.

Brooks, S. C. (1940). The intake of radioactive isotopes by living cells. *Cold Spring Harbor Symp. Quant. Biol.* 8: 171-180.
Brummer, S. B., Bradspies, J. I., Entine, G., Leung, C., and Lingertat, H. (1972). Polywater, an organic contaminant. *J. Phys. Chem.* 76: 457-458.

Bungenberg de Jong, H. (1932). Die Koazervation und ihre Bedeutung für die Biologie. *Protoplasma* 15: 110-173.

Buxbaum, R. E., Dennerll, T., Weiss, S., and Heidemann, S. R. (1987). F-actin and microtubule suspensions as indeterminate fluids. *Science* 235: 1511-1514.

C

Cameron, I. L. (1988). Ultrastructural observations on the transectioned end of frog skeletal muscles. *Physiol. Chem. Phys. Med. NMR.* 20: 221-225.

Cameron, I. L. , Kanal, K. M., Keener, C. R., and Fullerton, G. D. (1997). A mechanistic view of the non-ideal osmotic and motional behavior of intracellular water. *Cell Biol. Int'l.* 21(2): 99-113.

Cameron, I. L. Cook, K. R., Edwards, D., Fullerton, G. D., Schatten, G., Schatten, H., Zimmerman, A. M., and Zimmerman, S. (1987). *J. Cell. Physiol.* 133: 14-24.

Cameron, I. L., Cook, K. R., Edwards, D., Fullerton, G.D., Schatten, G., Schatten, H., Zimmerman, A. M., and Zimmerman, S. (1997). Cell cycle changes in water properties in sea urchin eggs. J. Cell Physiol. 133(1): 99-113.

Cameron, I. L., Fullerton, G. D., and Smith, N. K. R. (1988). Influence of cytomatrix proteins on water and on ions in cells. *Scanning Microscop.* 2(1): 275-288.

Cameron, I. L., Hardman, W. E., Fullerton, G. D., Miseta, A., Koszegi, T., Ludany, A., and Kellermayer, M. (1996). Maintenance of ions, proteins and water in lens fiber cells before and after treatment with non-ionic detergents. *Cell Bio. Int.* 20(2): 127-137.

Canny, M. J. (1998). Transporting water in plants. *Am. J. Sci.* 86: 152-159.

Carmeliet, E. (1992). A fuzzy subsarcolemmal space for intracellular Na^+ in cardiac cells? *Cardiovasc. Res.* 26: 433-442.

Casademont, J., Carpenter, S., and Karpati, G. (1988). Vacuolation of muscle fibers near sarcolemmal breaks represents T tubule dilatation secondary to enhanced sodium pump activity. *J. Neuropath. Exp. Neurol.* 47: 618-628.

Challice, C. E. (1965). Studies on the microstructure of the heart. *J. Roy. Microsc. Soc.* 85: 1-21.

Chanutin, A. and Hermann, E. (1969). The interaction of organic and inorganic phosphates with hemoglobin. *Arch. Biochem. Biophys.* 131: 180-184.

Chaudhury, M. K., and Whitesides, G. M. (1992). How to make water run uphill. *Nature* 256: 1539-1541.

Chen, W. T. (1981). Mechanism of retraction of the trailing edge during fibroblast movement. *J. Cell Biol.* 90(1): 187-200.

Cheng, Y.-P. (1971). The ultrastructure of the rat sino-arrial node. *Acta. Anat. Nippon.* 46: 339-358.

Choi, D. W. (1988). Glutamate neuortoxicity and diseases of the nervous system. *Neuron* 1: 623-34.

Chou, K. C. (1992). Energy-optimized structure of antifreeze protein and its binding mechanism. *J. Mol. Biol.* 223: 509-517.

Clarke, M. S. F., Caldwell, R. W., Miyake, K., and McNeil, P. L. (1995). Contraction-induced cell wounding and release of fibroblast growth factor in heart. *Circ. Res.* 76: 927-934.

Clegg, J. S. (1982). Interrelationships between water and cell metabolism in Artemia cysts. IX. *Evidence for organization of soluble cytoplasmic enzymes.* Cold Spring Harbor Symp. Quant. Biol. *46(Pt. 1): 23-37.*

Clegg, J. S. (1984). Intracellular water and the cytomatrix: some methods of study and current views. *J. Cell Biol.* 99: 167s-171s.

Clegg, J. S. (1988). On the internal environment of animal cells. *Microcompartmentation,* CRC Press, Boca Raton.

Clegg, J. S., and Drost-Hansen, W. (1991). On the biochemistry and cell physiology of water. *Biochem. and Mol. Biol. of Fishes,* Vol. 1, *ed.* Hochachka, and Mommsen, Elsevier, N. Y.

Clegg, J. S., and Jackson, S. A. (1988). Glycolysis in permeabilized L-929 cells. *Biochem. J.* 255: 335-344.

Cohn,W. E., and Cohn, E. T. (1939). Permeability of red corpuscles of the dog to sodium ion. *Proc. Soc. Exp. Biol.* 41: 445-449.

Collins, E. W., Jr. and Edwards, C. (1971). Role of Donnan equilibrium in the resting potentials in glycerol-extracted muscle. *Am. J. Physiol.* 22(4): 1130-1133.

Collins, K. D. (1995). Sticky ions in biological systems. *Proc. Nat'l. Acad. Sci.* 92: 5553-5557.

Cooke, R. (1997). Actomyosin interaction in striated muscle. *Physiol. Rev.* 77(3): 671-697.

Cope, F. W. (1969). Nuclear magnetic resonance evidence using D_20 for structured water in muscle and brain. *Biophys. J.* 9: 303-319.

Curran, M. J., and Brodwick, M. S. (1991). Ionic control of the size of the vesicle matrix of beige mouse mast cells. *J. Gen. Physiol.* 98: 771-790.

Cussler, E. L., Stokar, M. R., and Varberg, J. E. (1984). Gels as size selective extraction solvents. *AIChE Journal* 30(4): 578-582.

D

Damadian, R. (1971). Tumor detection by nuclear magnetic resonance. *Science* 171: 1151-1153.

Dayhoff, J., Hameroff, S., Lahoz-Beltra, R., and Swenberg, C. E. (1994). Cytoskeletal involvement in neuronal learning: a review. *Eur. Biophys. J.* 23: 79-93.

Dean, R. (1941). Theories of electrolyte equilibrium in muscle. *Biol. Symp.* 3: 331-348.

deBeer, E. L, Sontrop, A., Kellermayer, M. S. Z., and Pollack, G. H. (1998). Actin-filament motion in the *in vitro* motility assay is periodic. *Cell Motil. and Cytoskel.* 38: 341-350.

DeDuve, C. (1995). *Vital Dust,* Basic Books Press, New York.

Delay, M. J., Ishide, N., Jacobson, R. C., Pollack, G. H., and Tirosh, R. (1981). Stepwise sarcomere shortening: Analysis by high-speed cinemicrography. *Science* 213: 1523-1525.

Dempster, J. A., Van Hoek, A. N., and Van Os, C. H. (1992). The quest for water channels. *NIPS* 7: 172-176.

Derjaguin, B. V. (1966). Effects of lyophilic surfaces on the properties of boundary liquid films. *Disc. Frad. Soc.* 42: 109-119.

DeVries, A. L. (1982). Fish glycopeptide and peptide antifreezes: their interaction with ice and water. *Biophysics of Water, ed.* F. Franks, and S. Mathias, Wiley Interscience, N. Y.

Discher, E. E., Mohandas, N., and Evans, E. A. (1994). Molecular maps of red cell deformation: hidden elasticity and *in situ* connectivity. *Science* 266: 1032-1035.

Dos Remedios, C. G., and Moens, P. D. (1995). Actin and the actomyosin interface: a review. *Biochim. Biophys. Acta.* 1228(2-3): 99-124.

Doyle, D. A., Cabral, J. M., Pfuetzner, R. A., Kuo, A., Gulbis, J. M., Cohen, S. L., Chait, B. T., and MacKinnon, R. (1998). Structure of the potassium channel: molecular basis of K^+ conduction and selectivity. *Science* 280: 69-77.

Draper, M. H., and Weidmann, S. (1951). Cardiac resting and action potentials recorded with an intracellular electrode. *J. Physiol.* 115: 74-94.

Drummond, D. R., Peckham, M., Sparrow, J., and White, D. C. (1990). Alteration in kinetics caused by mutations in actin. *Nature* 348: 440-442.

Dujardin, F. (1835). Sur les prètendus estomacs des animalcules infusoires et sur une sustance appelée sarcode. part 2 of "Recherche sur les organismes inferieur." *Ann. De Sci. Natur, part. Zool.* 2d Ser., 4.

Dusek, K. and Patterson, D. (1968). A transition in swollen polymer networks induced by intramolecular condensation. *J. Polymer Sci. A-2* 6: 1209-1216.

E

Edelmann, L. (1978). Vizualization and x-ray microanalysis of potassium tracers in freeze-dried and plastic embedded frog muscle. *Microsc. Acta. Suppl.* 2: 166-174.

Edelmann, L. (1980). Preferential localized uptake of K^+ and Cs^+ over Na^+ in the A-band of freeze-dried embedded muscle section: detection by x-ray microanalysis and laser microprobe mass analysis. *Physiol. Chem. Phys.* 12(6): 509-514.

Edelmann, L. (1983). Electron probe x-ray microanalysis of K, Rb, Cs, and Ti in cryosections of striated muscle. *Physiol. Chem. Phys. Med. NMR.* 15(4): 337-344.

Edelmann, L. (1988). The cell water problem posed by electron microscopic studies of ion binding in muscle. *Scanning Microsc.* 2(2): 851-865.

Ehrenpries, S. (1967). Discussion on P. G. Waser: Receptor localization by autoradiographic techniques. *Ann. N.Y. Acad. Sci. (Discussion).* 144: 754(1).

Endo, (1972). Length dependence of activation of skinned muscle fibers by calcium. *Symp. on Quant. Biol.* Cold Spring Harbor, N.Y.

Endo, S., Sakai, H., and Matsumoto, G. (1979). Microtubules in squid giant axon. *Cell Struct. Funct.* 4: 285-293.

Ernst, E. (1970). Bound water in physics and biology. *Acta Biochim et Biophys. Acad. Sci. Hung.* 5(1): 57-69.

F

Fenn, W. O. (1953). Introduction to a symposium on the metabolism of potassium. *Lancet* 73: 163-166.

Fenn, W. O., and Cobb, D. M. (1934). The potassium equilibrium in muscle. *J. Gen. Physiol.* 17:629-656.

Fernandez, J. M., Villalon, M., and Verdugo, P. (1991). Reversible condensation of mast cell secretory products *in vitro*. *Biophys. J.* 59: 1022-1027.

Fischer, M. H., and Moore, G. (1907). On the swelling of fibrin. *Am. J. Physiol.* 20: 330-342.

Fisher, H. F. (1964). A limiting law relating the size and shape of protein molecules to their composition. Proc. Natl. Acad. Sci. USA. 51: 1285-1291.

Fisher, I. R., Gamble, R. A., and Middlehurst, J. (1981). The Kelvin equation and the condensation of water. Nature 290: 575-576.

Fishman, H. M., Tewari, K. P., and Stein, P. G. (1990). Injury-induced vesiculation and membrane redistribution in squid giant axon. *Biochim. Biophys. Acta.* 1023: 421-435.

Fleischer, S., Fleischer, B., and Stoeckenius, W. (1965). The structure of whole and fragmented mitochondria after lipid depletion. *Fed. Proc.* 24: 296-298.

Fletcher, N. H. (1970). *Chemical Physics of Ice,* Cambridge, London.

Frank, H. S. and Wen, W. Y. (1957). Structural aspects of ion-solvent interaction in aqueous solutions: a suggested pictrue of water structure. *Discuss. Faraday Soc.* 24: 133-140.

Frommer, M. A., and Lancet, D. (1972) Freezing and non-freezing water in cellulose acetate membranes. *J. Appl. Polymer Sci.* 16: 1295-1303.

Fushime, K., and Verkman, A. S. (1991). Low viscosity in the aqueous domain of cell cytoplasm measured by picosecond polarization microfluorimetry. *J. Cell Biol.* 112: 719-725.

G

Gabriel, C., Sheppard, R. J., and Grant, E. H. (1983). Dielectric properties of ocular tissues at 37 degrees C. *Phys. Med. Biol.* 28(1): 43-49.

Garamvolgyi, N. (1959). Kontraktion isolierter Muskelfibrillen. *Acta. Physiol. Acad. Sci. Hung.* 16: 139-146.

Garrigos, M., Morel, J. E., and Garcia de la Torre, J. (1983). Reinvestigation of the shape and state of hydration of the skeletal myosin subfragment 1 monomer in solution. *Biochem.* 22: 4961-4969.

Gary-Bobo, C. M., and Lindenberg, A. B. (1969). The behavior of nonelectroytes in gelatin gels. *J. Colloid and Interface Sci.* 29(4): 702-709.

Gershon, N. D., Porter, K. R., and Trus, B. L. (1985). The cytoplasmic matrix: its volumes and surface area and the diffusion of molecules through it. *Proc. Nat'l. Acad. Sci. USA.* 82(15): 5030-5034.

Ginzburg, B. Z., and Cohen, D. (1964). Calculation of internal hydrostatic pressure in gels from the distribution coefficients of nonelectrolytes between gels and solutions. *Trans. Faraday Soc.* 60: 185-189.

Glynn, I. M., and Karlish, S. J. D. (1975). The sodium pump. *Ann. Rev. Physiol.* 37:13-55.

Gomez, A. M., Kerfant, B. G., and Vassort, G. (2000). Microtubule disruption modulates $Ca2^+$ signaling in rat cardiac myocytes. *Circ. Res.* 86: 30-36.

Goodsell, D. S. (1991). Inside a living cell. *Trends Biochem. Sci.* 16(6): 206-210.

Graham. T. (1833). Title *Phil. Mag.* 2: 175, 269, 351.

Granick, S. (1991). Motions and relaxations of confined liquids. *Science* 253: 1374-1379.

Granzier, H. L. M., Myers, J. A., and Pollack, G. H. (1987). Stepwise shortening of muscle fiber segments. *J. Mus. Res. Cell Motil.* 8: 242-251.

H

Hagiwara, S., Chichibu, S., and Naka K. (1964). The effects of various ions on resting and spike potentials of barnacle muscle fibers. *J. Gen. Physiol.* 48: 163-179.

Hardy, W. (1932). Problems of the boundary state. *Phil. Trans. Roy. Soc. London Ser. A.* 230: 1-37.

Harrington, W. F. (1979). On the origin of the contractile force in skeletal muscle. *Proc. Natl. Acad. Sci.* 76(10): 5066-5070.

Hatano, S. (1972). Conformational changes of plasmodium actin polymers formed in the presence of Mg^{++}. *J. Mech. Cell Motil.* 1: 75-80.

Hatano, S., Totsuka, T., and Oosawa, F. (1967). Polymerization of plasmodium actin. *Biochim. Biophys. Acta.* 140: 109-122.

Hatori, K., Honda, H., and Matsuno, K. (1996a). ATP dependent fluctuations of single actin filaments *in vitro*. *Biophys. Chem.* 58: 267-272.

Hatori, K., Honda, H., and Matsuno, K. (1996b). Communicative interaction of myosins along an actin filament in the presence of ATP.

Biophys. Chem. 60: 149-152.

Hatori, K., Honda, H., Shimada, K., and Matsuno, K. (1998). Propagation of a signal coordinating force generation along an actin filament in actomyosin complexes. *Biophys. Chem.* 75: 81-85.

Hazlewood, C. F. (1979). A view of the significance and understanding of the physical properties of cell-associated water. *Cell-Associated Water, ed.* W. Drost-Hansen, and J. S. Clegg, Acad. Press, N. Y.

Hazlewood, C. F., Nichols, B. L., and Chamberlain, N. F. (1969). Evidence for the existance of a minimum of two phases of ordered water in a skeletal muscle. *Nature* 222: 747-750.

Hazlewood, C. F., Singer, D. B., and Beall, P. (1979). Electron microscope examination of common red cell ghosts: cytoplasmic contamination. *Physiol. Chem. Phys.* 11(2): 181-184.

Heidorn, D. B. (1985). *A Quasielectric Neutron Scattering Study of Water Diffusion in Frog Muscle,* PhD Thesis, Rice University, Houston, TX.

Heimburg, T., and Biltonen, R. L. (1996). A Monte Carlo simulation study of protein-induced heat capacity changes and lipid-induced protein clustering. *Biophys. J.* 70: 84-96.

Hennessey, E. S., Drummond, D. R., and Sparrow, J. C. (1993). Molecular genetics of actin function. *Biochem. J.* 282: 657-671.

Heppel, L. (1939). The electrolytes of muscle and liver in potassium-depleted rats. *Amer. J. Physiol.* 127:385-392.
Heppel, L. (1940). The diffusion of radioactive sodium into the muscles of potassium-depleted rats. *Amer. J. Physiol.* 128: 449-454.

Hille, B. (1973). The permeability of the sodium channel to metal cations in myelinated nerve. *J. Gen Physiol.* 59: 637-658.

Hille, B. (1984). *Ionic Channels of Excitable Membranes,* Sinauer, Sunderland, MA.

Hille, B. (1992). *Ionic Channels of Excitable Membranes,* Second edition, Sinauer, Sunderland, MA.

Hillman, H. (1994). New considerations about the structure of the membrane of the living animal cell. *Physiol. Chem. and Med. NMR.* 26:55-67

Hirokawa, N. (1998). Kinesin and dynein superfamily proteins and the mechanism of organelle transport. *Science* 279: 519-524.

Hodgkin, A. L, and Katz, B. (1949). The effect of sodium on the electrical conductivity of the giant axon of the squid. *J. Physiol.* 108: 37-77.

Hodgkin, A. L., and Horowicz, P. (1959a). Movements of Na and K in single muscle fibres. *J. Physiol.* 145: 405-432.

Hodgkin, A. L., and Horowicz, P. (1959b). The influence of potassium and chloride ions on the membrane potential of single muscle fibres. *J. Physiol.* 148: 127-160.

Hoenger, A., and Milligan, R. A. (1997). Motor domains of kinesin and *ncd* interact with microtubule protofilaments with the same binding geometry. *J. Mol. Biol.* 265(5): 553-564.

Hoeve, C. A. J., Willis, Y. A., and Martin, D. J. (1963). Evidence for a phase transition in muscle contraction. *Biochem.* 2: 282-286.

Hoffman, A. S. (1991). Conventional and environmentally-sensitive hydrogels for medical and industrial uses: a review paper. *Polymer Gels* 268(5): 82-87.

Holstein, T., and Tardent, P. (1984). An ultrahigh-speed analysis of exocytosis: nematocyst discharge. *Science* 223: 830-833.

Horn, R. G., and Israelachvili, J. (1981). Direct measurement of structural forces between two surfaces in a nonpolar liquid. *J. Chem. Phys.* 75(3): 1400-1411.

Howard, J. (1997). Molecular motors: structural adaptations to cellular functions. *Nature* 389: 561-567.

Hubley, M. J., Rosanske, R. C., and Moerland, T. S. (1995). Diffusion coefficients of ATP and creatine phosphate in isolated muscle: Pulse gradient ^{31}P NMR. of small biological samples. *NMR. in Biomed.* 8: 72-78.

Hutchings, B. L. (1969). Tetracycline transport in *Staphylococcus Aureus H. Biochim. Biophys. Acta.* 174: 734-748.

Huxley, A. F., and Niedergerke, R. (1954). Structural changes in muscle during contraction: Interference microscopy of living muscle fibres. *Nature* 173: 971-973.

Huxley, A. F., and Niedergerke, R. (1958). Measurement of the striations of isolated muscle fibres with the interference microscope. *J. Physiol.* 144: 403-425.

Huxley, H. E. (1996). A personal view of muscle and motility mechanisms. *Ann. Rev. Physiol.* 58: 1-19.

Huxley, H. E., and Hanson, J. (1954). Changes in the cross striations of muscle during contraction and stretch and their structural interpretation. *Nature* 173: 973-976.

Hyman, A. A., and Karsenti, E. (1996). Morphogenetic properties of microtubules and mitotic spindle assembly. *Cell* 84: 401-410.

I

Inoue, I., Kobatake, Y., and Tasaki, I. (1973). Excitability, instability, and phase-transition in squid axon membrane under internal perfusion with dilute salt solutions. *Biochem. Biophys. Acta.* 307: 471-477.

Inoue, S., and Salmon, E. D. (1995). Force generation by microtubule assembly/disassembly in mitosis and related movements. *Mol. Biol. of Cell* 6: 1619-1640.

Irving, M., St.-C., Allen, T., Sabido-David, C., Craik, J. S., Brandmeler, B., Kendrick-Jones, J., Corrie, J. E. T., Trentham, D. R., and Goldman, Y. E. (1995). Tilting of the light-chain region of myosin during step length changes and active force generation in skeletal muscle. *Nature* 375: 688-691.

Ishijima, A., Doi, T., Sakurada, K., and Yanagida, T. (1991). Subpiconewton force fluctuations of actomyosin *in vitro*. *Nature* 352: 301-306.

Israelachvili, J. N., and McGuiggan, P. M. (1988). Forces between surfaces in liquids. *Science* 241: 795-800.

Israelachvili, J., and Wennerström, H. (1996). Role of hydration and water structure in biological and colloidal interactions. *Nature* 379: 219-225.

Ito, T., Suzuki, A., and Stossel, T. P. (1992). Regulation of water flow by actin-binding protein-induced actin gelation. *Biophys. J.* 61: 1301-1305.

J

Jacobs, W. P. (1994). Caulerpa. *Sci. Amer.* 271(6): 100 –106.

Jhon, M. S., and Andrade, J. (1973). Water and hydrogels. *J. Biomed. Mater. Res.* 7: 509-522.

Jordan-Lloyd, D., and Shore, A. (1938). *The Chemistry of Proteins,* Second edition, J. A. Churchill, London.

Joseph, N. R., Engel, M. B., and Catchpole, H. R. (1961). Distribution of sodium and potassium in certain cells and tissues. *Nature* 4794: 1175-1178.

K

Kamitsubo, E. (1972). Motile protoplasmic fibrils in cells of the Characeae. *Protoplasma* 74: 53-70.

Kamiya, N. (1970). Contractile properties of the plasmodial strand. *Proc. Jpn. Acad.* 46: 1026-1031.

Käs, J., Strey, H., and Sackmann, E. (1994). Direct imaging of reptation for semi-flexible actin filaments. *Nature* 368: 226-229.

Kasturi, S. R., Hazlewood, C. F., Yamanashi, W. S., and Dennis, L. W. (1987). The nature and origin of chemical shift for intracellular water nuclei in *Artemia* cysts. *Biophys. J.* 52: 249-256.

Katayama, E. (1998). Quick-freeze deep-etch electron microscopy of the actin-heavy meromyosin complex during the *in vitro* motility assay. *J. Mol. Biol.* 278: 349-367.

Katayama, E., and Nonomura, Y. (1979). Electron microscopic analysis of tropomyosin paracrystals. *J. Biochem.* 86: 1511-1522.

Katchalsky, A., and Zwick, M. (1955). Mechanochemistry and ion exchange *J. Polymer Sci.* 16: 221-234.

Kellermayer, M. S. Z., and Pollack, G. H. (1996). Rescue of *in vitro* actin motility halted at high ionic strength by reduction of ATP to submicromolar levels. *Biochim. Biophys. Acta.* 1277: 107-114.

Kellermayer, M., Ladany, A., Jobst, K., Szucs, G., Trombitás, K., and Hazlewood, C. F. (1986). Cocompartmentation of proteins and K^+ within the living cell. *Proc. Natl. Acad. Sci. USA.* 83(4): 1011-1015.

Kellermayer, M., Ludany, A., Miseta, A., Koszegi, T., Berta, G., Bogner, P., Hazlewood, C. F., Cameron, I. L., and Wheatley, D. N. (1994). Release of potassium, lipids, and proteins from nonionic detergent treated chicken red blood cells. *J. Cell. Physiol.* 159: 197-204.

Kim, E., Bobkova, E., Miller, C. J., Orlova, A., Hegyi, G., Egelman, E. H., Muhlrad, A., and Reisler, E. (1998). Intrastrand cross-linked actin between Gln-41 and Cys-374. III. Inhibition of motion and force generation with myosin. *Biochem.* 37(51): 17801-17809.

Kiser, P. F., Wilson, G., and Needham, D. (1998). A synthetic mimic of the secretory granule for drug delivery. *Nature* 394: 459-462.

Kitamura, K., Tokanaga, M., Iwane, A. H., and Yanagida, T. (1999). A single myosin head moves along an acin filament with regular steps of 5.3 nanometers. *Nature* 397(6715): 129-134.

Klenchin, V. A., Sukharev, S. I., Serov, S. M., Chernomordik, L. V., and Chizmadzhev, Yu. A. (1991). Electrically induced DNA uptake by cells is a fast process involving DNA electrophoresis. *Biophys J.* 60: 804-811.

Klinger, R. G., Zahn, D. P., Brox, D. H., and Frundes, H. E. (1971). Interaction of the hemoglobin with ions. Binding of ATP to human hemoglobin under simulated *in vitro* conditions. *Europ. J. Biochem.* 18: 171-177.

Kojima, H., Muto, E., Higuchi, H., and Yanagida, T. (1997). Mechanics of single kinesin molecules measured by optical trapping nanometry. *Biophys. J.* 73: 2012-2022

Kolata, G. B. (1976). Water structure and ion binding: a role in cell physiology? *Science* 192: 1220-1222.

Kolberg, R. (1994). A membrane flip for a bacterial ion channel. *J. NIH Res.* 6: 35-36.

Korn, E. D. (1966). Structure of biological membranes. *Science* 153: 1491-1498.

Kraft, T., Messerli, M., Rothen-Rutishauser, B., Perriard, J.-C., Wallimann, T., and Brenner, B. (1995). Equilibration and exchange of fluorescently labeled molecules in skinned skeletal muscle fibers visualized by confocal microscopy. *Biophys. J.* 69: 1246-1258.

Krause, T. L., Fishman, H. M., Ballinger, M. L., and Bittner, G. D. (1984). Extent and mechanism of sealing in transected giant axons of squid and earthworms. *J. Neurosci.* 14: 6638-6651.

Krauskopf, K. B. (1967). *Introduction to Geochemistry,* McGraw-Hill, Appendix III.

Kühne, W. (1864). *Untersuchungen über das Protoplasma und die Contractilität*, Leipzig.

Kull, F. J., Sablin, E. P., Lau, R., Fletterick, R. J., and Vale, R. (1996). Crystal structure of the kinesin motor domain reveals a structural similarity to myosin. *Nature* 380: 550-555.

Kuroda, K. (1964). Behavior of naked cytoplasmic drops isolated from plant cells. *Primitive Motile Systems in Biology, ed.* R. D. Allen, and N. Kamiya, Acad. Press, N. Y.

Kushmerick, M. J., and Podolsky, R. J. (1969). Ionic mobility in muscle cells. *Science* 166(910): 1297-1298.

L

Lehninger, A. L. (1964). *The Mitochondrion,* Benjamin, Menlo Park, CA.

Lev, A. A., Korchev, Y. E., Rostovtseva, T. K., Bashford, C. L., Edmonds, D. T., and Pasternak, C. A. (1993). Rapid switching of ion current in narrow pores: implications for biological ion channels. *Proc. Roy. Soc. Lond. B.* 252: 187-192.

Lewis, A., Rousso, I., Khachatryan, E., Brodsky, I., Lieberman, K., and Sheves, M. (1996). Directly probing rapid membrane protein dynamics with an atomic force microscopic: a study of light-induced conformational alterations in bacteriorhodopsin. *Biophys. J.* 70: 2380-2384.

Ling, G. N. (1952). The role of phosphate in maintenance of the resting potential and selective ionic accumulation in frog muscle cells. Phosphorous Metabolism, 11: 749-795. ed. W. McElroy, and B. Glass, J. Hopkins Press, Baltimore.

Ling, G. N. (1955). New hypothesis for the mechanism of cellular resting potential. *Fed. Proc.* 14: 93-94.

Ling, G. N. (1962). *A Physical Theory of the Living State: the Association-Induction Hypothesis,* Blaisdell Publ. Co., Waltham, MA.

Ling, G. N. (1965). The physical state of water in living cell and model systems. *Proc. N. Y. Acad. Sci.* 125: 401-417.

Ling, G. N. (1973). How does ouabain control the levels of cell K^+ and Na^+? By interference with a Na pump or by allosteric control of K^+-Na^+ adsorption on cytoplasm protein sites. *Physiol. Chem. and Physics* 5: 295-311.

Ling, G. N. (1978). Maintenance of low sodium and high potassium levels investing muscle cells. *J. Physiol.* 280: 105-123.

Ling, G. N. (1984). *In Search of the Physical Basis of Life,* Plenum Publ. Co., New York.

Ling, G. N. (1988). A physical theory of the living state: application to water and solute distribution. *Scanning Microsc.* 2(2): 899-913.

Ling, G. N. (1992). *A Revolution in the Physiology of the Living Cell.* Krieger Pub. Co., Malabar, Fl.

Ling, G. N. (1997). Debunking the alleged resurrection of the sodium pump hypothesis. *Physiol. Chem. Phys. & Med NMR.* 29: 123-198.

Ling, G. N., and Ochsenfeld, M. M. (1973). Mobility of potassium ion in frog muscle cells, both living and dead. *Science* 181: 78-81.

Ling, G. N., and Walton, C. L. (1976). What retains water in living cells? *Science* 191: 293-295.

Ling, G. N., Kolebic, T., and Damadian, R. (1990). Low paramagnetic-ion content in cancer cells: its significance in cancer detection by magnetic resonance imaging. *Physiol. Chem. Phys. & Med. NMR.* 22: 1-14.

Ling, G. N., Niu, Z., and Ochsenfeld, M. (1993). Predictions of polarized multilayer theory of solute distribution confirmed from a study of the equilibrium distribution in frog muscle of twenty-one nonelectrolytes including five cryoprotectants. *Physiol. Chem. Phys. Med. NMR.* 25(3): 177-208.

Ling, G. N., Walton, C., and Bersinger T. J. (1980). Reduced solubility of polymer-oriented water for sodium salts, amino acids and other solutes normally maintained at low levels in living cells. *Physiol. Chem. Physics* 12: 111-138.

Lippincott, E. R., Cessac, G. L., Stromberg, R. R., and Grant, W. H. (1971). Polywater - a search for alternative explanations. *J. Colloid Interface Sci.* 36: 443-460.

Lippincott, E. R., Stromberg, R. R., Grant, W. H., and Cessac, G. (1969). Polywater. Vibrational spectra indicate unique stable polymeric structure. *Science* 164: 1482-1487.

López-Beltrán, E. A., Maté, M. J., and Cerdán, S. (1996). Dynamics and environment of mitochondrial water as detected by H NMR. *J. Biol. Chem.* 271(18): 10648-10653.

Lowey, S., and Luck, S. M. (1969). Equilibrium binding of andenosine diphosphate to myosin. *Biochem.* 8: 3195-3199.

Luby-Phelps, K., Mujumdar, S., Majumdar, R. B., Ernst, L. A., Galbraith, W., and Waggoner, A. S. (1993). A novel fluorescence ratiometric method confirms the low solvent viscosity of the cytoplasm. *Biophys. J.* 65: 236-242.

Luby-Phelps, K., Taylor, D. L., and Lanni, F. (1986). Probing the structure of cytoplasm. *J. Cell. Biol.* 102(6): 2015-2022.

Luck, W. A. P. (1976). Water in biological systems. *Topics Current Chem.* 64: 113-180.

M

Maniotis, A., and Schliwa, M. (1991). Microsurgical removal of centrosomes blocks cell reproduction and centriole generation in BSC-1 cells. *Cell* 67: 495-504.

Marsh, D. J., De Bruin, S. H., and Gratzer, W. B. (1977). An investigation of heavy meromyosin-ADP binding equilibria by proton release measurements. *Biochem.* 16: 1738-1742.

Matsumoto, G. (1984). A proposed membrane model for generation of sodium currents in squid giant axons. *J. Theor. Biol.* 107: 649-666.

Matsumoto, G., Ichikawa, I. M., Tasaki, A., Murofushi, H., and Sakai, H. (1984). Axonal microtubules neccessary for generation of sodium current in squid axons: I. Pharmacological study on sodium current by microtubule proteins and 260 K protein. *J. Membr. Biol.* 77: 77-91.

Matsumoto, G., Kobayashi, T., and Sakai, H. (1979). Restoration of the excitibility of squid giant axon by tubulin tyrosine ligase and microtubule proteins. *J. Biochem.* 86: 1155-1158.

Matsumoto, G., Murofushi, H., Endo, S., and Sakai, H. (1982). *Biological Functions of Microtubules and Related Structures, ed.* H. Sakai., H. Mohri, and G. G. Borisy, Academic Press, Tokyo.

Matsumoto, G., Murofushi, H., Endo, S., Kobayashi, T., and Sakai, H. (1983). *Structure and Function of Excitable Cells, ed.* D. C. Chang, I. Tasaki, W. J. Adelman, and R. H. Leuchtag, Plenum Press, N. Y.

Maughan, D., and Lord, C. (1988). Protein diffusivities in skinned frog skeletal muscle fibers. *Molecular Mechanism of Muscle Contraction, ed.* H. Sugi, and G. H. Pollack, Plenum Press, N.Y.

Maughan, D., and Recchia, C. (1985). Diffusible sodium, potassium, magnesium, calcium, and phosphorous in frog skeletal muscle. *J. Physiol* 368: 545-563.

McIntosh, J. R. (1973). The axostyle of *Saccinobaculus.* II. Motion of the microtubule bundle and a structural comparison of straight and bent axostyles. *J. Cell Biol.* 56: 324-339.

McLachlan, A. D., and Karn, J. (1982). Periodic charge distributions in the myosin rod amino acid sequence match cross-bridge spacings in muscle. *Nature* 299: 226-231.

McNeil, P. L., and Ito, S. (1990). Molecular traffic through plasma membrane disruptions of cells *in vivo.* *J. Cell Sci.* 96: 549-556.

McNeil, P. L., and Steinhardt, R. A. (1997). Loss, restoration, and maintenance of plasma membrane integrity. *J. Cell Biol.* 137(1): 1-4.

Meaves, H., and Vogel, W. (1973). Calcium inward currents in internally perfused giant axons. *J. Physiol.* (London) 235: 225-265.

Mehta, A. D., Rock, R. S., Rief, M., Spudich, J. A., Mooseker, M. S., and Cheney, R. E. (1999). Myosin-V is a processive actin-based motor. *Nature* 400(6744): 590-593.

Melcher, P., Webb, D., and Haase, A. (1994). High molecular weight organic compounds in the xylem sap of mangroves: implications for long-distance water transport. *Bot. Acta.* 107: 218-229.

Ménétret, J. F., Hoffmann, W., Schroeder, R. R., Rapp, G., and Goody, R. S. (1991). Time-resolved cryo-electron microscopic study of the dissociation of actomyosin induced by photolysis of photolabile nucleotides. J. Mol. Biol. 219(2): 139-144.

Mentré, P. (1995). *L'Eau dans la Cellule: Une Interface Dynamique et Hétérogène des Macromolécules,* Masson, Paris.

Mentré, P., and Debey, P. (1999). An unexpected effect of an ouabain-sensitive ATPase activity on the amount of antigen-antibody complexes formed *in situ.* *Cell and Mol. Biol.* 45(6): 781-791.

Metuzals, J., and Izzard, C. S. (1969). Spatial patterns of thread-like elements in the axoplasm of the giant nerve fiber of the squid (*Loligo pealii L.)* as disclosed by differential interference microscopy and by electron microscopy. *J. Cell Biol.* 43: 456-479.

Metuzals, J., and Tasaki, I. (1978). Subaxolemmal filamentous network in the giant nerve fiber of the squid *(Loligo pealei L.)* and its possible role in excitability. *J. Cell Biol.* 78: 597-621.

Miki, M., and Koyama, T. (1994). Domain motion in actin observed by fluorescence resonance energy transfer. *Biochem.* 33: 10171-10177.

Miller, D. J. (1979). Are cardiac muscle cells 'skinned' by EGTA or EDTA? *Nature* 277: 142-143.

Mimori, Y., and Miki-Nonumura, T. (1995). Extrusion of rotating microtubules on the dynein-track from a microtubule-dynein gamma-complex. *Cell Motil. And Cytoskel.* 30: 17-25.

Miyagishima, Y. (1975). Catecholamine in the myocardium: a fluorecence histochemical study. *Jpn. Circ. J.* 39: 357-375.

Miyano, M., and Osada, Y. (1991). Electroconductive organogel. 2. appearance and nature of current oscillation under electric field. *Macromol.* 24: 4755-4761.

Mond, R., and Amson, K. (1928). Über die Ionenpermeabilitaet des quergestreiften Muskels. *Pflüger's Arch. Ges. Physiol.* 220: 69-81.

Morel, J. E. (1991). The isometric force exerted per myosin head in a muscle fibre is 8 pN. Consequence on the validity of the traditional concepts of force generation. *J. Theor. Biol.* 151: 285-288.

Moriyama, Y., Yasuda, K., Ishiwata, S., and Asai, H. (1996). Ca (2+)-induced tension development in the stalks of glycerinated Vorticella convallaria. *Cell Motil. and Cytoskel.* 34(4): 271-278.

Muto, E., and Yanagida, T. (1997). Cooperative binding of kinesin molecules to a microtubule in the presence of ATP. *Biophys. J.* 72: A62.

N

Nanavati, C., and Fernandez, J. M. (1993). The secretory granule matrix: a fast-acting smart polymer. *Science* 259(5097): 963-965.

Natori, R. (1975). The electric potential change of internal membrane during propagation of contraction in skinned fibre of toad skeletal muscle. *Jpn. J. Physiol.* 25(1): 51-63.

Naylor, G. R., Bartels, E. M., Bridgman, T. D., and Elliott, G. F. (1985). Donnan potentials in rabbit *psoas* muscle in rigor. *Biophys. J.* 48(1): 47-59.

Naylor, W. G., and Merrillees, N. C. R. (1964). Some observations on the fine structure and metabolic activity of normal and glycerinated ventricular muscle of toad. *J. Cell Biol.* 22: 533-550.

Negendank, W. (1982). Studies of ions and water in human lymphocytes. *Biochim. Biophys. Acta.* 694: 123-161.

Neher, E., Sackmann, B., and Steinbach, J. H. (1978). The extracellular patch clamp: a method for resolving currents through individual open channels in biological membranes. *Pflüger's Archiv. Ges. Physiol.* 375: 219-228.

Nickels, J. (1970). Localisation of a microelectrode tip in muscle cell: A light and electron microscopic study. *Acta. Physiol. Scand.* 80: 360-369.

Nickels, J. (1971). A method for localisation of the muscle cell and the motor end plate after in vivo registration with a microelectrode. *Pflüger's Archiv. Ges. Physiol.* 330: 45-50.

Nicklas, R. B. (1997). How cells get the right chromosomes. *Science* 275: 632-637.

Noble, D., and Bett, G. (1993). Reconstructing the heart: a challenge for integrative physiology. *Cardiovasc. Res*: 27: 1701-1712..

Noda, C., and Yugari, Y. (1973). Effect of catecholamines in restoring the beating of cultured rat heart cells treated with reserpine. *Jpn. J. Pharmacol.* 23(6): 839-846.

O

Odelblad, E., Bhar, B. N., and Lindstrom, G. (1956). Proton magnetic resonance of human red blood cells in heavy-water exchange experiments. *Arch. Biochem. Biophys.* 63: 221- 225.

Ogata, M. (1996). Hydrodynamic properties of water in myoplasm in resting and active states. *Proc. Jpn. Acad.* Ser. B 72(6): 137-141.

Okada, Y., and Hirokawa, N. (2000). Mechanism of the single-headed processivity: Diffusional anchoring between the K-loop of kinesin and the C terminus of tubulin. *PNAS.* 97(2): 640-645.

Okuzaki, H., and Osada, Y. (1994). Electro-driven chemomechanical polymer gel as an intelligent soft material. *J. Biomater. Sci. Polymer Edn.* 5: 485-496.

Oosawa, F., Fujime, S., Ishiwata, S., and Mihashi, K. (1972). Dynamic property of F-actin and thin filament. *CSH Symposia on Quant. Biol.* XXXVII: 277-285.

Oparin, A. I. (1971). Routes for the origin of the first forms of life. *Sub-Cell Biochem.* 1: 75-81.

Oparin, A. I. (1964). *The Chemical Origin of Life, ed.* C. Thomas, Springfield, Ill.

Oplatka, A. (1996). The rise, decline, and fall of the swinging crossbridge dogma. *Chemtracts. Bioch. Mol. Biol.* 6: 18-60.

Oplatka, A. (1997). Critical review of the swinging crossbridge theory and of the cardinal active role of water in muscle contraction. *Crit. Rev. Biochem. Mol. Biol.* 32(4): 307-360.

Oplatka, A. (1998). Do the bacterial flagellar motor and ATP synthase operate as water turbines? *Biochem. Biophys. Res. Commun.* 249: 573-578.

Osada, Y., and Gong, J. (1993). Stimuli-responsive polymer gels and their application to chemomechanical systems. *Prog. Polym. Sci.* 18: 187-226.

Osada, Y., and Ross-Murphy, S. (1993). Intelligent gels. *Sci. Amer.* 268: 82-86.

Osterhout, W. J. V., and Hill, S. E., (1938). Calculations of bioelectric potentials. *J. Gen. Physiol.* 22: 139-146.

Overbeek, J. Th. G. (1956). The Donnan equilibrium. *Prog. Biophys Biophys Chem.* 6: 58-84.

P

Pashley, R. M., and Kitchener, J. A. (1979). Surface forces in adsorbed mmultilayers of water on quartz. *J. Colloid and Interface Sci.* 71: 491-500.

Pauling, L. (1945). The adsorption of water by proteins. *J. Am. Chem. Soc.* 67: 555-557.

Pauling, L. (1959a). A molecular theory of general anesthesia. *Science* 134: 15-21.

Pauling, L. (1959b). *Hydrogen Bonding, ed.* L. Hadzi, Pergamon Press, London.

Peachey, L. D. (1965). The sarcoplasmic reticulum and transverse tubules of the frog's sartorius. *J. Cell Biol.* 25: 209- 231.

Peters, R. (1984). Nucleo-cytoplasmic flux and intracellular mobility in single hepatocytes measured by fluorescence microphotolysis. *EMBO J.* 3: 1831-1836.

Petka, W. A., Harden, J. L., McGrath, K. P, Wirtz, D., and Tirrell, D. A. (1998). Reversible hydrogels from self-assembling artificial proteins. *Science* 281: 389-392.

Pfann, W. G. (1962). Zone melting. *Science* 135: 1101-1109.

Pfann, W. G. (1967). Zone refining. *Sci. Am.* 217: 63-71.

Pfeffer, W. F. (1877). *Osmotische Untersuchungen: Studien zur Zell-Mechanik,* Engelmann, Leipzig.

Pinto da Silva, P., and Branton, D. (1970). Membrane splitting in freeze-ethching. Covalently bound ferritin as a membrane marker. *J. Cell Biol.* 45: 598-605.

Pissis, P., Anagnostopoulou-Konsta, A., and Apekis, L. (1987). A dielectric study of the state of water in plant stems. *J. Exptl. Bot.* 38(194): 1528-1540.

Plumb, R. C., and Bridgman, W. B. (1972). Ascent of sap in trees. *Science* 176: 1129-1131.

Pollack, G. H. (1977). Cardiac pacemaking: an obligatory role of catecholamines? *Science* 196: 731-738.

Pollack, G. H. (1983). The sliding filament / cross-bridge theory. *Physiol. Rev.* 63: 1049-1113.

Pollack, G. H. (1990). *Muscle & Molecules: Uncovering the Principles of Biological Motion,* Ebner and Sons, Seattle, WA.

Pollack, G. H. (1996). Phase transitions and the molecular mechanism of contraction. *Biophs. Chem.* 59: 315-328.

Pollack, G. H., Iwazumi, T., ter Keurs, H. E. D. J., and Shibata, E. F. (1977). Sarcomere shortening in striated muscle occurs in stepwise fashion. *Nature* 268: 757-759.

Pollard, T. (1984). Actin-binding protein evolution. *Nature* 312(5993): 403.

Porter, K. R., Beckerle, M., and McNivan, M. (1983). The cytoplasmic matrix. *Mod. Cell Biol.* 2: 259-302.

Prausnitz, M. R., Milano, C. D., Gimm, J. A., Langer, R., and Weaver, J. C. (1994). Quantitative study of molecular transport due to electroporation: uptake of bovine serum albumin by erythrocyte ghosts. *Biophys. J.* 66(5): 1522-1530.

Prochniewicz, E., Zhang, Q., Janmey, P. A., and Thomas, D. D. (1996). Cooperativity in F-actin: binding of gelsolin at the barbed end affects structure and dynamics of the whole filament. *J. Mol. Biol.* 260(5): 756-766.

Prulière, G., and Douzou, P. (1989). Sol-gel processing of actin to obtain homogeneous glasses at low temperatures. *Biophys. Chem.* 34: 311-315.

Purcell, E. M. (1977). Life at low Reynolds number. *Am. J. Physics* 45(1): 3-11.

R

Rabenstein, D. L., Ludowyke, R., and Lagunoff, D. (1987). Proton nuclear magnetic resonance studies of mast cell histamine. *Biochemistry* 26: 6923-6926.

Rao, P. N., Hazlewood, C. F., and Beall, P. T. (1982). Cell cycle phase-specific changes in relaxation times and water content in HeLa cells. *Cell Growth, ed.* C. Nicolini, NATO Adv. Study Inst. Series, Ser. A38: 535-547.

Rieder, C. L, and Salmon, E. D. (1994). Motile kinetochores and polar ejection forces dictate chromosome position on the vertebrate mitotic spindle. *J. Cell Biol.* 124(3): 223-233.

Rief, M., Gautel, M., Oesterhelt, F., Fernandez, J. M., and Gaub, H. E. (1997). Reversible unfolding of individual titin immunoglobin domains by AFM. *Science* 276: 1109-1112.

Robinson, G. W., and Cho, C. H. (1999). Role of hydration water in protein unfolding. *Biophys. J.* 77: 3311-3318.

Robinson, G. W., Zhu, S.-B., Singh, S., and Evans, M. W. (1996). *Water in Biology, Chemistry, and Physics,* World Scientific, London.

Rorschach, H. E., Bearden, D. W., Hazlewood, C. F., Heidorn, D. B., and Nicklow, R. M. (1987). Quasi-elastic scattering studies of water diffusion. *Scanning Microsc.* 1(4): 2043-2049.

Rorschach, H. E., Lin, C., and Hazlewood, C. F. (1991). Diffusion of water in biological tissues. *Scanning. Microsc.* Suppl. 5: S1-S10.

Rozycka, M., Gonzalez-Serratos, H., and Goldman, W. (1993). Non-homogeneous Ca release in isolated frog skeletal muscle fibres. *J. Mus. Res. Cell Motil.* 14: 527-532.

Ruch, T. C., and Patton, H. D. (1965). *Physiology and Biophysics,* Saunders, Philadelphia.

S

Sachs, F., and Qin, F. (1993). Gated, ion-selective channels observed with patch pipettes in the absence of membranes: novel properties of a gigaseal. *Biophys. J.* 65(3): 1101-1107.

Sakahibara, H., Kojima, H., Sakai, Y., Katayama, E., and Oiwa, K. (1999). Inner-arm dynein c of *Chlamydomonas* flagella is a single-headed processive motor. *Nature* 400(6744): 586-590.

Sato, H., Tasaki, I., Carbone, E., and Hallett, M. (1973). Changes in axon birefringence associated with excitation: implications for the structure of the axon membrane. *J. Mechanochem. Cell Motil.* 2: 209-217.

Sato, M., Wong, T. Z., Brown, T., and Allen, R. D. (1984). Rheological properties of living cytoplasm: a preliminary investigation of squid axoplasm (*Loligo pealei*). *Cell Motil.* 4: 7-23.

Sawahata, K., Gong, J. P., and Osada, Y. (1995). Soft and wet touch-sensing system made of hydrogel. *Macromol. Rapid Commun.* 16: 713-716.

Schliwa, M. (1986). *The Cytoskeleton: An Introductory Survey,* Springer, N. Y.

Schültze, M. (1861). *Müller's Arch. für Anatomie und Physiologie und für Wissenschaftliche Medicin,* Berlin.

Schültze, M. (1863). *Das Protoplasma des Rhizopoden und der Pflanzenzellen. Ein Beitrag zur Theorie der Zelle,* Leipzig.

Schutt, C. E., and Lindberg, U. (1992). Actin as the generator of tension during muscle contraction. *Proc. Natl. Acad. Sci. USA.* 89(1): 319-333.

Schutt, C. E., and Lindberg, U. (1993). A new perspective on muscle contraction. *FEBS.* 325: 59-62.

Schutt, C. E., and Lindberg, U. (1998). Muscle contractions as a Markov process I: energetics of the process. *Acta. Physiol. Scan.* 163: 307-324.

Schwann, T. (1839). *Mikroskopische Untersuchungen über die Übereinstimmung in der Struktur und dem Wachstum der Thiere und Pflanzen,* Berlin.

Schwister, K., and Deuticke, B. (1985). Formation and properties of aqueous leaks induced in human erythrocytes by electrical breakdown. *Biochim. Biophys. Acta.* 816: 332-348.

Schwyter, D. H., Kron, S. J., Toyoshima, Y. Y., Spudich, J. A., and Reisler, E. (1990). Subtilisin cleavage of actin inhibits in vitro sliding movement of actin filaments over myosin. *J. Cell Biol.* 111: 465-470.

Serpersu, E. H., Kinosita, K., Jr. and Tsong, T. Y. (1985). Reversible and irreversible modification of erythrocyte membrane permeability by electric field. *Biochim. Biophys. Acta.* 812: 779-785.

Slatin, S. L., Qiu, X-Q, Jakes, K. S., and Finkelstein, A. (1994). Identification of a translocated protein segment in a voltage-dependent channel. *Nature* 371: 158-161.

Solomon, A. K. (1960). Red cell membrane structure and ion transport. *J. Gen. Physiol.* 43: 1-15.

Somlyo, A. V., Gonzalez-Serratos, H. G., Shuman, H., McClellan, G., and Somlyo, A. P. (1981). Calcium release and ionic charges in the sarcoplasmic reticulum of tetanized muscle: an electron probe study. *J. Cell Biol.* 90(3): 577-594.

Spira, M. E., Benbassat, D., and Dormann, A. (1993). Resealing of the proximal and distal cut ends of transected axons: electrophsiological and ultrastructural analysis. *J. Neurobiol.* 24: 300-316.

Spudich, J. A. (1994). How molecular motors work. *Nature* 372: 515-518.

Spyropolous, C. S. (1961). Initiation and abolotion of electric response by thermal and chemical means. *Am. J. Physiol.* 200: 203-208.

Stebbings, H., and Hunt, C. (1982). The nature of the clear zone around microtubules. *Cell and Tissue Res.* 227: 609-617.

Stebbings, H., and Willison, J. H. M. (1973). Stucture of microtubules: a study of freeze-etched and negatively stained microtubules from the ovaries of *Notonecta. Z. Zellforsch u. Mikrosc. Anat.* 138(3): 387-396.

Stein, W. D. (1990). *Channels, Carriers, and Pumps: An Introduction to Membrane Transport,* Acad. Press, San Diego.

Steinbach, B. (1940). Sodium and potassium in frog muscle. *J. Biol. Chem.* 133: 695-701.

Steinberg, I. Z., Oplatka, A., and Katchalsky, A. (1966). Mechanochemical engines. *Nature* 210: 568-571.

Stephenson, D. G., Wendt, I. R., and Forrest, Q. G. (1981). Non-uniform ion distributions and electrical potentials in sarcoplasmic regions of skeletal muscle fibres. *Nature* 289(5799): 690-692.

Stillinger, F. A. (1980). Water revisited. *Science* 209: 451-457.

Sugitani, M, Kobayashi, T., and Tanaka, T. (1987). *Polym. Preprints Jpn.* 36: 2876-2878.

Suzuki, M. (1994). A new concept of a hydrophobicity motor based on local hydrophobicity transition of functional polymer substrate for micro/nano machines. *Polym. Gels and Ntwks.* 2: 279-287.

Szasz, A., van Noort, D., Scheller, A., and Douwes, F. (1994). Water states in living systems. I. Structural aspects. *Physiol. Chem. Phys. & Med. NMR.* 26: 299-322.

Szent-Györgyi, A. (1951). *Chemistry of Muscular Contraction,* Acad. Press, N. Y.

Szent-Györgyi, A. (1960). *Introduction to a Submolecular Biology,* Acad. Press, N. Y., London.

Szent-Györgyi, A. (1972). *The Living State. With Observations on Cancer,* Acad. Press, N. Y.

T

Tabcharani, J. A., Jensen, T.J., Riordan, J.R., and Hanrahan, J.W. (1989). Bicarbonate permiability of the outwardly rectifying anion channel. *J. Memb. Biol.* 112(2): 109-122.

Tanaka, T. (1981). Gels. *Sci. Amer.* 244: 110-113.

Tanaka, T., Annaka, M., Franck, I., Ishii, K., Kokufuta, E., Suzuki, A., and Tokita, M. (1992). Phase transitions of gels.
Mechanics of Swelling. *NATO ASI Series Vol H 64,* Springer-Verlag, Berlin.

Tang, J., and Janmey, P. A. (1998). Two distinct mechanisms of actin bundle formation. *Biol. Bull.* 194: 406-408.

Tang, J., and Janmey, P. A. (1996). Polyelectrolyte nature of F-actin and the mechanism of actin bundle formation. *J. Biol. Chem.* 271(15): 8556-8563.

Taniguchi, Y., and Horigome, S. (1975). The states of water in cellulose acetate membranes. *J. Appl. Polymer Sci.* 19: 2743-2748.

Tasaki, I, Byrne, P. M., and Masumura, M. (1987). Detection of thermal responses of the retina by use of polyvinylidene fluoride multilayer detector. *Jpn. J. Physiol.* 37: 609-619.

Tasaki, I. (1988). A macromolecular approach to excitation phenomena: mechanical and thermal changes in nerve during excitation. *Physiol. Chem. and Phys. and Med. NMR.* 20: 251-268.

Tasaki, I. (1982). *Physiology and Electrochemistry of Nerve Fibers,* Acad. Press, NY.

Tasaki, I. (1998). Repetitive mechanical responses of the amphibian skin to adrenergic stimulation. *Jpn. J. Physiol.* 48: 297-300.

Tasaki, I. (1999a). Rapid structural changes in nerve fibers and cells associated with their excitation processes. *Jpn. J. Physiol.* 49(2): 125-138.

Tasaki, I. (1999b). Evidence for phase transition in nerve fibers, cells and synapses. *Ferroelectrics* 220: 305-316.

Tasaki, I., and Hagiwara, S. (1957). Demonstration of two stable potential states in the squid giant axon under tetraethylammonium chloride. *J. Gen. Physiol.* 40: 859-885.

Tasaki, I., and Byrne, P. M. (1992). Discontinuous volume transitions in ionic gels and their possible involvement in the nerve excitation process. *Biopolymers* 32: 1019-1023.

Tasaki, I., and Byrne, P. M. (1994). Discontinuous volume transitions induced by calcium-sodium ion exchange in anionic gels and their neurobiological implications. *Biopolymers* 34: 209-215.

Tasaki, I., and Iwasa, K. (1981). Temperature changes associated with nerve excitation: detection by using polyvinylidene fluoride film. *Biochem. Biophys. Res. Commun.* 101: 172-176.

Tasaki, I., Kusano, K., and Byrne, P. M. (1989). Rapid mechanical and thermal changes in the garfish olfactory nerve associated with a propagated impulse. *Biophys. J.* 55: 1033-1040.

Tasaki, I., Singer, I., and Takenada, T. (1965). Effects of internal and external ionic environment on excitability of squid giant axon: A macromolecular appraoach. *J. Gen. Physiol.* 48: 1095-1123.

Taylor, S. R., Shlevin, H. H., and Lopez, J. R. (1975). Calcium in excitation-contraction coupling of skeletal muscle. *Biochem. Soc. Transact.* 7: 759-764.

Terakawa, S., Nagano, M., and Watanabe, A. (1977). Intracellular divalent cations and plateau duration of squid giant axons treated with tetraethylammonium. *Jpn. J. Physiol.* 27: 785-800.

Thomas, D. D. (1987). Spectroscopic probes of muscle cross-bridge rotation. *Ann. Rev. Physiol.* 49: 641-709.

Tigyi, J., Kellermayer, M., and Hazlewood, C. F. (1991). *The Physical Aspect of the Living Cell,* Akad. Kiado, Budapest.

Tirosh, R., and Oplatka, A. (1982). Active streaming against gravity in glass microcapillaries of solutions containing acto-heavy meromyosin and native tropomyosin. *J. Biochem.* 91: 1435-1440.

Tokita, M., and Tanaka, T. (1991). Reversible decrease of gel-solvent friction. *Science* 253: 1121-1123.

Toney, M. F., Howard, J. N., Richer, J., Borges, G. L., Gordon, J. G., Melroy, O. R., Wiesler, D. G., Yee, D., and Sorensen, L. B. (1994). Voltage-dependent ordering of water molecules at an electrode-electrolyte interface. *Nature* 368: 444-446.

Trantham, E.C., Rorschach, H. E., Clegg, J. C., Hazlewood, C. F., Nicklow, R. M., and Wakabayashi, N. (1984). Diffusive properties of water in Artemia cysts as determined from quasi-elastic neutron scattering spectra. *Biophys. J.* 45(5): 927-938.

Trautwein, W., and Uchizono, K. (1963). Electron microscopic and electrophysiologic study of the pacemaker in the sino-atrial node of the rabbit heart. *Z. Zellforsch.* 61: 96-109.

Trentham, D. R., Eccleston, J. F., and Bagshaw, C. R. (1967). Kinetic Analysis of ATPase Mechanisms. *Quart. Rev. Biophys.* 9: 218-281.

Trombitás, K., and Pollack, G. H. (1995). Visualization of the transverse cytoskeletal network in insect-flight muscle by scanning-electron microscopy. *Cell Motil. and Cytoskel.* 32: 226-232.

Trombitás, K., and Tigyi-Sebes, A. (1979). The continuity of thick filaments between sarcomeres in honey-bee flight muscle. *Nature* 281(5729): 319-320.

Trombitás, K., Baatsen, P., Schreuder, J., and Pollack, G. H. (1993). Contraction-induced movements of water in single fibres of frog skeletal muscle. *J. Mus. Res. Cell Motil.* 14: 573-584.

Troshin, A. (1948). Salt currents in the complex coascervate: gelatine and gum arabic. Izv. Akad Nauk SSSR ser biol. No 4: 180-185.

Troshin, A. (1966). *Problems of Cell Permeability,* Pergamon Press, Oxford.

Troshin, A. S. (1956). *Problems of Cell Permeability,* English edition: Pergammon Press, Oxford, Translated by W. F. Widdas.

Troyer, D. (1975). Possible involvement of the plasma membrane in saltatory particle movement in heliozoan axopods. *Nature* 254: 696-698.

Tskhovrebova, L., Trinick, J., Sleep, J. A., and Simmons, R. M. (1997). Elasticity and unfolding of single molecules of the giant muscle protein titin. *Nature* 387: 308-312.

Tsukita, S., Tsukita, S., Kobayashi, T., and Matsumoto, G. (1986). Subaxolemmal cytoskeleton in squid giant axon. II. Morphological identification of microtubule- and microfilament-asociated domains of axolemma. *J. Cell Biol.* 102: 1710-1725.

Tuganowski, W., Krause, M., and Dorczak, K. (1973). The effect of dibutyryl 3'5'-cylic AMP on the cardiac pacemaker, arrested with reserpine and alpha-methyl-tyrosine. *Naunyn-Schmiedebergs Arch. Pharmakol.* 280: 63-70.

U

Urry, D. W. (1993). Molecular Machines: How motion and other functions of living organisms can result from reversible chemical changes. *Angewandte Chemie Intl. Ed. Engl.* 32: 819-841.

Uvnas, B., Aborg, C. H., Lyssarides, L., and Thyberg, J. (1985). Cation exchanger property of isolated rat peritoneal mast cell granules. *Acta. Physiol. Scand. Suppl.* 125: 25-31.

Uvnas, B., and Aborg, C. H. (1977). On the cation exchanger properties of rat mast cell granules and their storage of histamine. *Acta. Physiol. Scand.* 100: 309-314.

V

Vale, R. D., and Toyoshima, Y. Y. (1988). Rotation and transition of microtubules *in vitro* induced by dyneins from *Terahymena Cilia.* *Cell* 52: 459-469.

Vale, R. D., Funatsu, T., Pierce, D. W., Romberg, L., Harada, Y., and Yanagida, T. (1996). Direct observation of single kinesin molecules moving along microtubules. *Nature* 380: 451–453.

Valentijn, K., Valentijn, J. A., and Jamieson, J. D. (1999). Role of actin in regulated exocytosis and compensatory membrane retrieval: insights from an old acquaintance. *Biochem. Biophys. Res. Commun.* 266: 652-661.

van Iterson, W. (1965). Symposium on the fine structure and replication of bacteria and their parts. *Bact. Rev.* 29: 299-325.

Verdugo, P. (1990). Goblet cells secretion and mucogenesis. *Ann.. Rev. Physiol.* 52: 157-176.

Verdugo, P., and Orellana, M. V. (1995). The secretory granule as a biomimetic model for drug delivery. *Proc. Intern. Symp. Control Rel. Bioact. Mater.* 22: 25-33.

Verdugo, P., Deyrup-Olsen, I., Martin, A. W., and Luchtel, D. L. (1992). Polymer gel-phase transition: the molecular mechanism of product release in mucin secretion? *NATO ASI Series Vol. H64, Mechanics of Swelling, ed.* T. K. Karalis, Springer, Berlin.

Verdugo, P., Orellana, M. V., and Freitag, C. (1995). The secretory granule as a biomimetic model for drug delivery. *Proc. Intern. Symp. Control Rel. Bioact. Mater.* 22: 25.

Visscher, K., Schnitzer, M. J., and Block, S. M. (1999). Single kinesin molecules studied with a molecular force clamp. *Nature* 400(6740): 184-189.

Vogler, E. A. (1998). Structure and reactivity of water at biomaterial surfaces. *Adv. Colloid and Interface Sci.* 74: 69-117.

Von Hippel, P. H., and Wong, K-Y. (1964). Neutral salts: the generality of the stability of macromolecular conformation. *Science* 145: 577-580.

Von Zglinicki, T. (1988). Monovalent ions are spatially bound within the sarcomere. *Gen. Physiol. Biophy.* 7: 495-504.

W

Walczak, E. C., Mitchison, T. J., and Desai, A. (1996). XKCM1: a Xenopus kinesin-related protein that regulates microtubule dynamics during mitotic spindle assembly. *Cell* 84: 37-47.

Walker, M. and Trinick, J. (1986). Electron microscope study of the effect of temperature on the length of the tail of the myosin molecule. *J. Mol. Biol.* 192: 661-667.

Walker, R. A., Salmon, E. D., and Endow, S. A. (1990). The *Drosophila* claret segregation protein is a minus-end directed motor molecule. *Nature* 347: 780-782.

Walzthöny, D., and Eppenberger, H. M. (1986). Melting of myosin rod as revealed by electron microscopy. II. Effects of temperature and pH on length and stability of myosin rod and its fragments. *Eur. J. Cell Biol.* 41: 38-43.

Walzthöny, D., Eppenberger, H. M., and Walliman, T. (1986). Melting of myosin rod as revealed by electron microscopy. I. Effects of glycerol and anions on length and stability of myosin rod. *Eur. J. Cell Bio.* 41: 33-37.

Wang, C., Stewart, R. J., and Kopecek, J. (1999). Hybrid hydrogels assembled from synthetic polymers and coiled-coil protein domains. *Nature* 397: 417-420.

Wang, N., Butler, J. P., and Ingber, D. E. (1993). Mechanotransduction across the cell surface and through the cytoskeleton. *Science* 260: 1124-1127.

Warner, F. D., and Mitchell, D. R. (1981). Polarity of dynenin-microtubule interactions *in vitro*: cross-bridging between parallel and anti-parallel microtubules. *J. Cell Biol.* 89: 35-44.

Washabaugh, M. W., and Collins, K. D. (1986). The systematic characterization by aqueous column chromatography of solutes which affect protein stability. *J. Biol. Chem.* 261: 12477-12485.

Watterson, J. G. (1991). The role of water in cell function. *Biofizika.* 36(1): 5-30.

Watterson, J. G. (1997). The pressure pixel-unit of life? *BioSytems* 41: 141-152.

Weisenberg, R. C., and Cianci, C. (1984). ATP-induced gelation-contraction of microtubules assembled in vitro. *J. Cell Biol.* 99: 1527-1533.

Weiss, R. M., Lazarra, R., and Hoffman, B. F. (1967). Potentials measured from glycerinated cardiac muscle. *Nature* 215: 1305-1307.

Wells, A. L., Lin, A. W., Chen, L.-Q., Safer, D., Cain, S. M., Hasson, T., Carragher, B. O., Milligan, R. A., and Sweeney, H. L. (1999). Myosin VI is an actin-based motor that moves backwards. *Nature* 401: 505-508.

Whittam, R. (1961). Active cation transport as a pace-maker of respiration. *Nature* 19: 603-604.

Wiggins, P. M. (1990). Role of water in some biological processes. *Microbiol. Rev.* 54(4): 432-449.

Wiggins, P. M., and van Ryn, R. T. (1990). Changes in ionic selectivity with changes in density of water in gels and cells. *Biophys. J.* 58: 585-596.

Wohlfarth-Botterman, K. E. (1964). Differentiations of the ground cytoplasm and their significance for the generation of the motive force of ameboid movement. *Primitive Motile Systems in Cell Biology, ed.* R. D. Allen, and N. Kamiya, Acad. Press, N. Y.

Wojcieszyn, J. W., Schlegel, R. A., Wu, E.-S., and Jacobson, K. A. (1981). Diffusion of injected macromolecules within cytoplasm of living cells. *Proc. Nat'l. Acad. Sci. USA* 78(7): 4407-4410.

Woodbury, D. (1989). Pure lipid vesicles can induce channel-like conductances in planar bilayers. *J. Memb. Biol.* 109(2): 145-150.

Wordeman, L., and Mitchison, T. J. (1995). Identification and partial characterization of mitotic centromere-associated kinesin, a kinesin-related protein that associates with centromeres during mitosis. *J. Cell Biol.* 128: 95-104.

X

Xie, T-D, Sun, L., and Tsong, T. Y. (1990). Studies of mechanisms of electric field-induced DNA transfection. *Biophys. J.* 58: 13-19.

Xu, X.-H. N, and Yeung, E. S. (1998). Long-range electrostatic trapping of single-protein molecules at liquid-solid interface. *Science* 281: 1650-1653.

Y

Yamada, T. (1998). ^{1}H-NMR spectroscopy of the intracellular water of resting and rigor frog skeletal muscle. *Mechanisms of Work Production and Work Absorption in Muscle, ed.* H. Sugi, and G. H. Pollack, Plenum Press, N. Y.

Yanagida, T., and Oosawa, F. (1978). Polarized fluorescence from epsilon-ADP incorporated into F-actin in a myosin-free single fiber: conformation of F-actin and changes induced in it by heavy meromyosin. *J. Mol. Biol.* 126: 507-524.

Yanagida, T., Nakase, M., Nishiyama, K., and Oosawa, F. (1984). Direct observation of motion of single F-actin filaments in the presence of myosin. *Nature* 307: 58-60.

Yang, P., Tameyasu, T., and Pollack, G. H. (1998). Stepwise shortening in single sarcomeres of single myofibrils. *Biophys. J.* 74: 1473-1483.

Yano, M., Yamada, T., and Shimizu, H. (1978). Studies of the chemo-mechanical conversion in artificially produced streamings. *J. Biochem.* 84: 277-283.

Yano, M., Yamamoto, Y., and Shimizu, H. (1982). An actomyosin motor. *Nature* 299: 557-559.

Yasunaga, H., and Ando, I. (1993). Effect of cross-linking on the molecular motion of water in polymer gel as studied by pulse ^{1}H NMR and PGSE ^{1}H NMR. *Polym Gels and Ntwks* 1: 267-274.

Yasunaga, H., Kobayashi, M., Matsukawa, S., Kurosu, H., and Ando, I. (1997). Structures and dynamics of polymer gel systems viewed from NMR. spectroscopy. *NMR. in Polymer Science,* Acad. Press, London.

Yawo, H., and Kuno, M. (1985). Calcium dependence of membrane sealing at the cut end of the cockroach giant axon. *J. Neurosci.* 5: 1626-1632.

Yen, T. J., Li, G., Cshaar, B. T., Szilak, I., and Cleveland, D. W. (1992). CENP-E is a putative kinetochore motor that accumulates just before mitosis. *Nature* 359: 536-539.

Yoshizaki, K., Seo, Y., Nishikawa, H., and Morimoto, T. (1982). Application of pulsed-gradient 31P NMR on frog muscle to measure the diffusion rates of phosphorous compounds in cells. *Biophys. J.* 38: 209-211.

Yuan, S., and Hoffman, A. S. (1995). Synthetic sulfonated microspheres as drug delivery carriers. Proc. Intern. Symp. Control Rel. Bioact. Mater. 22: 26-27.

Z

Zimmermann, U., Zhu, J. J., Meinzer, F. C., Goldstein, G., Schneider, H., Zimmermann, G., Benkert, R., Thurmer, F., Melcher, P., Webb, D., and Haase, A. (1994). High molecular weight organic compounds in the xylem sap of mangroves: implications for long-distance water transport. *Bot. Acta.* 107: 218-229.

Index

A

B

C

D

E

F

Q

R

S

T

U

V

W

X

Z